Clifford P Shearman
June 1991

D1758334

DOPPLER ULTRASOUND

DOPPLER ULTRASOUND

Physics, Instrumentation, and Clinical Applications

D H Evans, PhD, FInstP
Department of Medical Physics and Clinical Engineering,
Leicester Royal Infirmary,
Leicester, England

W N McDicken, PhD
Department of Medical Physics and Medical Engineering,
The Royal Infirmary,
Edinburgh, Scotland

R Skidmore, PhD, FInstP
Medical Physics Department,
Bristol General Hospital,
Bristol, England

J P Woodcock, PhD, FInstP
Department of Medical Physics and Bioengineering,
University Hospital of Wales,
Cardiff, Wales

A Wiley Medical Publication

JOHN WILEY & SONS
Chichester · New York · Brisbane · Toronto · Singapore

© 1989 John Wiley & Sons Ltd

No part of this book may be reproduced by any means, or transmitted, or translated into a machine language without the written permission of the publisher.

Library of Congress Cataloging-in-Publication Data
Doppler ultrasound: physics, instrumentation, and clinical
 applications/D.H. Evans . . . [et al.].
      p.   cm.   — (A Wiley medical publication)
   Includes bibliographies and index.
   ISBN 0 471 91489 4
   1. Doppler ultrasonography. I. Evans, D. H., Ph.D.
   II. Series.
   [DNLM: 1. Ultrasonic Diagnosis—instrumentation.
   2. Ultrasonic Diagnosis—methods. 3. Ultrasonics.
   WB 289 D692]
 RC78.7.U4D66 1989
 616.07′543—dc 19
 DGPO/DLC
 for Library of Congress   88-12230
                                                      CIP

British Library Cataloguing in Publication Data

Evans, D. H.
   Doppler ultrasound: physics, instrumentation and clinical
   applications
   1. Man. Diagnosis. Doppler ultrasonography
   I. Title
   616.07′543

   ISBN 0 471 91489 4

Typeset by Cotswold Typesetting Ltd, Gloucester
Printed and bound by Anchor Press Ltd, Tiptree, Colchester

To Jennifer Lucy Evans, who had the misfortune to be born at the same time as this book was conceived

PREFACE

Doppler ultrasound has come of age in the last ten years. No longer is it a technique confined to research laboratories and specialist centres, it is now in widespread clinical use, and its use continues to grow at a rapid pace. In concert with its rapid growth a great many books have been published that deal with the clinical aspects of the subject. Such books often contain a brief technical introduction to Doppler ultrasound, but these are invariably superficial in nature, and there are virtually no books that deal primarily with the physics and instrumentation of the method. This is unfortunate because in order to obtain the most from Doppler ultrasound it is important that the user has a firm grasp of the underlying physical principles.

The purpose of this book is to provide an up-to-date introduction to the more technical aspects of Doppler ultrasound. We have attempted to address it to as wide a readership as possible and we hope that it will be suitable as a companion to all Doppler enthusiasts whether their primary discipline be medical, scientific or technical. In order to achieve this we have, wherever possible, avoided the use of complex equations, and leave the more mathematically inclined to pursue these in the cited literature. Most of the included mathematics can be avoided during an initial study of the text.

In order to satisfy the needs of as wide a range of interests as possible we have divided the text into two parts which could be described as 'essential' and 'advanced'. The latter is indicated by a vertical rule in the left-hand margin. We have confined less important technical material and all but the most important mathematics to the advanced sections, and have tried to arrange the separation of the two parts in such a way that the essential material is complete in itself. We hope that this arrangement will appeal to physicists and medical personnel alike and that this structure will tempt *all* readers to dip into the advanced material of special relevance to themselves.

The final part of the book is a short section on clinical applications. It has already been mentioned that there are many books available that deal with this facet of Doppler ultrasound and it is clear that a book such as this cannot hope to compete with these in terms of depth or breadth of coverage. We have included this material since it both illustrates the applications of the physical principles described earlier, and serves as an entry point into the clinical literature for those who have not approached the subject from a medical background.

We have tried to maintain notational consistency throughout the book, but this has not always been easy since, for example, the 'standard notation' of haemodynamics often clashes with that of ultrasonic wave propagation. We have therefore arrived at something of a compromise: we have attempted to use a given symbol for a given quantity throughout the book, but the same symbol may indicate different quantities in different chapters. Symbols are usually defined on the first occasion they appear in each chapter, and Chapters 2, 3, 8 and 10 each contain notation lists.

Many people have contributed to the production of this book directly or indirectly; they are too numerous to mention individually, but to them we offer our heartfelt thanks. We are also indebted to a number of individuals and publishers for allowing us to reproduce tables and illustrative material. We are grateful to Oxford Sonicaid Ltd for their generous support in the production of some of the illustrations.

D.H.E.
W.N.McD.
R.S.
J.P.W.

CONTENTS

x   Contents

CHAPTER 10 – Waveform analysis and pattern recognition

CHAPTER 11 – Volumetric blood flow measurements

CHAPTER 12 – Doppler test phantoms and quality control

CHAPTER 13 – Safety considerations in Doppler ultrasound

Chapter 1

INTRODUCTION

Doppler ultrasound is an important technique for non-invasively detecting and measuring the velocity of moving structures, and particularly blood, within the body. Although first used for this purpose some 30 years ago, it is only in the last decade or so that Doppler instrumentation has advanced to a stage where it is suitable for routine use outside the hospital research laboratory. The correct interpretation of the results obtained from Doppler equipment depends to a considerable extent on the understanding of both the physical mechanisms and signal processing methods that result in the Doppler signal. The aim of this book is to describe these processes and the effects they have on clinical studies.

The text is addressed to a wide readership, including medical, paramedical and scientific staff using, or interested in Doppler ultrasound. It is realized that some readers will require a more rigorous discussion of the subject matter than others and therefore the authors have distinguished between what might be classed as 'essential reading' and 'advanced reading', the latter sections being indicated by a vertical rule in the left-hand margin. Less mathematically inclined readers who have a firm grasp of the basic principles may wish to study the 'advanced reading' on subsequent occasions or in collaboration with a biomedical engineering colleague. This will give worthwhile access to additional aspects of Doppler techniques.

The remainder of this chapter consists of a very brief introduction to the Doppler effect and its use in medicine.

1.1 THE DOPPLER EFFECT

When an observer is moving relative to a wave source, the frequency he measures is different from the emitted frequency. If the source and observer are moving towards each other, the observed frequency is higher than the emitted frequency; if they are moving apart the observed frequency is lower. This simple but important assertion is known as the Doppler effect after the Austrian physicist Christian Doppler (1803–1853) who first postulated it in a paper given before the Royal Bohemian Society of Learning in 1842 (White 1982), and which was published in the proceedings of that society during the following year (Doppler 1843).

The details of Doppler's famous paper, and the subsequent experimental work of Buys Ballot in 1845 using the Amsterdam–Utrecht railway (originally intended to disprove Doppler's theory) have been described by a number of authors (Jonkman 1980, White 1982, Eden 1986) and make fascinating reading.

2   Doppler ultrasound

In his original paper Doppler made a number of incorrect observations concerning the velocities of the stars, and therefore it seems ironical that one of the most important applications of the Doppler effect has been for making velocity measurements in astronomy (using the absorption lines present in the light spectra from stars and galaxies). The Doppler effect has also been widely used in terrestrial applications, and it is only relatively recently that it has become important in medicine. Both ultrasound and laser Dopplers have found a place for detecting motion within the body, but the ultrasound technique is much more widely used and is the subject of this book.

1.2   ULTRASOUND

Ultrasound is sound that has a frequency above the audible range of man, that is greater than 20 kHz. In both imaging and Doppler applications in medicine the usual range of frequencies used is between about 1 MHz and 10 MHz. The lower limit is determined by wavelength considerations (the longer the wavelength the poorer the spatial resolution – both axial and transverse), and the upper limit by acceptable power levels (attenuation rises very rapidly with frequency and so a very small proportion of the transmitted power is returned to the transducer at high frequencies).

Both imaging and Doppler devices function by transmitting a beam of ultrasound into the body, and collecting and analysing the returning echoes. Imaging devices are able to calculate the coordinates from which echoes originated, and these together with knowledge of the echo amplitudes allow cross-sectional images of the body to be constructed. Doppler devices may also be able to determine the position of the source of echoes, but their fundamental concern is the frequency of the returning echoes, and whether there has been a Doppler shift as a result of interaction with a moving target.

1.3   DOPPLER ULTRASOUND IN MEDICINE

The method of utilizing the Doppler principle in medicine varies slightly from the 'classical' Doppler method in that the targets do not spontaneously emit a radiation, and it is therefore necessary to transmit a signal into the body and to observe the changes in frequency that occur when it is reflected or scattered from the targets.

It can be shown that under these conditions (see Section 3.4) there is a shift in the ultrasound frequency given by:

$$f_d = f_t - f_r = 2f_t v \cos \theta / c \qquad\qquad 1.1$$

where f_t and f_r are the transmitted and received ultrasound frequencies respectively, v the velocity of the target, c the velocity of sound in the medium, and θ the angle between the ultrasound beam and the direction of motion of the target. The velocity, c, and the transmitted frequency, f_t, are known in any given situation, and therefore the velocity of a target can be found from the expression:

$$v = K f_d / \cos \theta \qquad\qquad 1.2$$

where K is a known constant given by $c/2f_t$. This equation may be used to monitor changes in velocity, and if the angle θ can be determined absolute velocity may be calculated.

In practice there is unlikely to be just a single target contributing to the Doppler shift, and furthermore multiple targets are unlikely all to have the same velocity, and so the

Doppler shift signal will contain not a single frequency, but a spectrum of frequencies, and it is this spectrum that must be interpreted if full use is to be made of the Doppler shift signal.

Early Doppler units were continuous wave (CW) non-directional devices which presented the Doppler signal either as an audible signal or as a simple envelope signal related to the instantaneous average frequency, but since then a number of developments have taken place (Table 1.1) which have made the Doppler technique both sophisticated and widely applicable.

Table 1.1   Major developments in Doppler ultrasound

Development	Approximate year	Some early key references*
Doppler effect described	1842	Doppler (1843)
Doppler ultrasound used in medical applications	1957	Satomura (1959) Franklin et al (1961, 1963) Stegell et al (1966)
Directional Doppler	1966	McLeod (1967) Cross and Light (1971) Nippa et al (1975)
Pulsed wave Doppler	1967	Wells (1969) Peronneau and Leger (1969) Baker (1970)
Multigate and infinite gate systems	1970/75	Baker (1970) Keller et al (1976) Brandestini (1978) Nowicki and Reid (1981)
Doppler imaging	1971	Mozersky et al (1971) Reid and Spencer (1972) Fish (1975)
Duplex echo-Doppler systems	1974	Barber et al (1974) Phillips et al (1980)
Real-time colour flow mapping	1981	Eyer et al (1981) Namekawa et al (1982) Kasai et al (1985)

* Most of these references have been chosen because they are full papers written in English. In many cases the earliest reports in the literature are in less complete and less accessible conference proceedings.

Almost without exception Doppler units are now directional and are able to deal simultaneously with motion both towards and away from the transducer. Many units include real-time spectrum analysis hardware and this has greatly facilitated the interpretation of the Doppler signal. Pulsed Doppler units are also widely used and allow the operator to select signals from a particular range by gating the returning signal. Both conventional ultrasound imaging and Doppler imaging have added a new dimension to quantitative Doppler studies by allowing the operator to place the Doppler sample volume at a known point in the body or at a point of particular haemodynamic interest, whilst the Doppler images themselves may be of great value in some applications.

Doppler techniques are now used in a multitude of clinical applications and it seems that there are few, if any, branches of medicine that will in the future not benefit from such methods. It is hoped that this book will help to improve the understanding of the underlying principles of Doppler measurements.

1.4 SUMMARY

Doppler ultrasound is a powerful technique for the non-invasive measurement of velocity within the body. The Doppler signal, however, is influenced by a variety of factors which need to be appreciated if the maximum benefit is to be gained from the method. This book describes the physical mechanisms involved in the generation of Doppler signals, and the signal processing methods that are used to extract velocity signals from them. A number of clinical applications are reviewed to illustrate the power and versatility of the technique.

1.5 REFERENCES

Baker DW (1970) Pulsed ultrasonic Doppler blood-flow sensing. IEEE Trans Sonics Ultrasonics SU-17, 170–185.

Barber FE, Baker DW, Nation AWC, Strandness DE, Reid JM (1974) Ultrasonic duplex echo-Doppler scanner. IEEE Trans Biomed Eng BME-21, 109–113.

Brandestini M (1978) Topoflow – A digital full range Doppler velocity meter. IEEE Trans Sonics Ultrasonics SU-25, 287–293.

Cross G, Light LH (1971) Direction resolving Doppler instrument with improved rejection of tissue artifacts for transcutaneous aortovelography. J Physiol Lond 217, 5p–7p.

Doppler C (1843) Ueber das farbige Licht der Doppelsterne und einiger anderer Gestirne des Himmels. Abhandl Konigl Bohm Ges Ser 2, 465–482.

Eden A (1986) The beginnings of Doppler. In: Transcranial Doppler sonography (Ed. R Aaslid), pp 1–9, Springer-Verlag, New York.

Eyer MK, Brandestini MA, Phillips DJ, Baker DW (1981) Color digital echo/Doppler image presentation. Ultrasound Med Biol 7, 21–31.

Fish PJ (1975) Multichannel, direction-resolving Doppler angiography. In: Proc 2nd Eur Cong Ultrasonics Med (Eds E Kazner, M de Vlieger, HR Muller, VR McCready) pp 153–159. Excerpta Medica, Amsterdam.

Franklin DL, Schlegel W, Rushmer RF (1961) Blood flow measured by Doppler frequency shift of back-scattered ultrasound. Science 134, 564–565.

Franklin DL, Schlegel WA, Watson NW (1963) Ultrasonic Doppler shift blood flowmeter: circuitry and practical applications. Biomed Sci Instrum 1, 309–311.

Jonkman EJ (1980) Doppler research in the nineteenth century. Ultrasound Med Biol 6, 1–5.

Kasai C, Namekawa K, Koyano A, Omoto R (1985) Real-time two-dimensional blood flow imaging using an autocorrelation technique. IEEE Trans Sonics Ultrasonics SU-32, 458–464.

Keller HM, Meier WE, Anliker M, Kumpe DA (1976) Non-invasive measurement of velocity profiles and blood flow in the common carotid artery by pulsed Doppler ultrasound. Stroke 7, 370–377.

McLeod FD (1967) A directional Doppler flowmeter. Digest 7th Int Conf Med Biol Eng, pp 213.

Mozersky DJ, Hokanson DE, Baker DW, Sumner DS, Strandness DE (1971) Ultrasonic arteriography. Arch Surg 103, 663–667.

Namekawa K, Kasai C, Tsukamoto M, Koyano A (1982) Realtime bloodflow imaging system utilizing autocorrelation techniques. In: Ultrasound '82 (Ed. RA Lerski, P Morley), pp 203–208, Pergamon Press, New York.

Nippa JH, Hokanson DE, Lee DR, Sumner DS, Strandness DE (1975) Phase rotation for separating forward and reverse blood velocity signals. IEEE Trans Sonics Ultrasonics SU-22, 340–346.

Nowicki A, Reid JM (1981) An infinite gate pulse Doppler. Ultrasound Med Biol 7, 41–50.

Peronneau PA, Leger F (1969) Doppler ultrasonic pulsed blood flowmeter. Proc 8th Int Conf Med Biol Eng, 10–11.

Phillips DJ, Powers JE, Eyer MK, Blackshear WM, Bodily KC, Strandness DE, Baker DW (1980) Detection of peripheral vascular disease using the duplex scanner III. Ultrasound Med Biol 6, 205–218.

Reid JM, Spencer MP (1972) Ultrasonic Doppler technique for imaging blood vessels. Science 176, 1235–1236.

Satomura S (1959) Study of the flow patterns in peripheral arteries by ultrasonics. J Acoust Soc Jap 15, 151–158.

Stegall HF, Rushmer RF, Baker DW (1966) A transcutaneous ultrasonic blood-velocity meter. J Appl Physiol 21, 707–711.

Wells PNT (1969) A range-gated ultrasonic Doppler system. Med Biol Eng 7, 641–652.

White DN (1982) Johann Christian Doppler and his effect – a brief history. Ultrasound Med Biol 8, 583–591.

Chapter 2

BLOOD FLOW

2.1 INTRODUCTION

Although Doppler ultrasound may be used for the study of various types of motion within the body, its major use remains the detection and quantification of flow in the heart, arteries and veins. Doppler signals from these sources contain a great deal of information about flow, but before these signals can be interpreted it is essential to understand the basics of haemodynamics.

Blood flow in arteries is complex. The flow is pulsatile, the blood is an inhomogeneous non-Newtonian fluid, and the tethered viscoelastic arteries branch, curve and taper. The disease process only adds further complication. Despite this, useful insight into the way in which blood flows may be gained from the study of relatively simple models.

In this chapter the major emphasis is on blood flow and blood velocity profiles as these have such an important bearing on the Doppler signal, but flow and pressure are inseparably linked and the latter cannot be completely ignored. Indeed the detection or measurement of blood velocity may allow pressure information to be derived as in the case of the quantification of the pressure drop across a stenosed heart valve (Sections 2.6.1 and 16.3.2.1) or the inference of pressure changes from changes in pulse wave velocity (Section 2.5.1).

2.2 BASIC CONCEPTS AND DEFINITIONS

2.2.1 Viscosity

When one layer of a fluid moves with respect to an adjacent layer a frictional force arises between them due to viscosity. Put more formally, viscosity is that property of a fluid whereby it offers resistance to shear. Treacle and engine oils are examples of highly viscous fluids, while water and ethanol are examples of liquids with relatively low viscosities.

For most simple fluids (those known as Newtonian fluids) viscosity is independent of shear rate (i.e. the rate at which the adjacent layers slip over each other), but many fluids exhibit a change in viscosity with shear rate. Blood is not a simple fluid, it is essentially a suspension of red blood cells in plasma, and exhibits anomalous viscous properties at low shear rates (where it appears to have a finite yield stress), and in small tubes (where the size of the blood cells becomes significant in comparison to the size of the tube). Fortunately neither effect is really significant in blood vessels with diameters of 0.1 mm or more and it is possible to make use of the 'asymptotic' viscosity (the value to which the apparent viscosity tends at high shear rates) in calculations of blood flow in major arteries. The asymptotic viscosity of blood is dependent on both haematocrit and temperature, but is roughly four times as great as that of pure water (i.e. 0.04 P (poise) or 0.004 kg m^{-1} s^{-1}).

In addition to the 'absolute' or 'dynamic' viscosity introduced above, the 'kinematic viscosity', which is equal to absolute viscosity divided by density, is often used in equations describing fluid behaviour. The kinematic velocity of blood is approximately 0.038 St (stokes) or 3.8×10^{-6} m^2 s^{-1}.

The viscous behaviour of blood has been reveiwed in detail by McDonald (1974a).

2.2.2 Laminar, turbulent and disturbed flow

There are two distinctly different types of fluid flow. The first type is known as laminar or streamline flow, the second as turbulent flow. In laminar flow the fluid particles move along smooth paths in layers (or laminae) with every layer sliding smoothly over its neighbour. Laminar flow becomes unstable at high velocities (see Section 2.3.3) and breaks down into turbulent flow. In turbulent flow the particles follow very irregular and erratic paths, their velocity vectors varying continually both in magnitude and direction. The term turbulence is often rather loosely used in the vascular literature to indicate any non-laminar flow, but vortices and irregular movements of large bodies of fluid which can be traced to some obvious source of disturbance do not constitute turbulence, and are properly referred to as disturbed flow.

2.2.3 Steady flow

A steady flow is one in which all conditions (such as velocity and pressure) at any point in a stream remain constant with respect to time. Strictly therefore turbulent flow may never be described as steady as there are continual fluctuations in both velocity and pressure at every point. However, the definition is usually expanded to include flow in which the conditions fluctuate equally on both sides of a constant average value.

2.3 STEADY FLOW IN RIGID TUBES

The steady flow of Newtonian fluids in rigid tubes is well understood and serves as a convenient starting point in the discussion of blood flow in arteries. It is also relevant because pulsatile flow can be considered to be the sum of a steady component and a number of oscillatory components which interact neither with each other nor with the steady component. (This assumes that the system is linear, a concept which will be reintroduced in Section 2.4.1.)

2.3.1 Poiseuille flow

Established steady laminar flow in long cylindrical pipes is sometimes referred to as Poiseuille flow. In this, most basic, type of flow the fluid (blood) moves in a series of concentric shells such that the velocity profile across the vessel is parabolic, with the blood in the centre of the vessel moving most rapidly and the blood in contact with the vessel wall not at all.

The velocity of any lamina at radius r from the centre of the vessel may be written:

$$v(r) = v_{max}(1 - r^2/R^2) \qquad 2.1$$

where v_{max} is the velocity of the centre stream and R the radius of the vessel. This type of flow is illustrated in Fig. 2.1a.

> An interesting feature of parabolic flow is that the number of blood cells with velocities in the range v to $v + \delta v$ is independent of v. This is because, although dv/dr is proportional to r, the cross-sectional area of equally spaced laminae also rises in proportion to the distance from the vessel axis, and these two effects exactly cancel. The velocity distribution histogram corresponding to the profile shown in Fig. 2.1a is illustrated in Fig. 2.1b.

The mean velocity is of special importance because when multiplied by the cross-sectional area of the vessel it yields the volumetric flow, and it follows from the shape of the velocity histogram illustrated in Fig. 2.1b that the mean velocity in a vessel containing a parabolic velocity profile is exactly one half of the maximum (centre stream) velocity.

> This can be shown more formally by integrating the velocity given by eqn 2.1 over the vessel and dividing by the total area:
>
> $$\bar{v} = 1/\pi R^2 \int_{r=0}^{R} v_{max}\,(1 - r^2/R^2) \times 2\pi r\ dr \qquad 2.2$$
>
> i.e.
>
> $$\bar{v} = v_{max}/2 \qquad 2.3$$

The relationship between flow and pressure drop in a section of tube containing steady parabolic lamina flow is particularly simple and is named after Poiseuille in honour of his detailed measurements of flow through capillary tubes published in 1840. In essence Poiseuille's law states that the pressure drop across a section of tube; Δp, is proportional to the volumetric flow, the fluid viscosity, and the length of the tube, and is inversely

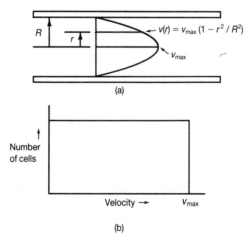

Figure 2.1   (a) Parabolic velocity profile found in steady laminar flow in a long cylindrical pipe. (b) Histogram illustrating the number of cells moving with a given velocity in the profile shown in (a). From this it follows that for this type of profile the maximum velocity is twice the mean velocity.

proportional to the fourth power of the radius. Mathematically it may be expressed as:

$$\Delta p = 8\mu Q L/\pi R^4 \qquad\qquad 2.4$$

where μ is the (absolute) viscosity, Q the volumetric flow rate, L the length of tube, and R its radius.

2.3.2 Entrance effects and the inlet length (steady laminar flow)

When flow enters a tube it must travel for some distance before it achieves its steady-state velocity profile. This is because the viscous drag exerted by the walls of the tube can only be transmitted to the central part of the tube by a progressive growth of the region of shear stress or boundary layer. This process is illustrated in Fig. 2.2.

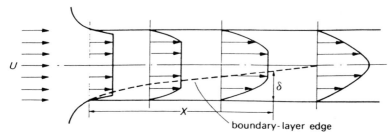

Figure 2.2   The development of the steady-state parabolic velocity profile at the entrance to a tube. The initially flat profile becomes progressively modified as the boundary layer grows with distance from the inlet. (Reprinted with permission from Caro et al (1978), Oxford University Press.)

Initially only the fluid directly in contact with the wall is stationary, and a large velocity gradient is present at this point. This gradient becomes reduced as more of the fluid in the core becomes sheared. Since the volumetric flow through any cross-section of the pipe must be the same, the fluid at the centre of the tube must accelerate to compensate for the deceleration at the periphery. The boundary layer thickness, δ, is initially proportional to the square root of the distance from the entrance of the tube, x, and may be written:

$$\delta \propto (vx/U)^{\frac{1}{2}} \qquad\qquad 2.5$$

where v is the kinematic viscosity and U is the free stream velocity (i.e. the velocity of the unsheared central core) (Caro et al 1978).

The distance required for the profile to achieve its final steady-state shape is known as the inlet length, and has been found experimentally to be approximately:

$$X = 0.03D(Re) \qquad\qquad 2.6$$

where D is the diameter of the pipe and Re the Reynolds number.

Because of the additional energy required to accelerate the flow in the inlet length, the pressure gradient in this region is much greater than where the flow is fully established.

2.3.3 Turbulent flow

Laminar flow becomes unstable at high velocities and breaks down into turbulent flow.

The point at which the transition between the two flow regimes occurs cannot be predicted exactly, but is largely determined by the Reynolds number, Re, which for a circular pipe is defined as:

$$Re = \bar{v}D/v \qquad 2.7$$

where $\bar{v}$ is the average velocity of flow across the pipe, D the diameter of the pipe, and v the kinematic viscosity. The critical Reynolds number, Re_{crit} (i.e. the Re at which the flow becomes turbulent) can, depending on geometry and the extent of upstream disturbances, be anywhere between about 2000 and 50 000. For values of less than 2000 it is practically impossible for turbulent flow to persist in a straight smooth pipe, and for values of greater than 10 000 the flow is inherently unstable, and the least disturbance will transform it into turbulent flow immediately. The critical Reynolds number is affected by vessel geometry and is greater in converging vessels, and less in curved, diverging, and rough vessels where it may be as low as 1000. For most practical purposes Re_{crit} is taken to be between 2000 and 2500.

Both the average velocity profile and the pressure gradient observed in turbulent flow differ from those in laminar flow. (Because of the erratic paths followed by individual particles it is necessary to distinguish between the instantaneous velocity profile and the average velocity profile even in 'steady' turbulent flow.) The average velocity profile found in turbulent flow is much flatter than that in steady laminar flow, so that the maximum velocity at the centre of the vessel is only about 20% higher than the average velocity (this compares with a value of 100% for laminar flow). The pressure gradient in a given tube is also much larger, and the pressure drop is proportional not to the flow rate but rather to the flow rate squared.

Fortunately, from the standpoint of the interpretation of Doppler signals, turbulence is rarely seen in the healthy circulation (although it is possible that the flow in the proximal aorta can sometimes be turbulent), and its very presence is usually a clue that an abnormality is present.

2.3.4   Entrance effects (turbulent flow)

The entrance length for turbulent flow is less than that for laminar flow and may be estimated from eqn 2.8 (Caro et al 1978):

$$X_{turbulent} = 0.69D(Re)^{\frac{1}{4}}. \qquad 2.8$$

2.4   PULSATILE FLOW IN RIGID TUBES

So far we have considered only steady flow. Arterial blood flow is pulsatile and in this section we show how such a flow may be analysed in terms of a steady flow together with a number of superimposed sinusoidal components.

2.4.1   Fourier analysis

In essence Fourier's theorem states that any periodic waveform may be broken down into (and resynthesized from) a series of sinusoidal waveforms with frequencies that are integral multiples of the repetition frequency of the original waveform. For any given waveform the amplitudes and phases of the sinusoidal components are unique and can be simply calculated. The way in which a waveform may be reconstructed from its Fourier components is illustrated in Fig. 2.3. The reason why Fourier analysis is so important is that the response of most physical systems to a signal of a single frequency is independent of the presence and magnitude of any other frequency components. Such systems are known as linear systems because a change in the size of the input signal is accompanied by the same proportional change in the output signal.

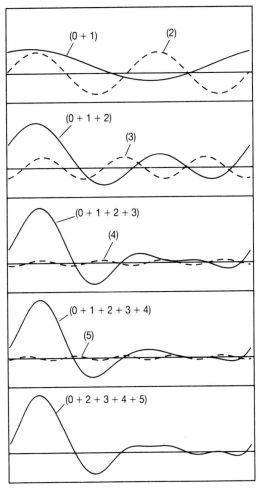

Figure 2.3 Synthesis of a common femoral flow waveform from its Fourier components. The numbers in the brackets refer to the harmonics used to construct each of the waveforms. The coefficients used to synthesize this waveform can be found in the first half of Table 2.2.

Thus, for example, if the relationship between the blood pressure gradient and blood flow is a linear one (as we assume in normal peripheral arteries) and we know the relationship between the pressure gradient and flow at any given frequency, we can predict the flow due to a complex pressure gradient in three simple steps. Firstly the pressure gradient is split up into its sinusoidal (Fourier) components; secondly the flow component (both its amplitude and phase) at each frequency is calculated from the pressure gradient at the same frequency; thirdly the derived flow components are added together to give the complete flow waveform. In Section 2.4.3 an analogous method will be used to derive the velocity profiles seen in pulsatile flow.

Mathematically the Fourier series for a periodic function $F(t)$ may be written as

$$F(t) = F_0 + F_1 \cos(\omega t - \phi_1) + F_2 \cos(2\omega t - \phi_2) + \cdots \qquad 2.9$$

or

$$F(t) = F_0 + \sum_{p=1}^{\infty} F_p \cos(p\omega t - \phi_p) \qquad 2.10$$

F_0 is the amplitude of the mean term (or the zeroth component), and F_p and ϕ_p the amplitudes and phases of the pth components. ω is the angular frequency and is equal to $2\pi f$ where f is the frequency of repetition of the waveform. The coefficients F_p and ϕ_p may be calculated using the following equations and there are now many computer routines which will do this swiftly and accurately.

$$F_p = (a_p^2 + b_p^2)^{\frac{1}{2}} \qquad 2.11$$

$$\phi_p = -\tan^{-1}(b_p/a_p) \qquad 2.12$$

where

$$a_p = 2/T \int_0^T F(t)\cos(p\omega t)\,\mathrm{d}t \qquad 2.13$$

and

$$b_p = 2/T \int_0^T F(t)\sin(p\omega t)\,\mathrm{d}t \qquad 2.14$$

2.4.2 Sinusoidal flow

There are two important aspects of sinusoidal flow to consider. The first is the relationship between pressure gradient and volumetric flow (i.e. the sinusoidal version of Poiseuille's law), and the second the relationship between volumetric flow and velocity profile. For both these it is necessary to refer to the results obtained by Womersley (1955). Womersley showed that Poiseuille's equation (eqn 2.4) can be modified to describe the relationship between pulsatile flow and pressure, and can be written thus:

$$Q = \frac{M'_{10}}{\alpha^2} \cdot \frac{\pi R^4}{\mu L} \cdot \Delta P \sin(\omega t - \phi + \varepsilon'_{10}) \qquad 2.15$$

where the pressure difference across the section of tube under consideration is $\Delta P \cos(\omega t - \phi)$. M'_{10}/α^2 and ε'_{10} are both functions of the non-dimensional parameter, α:

$$\alpha = R(\omega/v)^{\frac{1}{2}} \qquad 2.16$$

(where v is the kinematic viscosity) which characterizes kinematic similarities in liquid motion. The definitions of M'_{10} and ε'_{10} are given in the appendix (eqns A.1 and A.2) and plotted in Fig. 2.4. It can be seen from this figure that M'_{10}/α^2 tends to 1/8 and ε'_{10} tends to $90°$ when ω tends to zero, and therefore eqn 2.15 reduces to eqn 2.4 for very low values of α. As α increases (i.e. the frequency increases) and inertial effects tend to dominate viscous effects, the amplitude of flow oscillations in response to given pressure oscillations become smaller, and lag further and further behind the pressure gradient, and the artery effectively acts as a low pass filter. Some typical values of α found in the human circulation are given in Table 2.1.

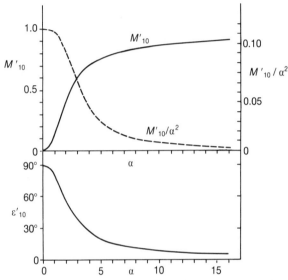

Figure 2.4   Womersley's parameters M'_{10}, M'_{10}/α^2, and ε'_{10} plotted as a function of α. (Reprinted with permission from Milnor (1982), © The Williams & Wilkins Co, Baltimore.)

Perhaps more important than the pressure–flow relationship is the flow–velocity profile relationship, since it is the velocity profile that will determine the Doppler signal we detect from the artery. We saw that for steady flow in a long rigid tube the velocity profile is parabolic (eqn 2.1). For oscillatory flow, however, the profile is a function of the parameter α. By manipulating Womersley's equations it is possible to show (Evans 1982) that the velocity profile resulting from a sinusoidal flow $Q\cos(\omega t - \phi)$ is given by:

$$U(y) = \frac{1}{\pi R^2} \cdot Q|\psi|\cos(\omega t - \phi + \chi) \qquad 2.17$$

where y is the non-dimensional radial coordinate, r/R, and $|\psi|$ and χ are the amplitude and phase of a complex function ψ, which is defined in the appendix (eqn A.5). ψ is a function of both y and α, and its modulus and phase are plotted in Fig. 2.5. For very low values of α the velocity profile is parabolic and all the laminae move with the same phase, but as α becomes large the profile becomes quite blunt and the motion of the

Table 2.1   Some typical values of haemodynamic parameters measured from or calculated for the human circulation

Vessel	Diameter (mm)	Flow (ml min^{-1})	Velocity (cm s^{-1})	Reynolds number	Fundamental α	Entrance length (cm)
Ascending aorta	(31)	6400†	18‡	1500‡	21‡	(140)
Abdominal aorta	18*	(2000)	14‡	640‡	12‡	(34)
Common carotid	5.9§	387§	14§	(217)	(4.2)	(3.8)
Renal artery	(6.2)	(725)	40‡	700‡	4‡	(13)
External iliac	8.2*	380¶	(12)	(260)	(5.8)	(5.0)
Superficial femoral	6.4*	150¶	12‡	200‡	4‡	(3.8)
Posterior tibial	3.8*	10¶	3.5¶	(35)	(2.7)	(0.4)

Values in brackets are derived values. Wherever possible measured figures are quoted, and therefore values for the same site are not always totally consistent with each other.
* Callum et al (1983).
† Ganong (1971).
‡ Milnor (1982).
§ Payen et al (1982).
¶ Strandness and Sumner (1975).

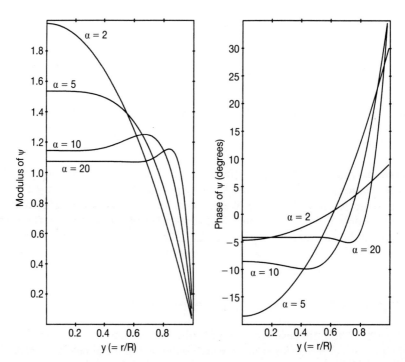

Figure 2.5   The modulus and phase of the function ψ which relates the shape of the velocity profile to sinusoidal laminar flow, plotted for α values of 2, 5, 10 and 20. (Reprinted with permission from Evans (1982), © Pergamon Journals Ltd.)

outside laminae leads that of those nearer the centre. This can be seen in Fig. 2.6 which shows the velocity profiles that develop in a tube for a variety of values of α. Note that these shapes differ from those in Fig. 2.5a because of the phase differences across the vessel.

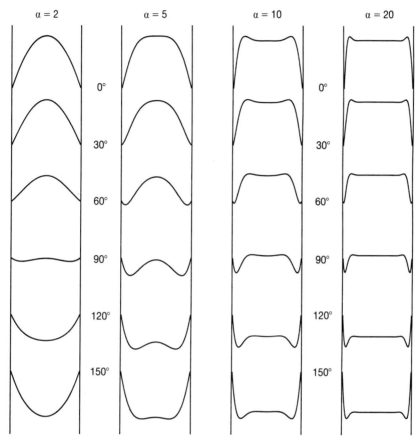

Figure 2.6   The velocity profiles that occur during sinusoidal flow with α values of 2, 5, 10 and 20 plotted at 30° intervals. Each series has been normalized to have the same maximum velocity at 0°, but note that in reality the flow oscillations in response to a given pressure oscillation become smaller as frequency increases. Only half of the cycle has been illustrated as the profiles in the other half have the same shape but are of opposite sign. These shapes differ from those in Fig. 2.5a because of the phase differences across the vessel. The wall velocities are always zero.

2.4.3   Pulsatile flow

We have discussed steady flow and sinusoidal flow and the way in which Fourier's theorem allows us to treat pulsatile flow as a combination of a steady flow and a series of sinusoidal components. These principles are now used to calculate the types of velocity profiles to be found in healthy common femoral and common carotid arteries in man. (We shall assume that entrance effects can be ignored.)

The starting points for these calculations are the mean velocity waveforms illustrated in Fig. 2.7. These waveforms were actually derived from ultrasound Doppler signals. The first step is to express the velocity signals as Fourier series:

$$V(t) = V_0 + \sum_{p=1}^{\infty} V_p \cos(p\omega t - \phi_p).$$
2.18

The values of V_0, V_p and ϕ_p, $p=1$ to 8, which have been calculated from the waveforms in Fig. 2.7 are given in Table 2.2. It is also necessary to find the values of α corresponding to each of the harmonics and these have been calculated from eqn 2.16 and are also to be found in Table 2.2. The shape of the velocity profile at any phase of the cardiac cycle may now be calculated by summing the contributions from the velocity profiles arising from each harmonic, i.e.

$$U(y) = \left\{ 2V_0(1 - y^2) + \sum_{p=1}^{\infty} V_p |\psi|_p \cos(p\omega t - \phi_p + \chi_p) \right\}$$
2.19

In practice the magnitude of V_p decreases rapidly with increasing p and the summation can be truncated after only a few terms.

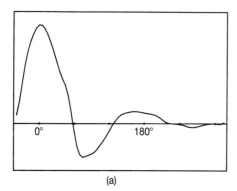
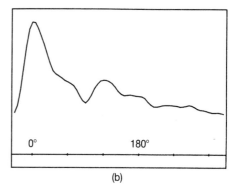

(a)   (b)

Figure 2.7 Mean velocity waveforms recorded from (a) the common femoral and (b) the common carotid arteries of a normal volunteer. The scale of the y-axes is arbitrary. The Fourier components of these waveforms are given in Table 2.2 and the calculated velocity profiles in Fig. 2.8.

The results obtained from this calculation are shown in Fig. 2.8. The velocity profiles that develop in the two arteries are clearly very different from each other and this will have a major influence on the type of Doppler signal that will be obtained from the two arteries and has implications for sampling and signal processing techniques. These velocity profiles will be discussed further in Chapter 8, and further information on velocity profiles in human arteries is to be found in Section 2.7.

2.4.4 Entrance effects in pulsatile flow

The mean velocity profile in pulsatile flow develops, at the entrance to a tube, just as it does for steady flow (Section 2.3.2). It will, however, have the pulsatile components of the profile superimposed on it, and these too require some distance to develop fully. In

Table 2.2   Fourier components and corresponding values of the non-dimensional parameter α for flow waveforms recorded from the common femoral and common carotid arteries of a healthy young subject. The values of V_p have been normalized to V_0, and the angle ϕ_p is given in degrees from an arbitrary starting point

Common femoral Diameter* = 8.4 mm Heart rate† = 62 bpm Viscosity = 0.038 St					Common carotid Diameter* = 6.0 mm Heart rate† = 62 bpm Viscosity = 0.038 St				
Harmonic	Frequency	α	V_p	ϕ_p	Harmonic	Frequency	α	V_p	ϕ_p
0	—	—	1.00	—	0	—	—	1.00	—
1	1.03	5.5	1.89	32	1	1.03	3.9	0.33	74
2	2.05	7.7	2.49	85	2	2.05	5.5	0.24	79
3	3.08	9.5	1.28	156	3	3.08	6.8	0.24	121
4	4.10	10.9	0.32	193	4	4.10	7.8	0.12	146
5	5.13	12.2	0.27	133	5	5.13	8.7	0.11	147
6	6.15	13.4	0.32	155	6	6.15	9.6	0.13	179
7	7.18	14.5	0.28	195	7	7.18	10.3	0.06	233
8	8.21	15.5	0.01	310	8	8.21	12.4	0.04	218

* Estimated values.
† These waveforms were recorded from a healthy young subject with a relatively low heart rate.

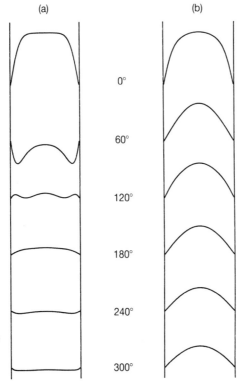

Figure 2.8   Velocity profiles for (a) a common femoral artery and (b) a common carotid artery, calculated from the mean waveforms shown in Fig. 2.7, plotted at 60° intervals. The two sets of profiles have been normalized to have the same maximum velocity at 0°.

general, however, the pulsatile components become fully established much more rapidly, and the unsteady entrance length $X_{unsteady}$ may be written:

$$X_{unsteady} \simeq 3.4 v_{core}/\omega \qquad\qquad 2.20$$

where v_{core} is the core velocity (Caro et al 1978).

2.5   PULSATILE FLOW IN ELASTIC AND VISCOELASTIC TUBES

A rigid tube model is adequate under most circumstances for predicting the velocity profiles which develop within arteries, and to a certain extent for predicting the relationship between pressure gradient and flow. However, it cannot give any information about arterial pulsations or the velocity of the pulse wave, and therefore it is necessary to develop other models to describe these aspects of haemodynamics.

2.5.1   Pulse wave velocity

It is important to distinguish between two types of velocity in the artery. The first is the velocity of the blood itself, the second (the pulse wave velocity, PWV) the velocity with which the blood flow waveform propagates. The distinction is particularly easy to envisage in the flow of an incompressible fluid through a rigid pipe. If a force is applied to the fluid at the entrance to the tube it will move with a finite velocity; however, the fluid at every point along the tube will also start to move with the same velocity at exactly the same time and the PWV will therefore be infinite. The PWV is finite in the case of elastic tubes but is still considerably higher than the blood flow velocity.

A further point to consider is exactly what is meant by PWV. If the pressure (or flow) wave did not alter in shape as it propagated this would be easy; it would simply be the distance travelled by any one feature of the waveform in unit time. In the arterial system, however, the pulse wave changes its shape as it propagates due to the effects of wave reflection and because both attenuation and (phase) velocity are frequency dependent. It has been usual in the past to measure the velocity of the 'foot' of the wave (the point at the end of diastole where the steep rise of the wavefront begins) since this point is easy to identify and is little affected by reflections, but this is a question of expediency.

The equation that predicts the velocity of pulse propagation in a thin-walled elastic tube filled with a liquid that is incompressible and has no viscosity was derived over a century ago and is known as the Moens–Korteweg equation after two Dutch scientists. The wave velocity c_0 is given by:

$$c_0 = (Eh/\rho D)^{\frac{1}{2}} \qquad\qquad 2.21$$

where E is Young's modulus of elasticity in the circumferential direction, h is the wall thickness, ρ is the density of the liquid, and D is the diameter of the tube. There have been a number of more sophisticated equations derived which take account of such parameters as the viscosity of the fluid, but eqn 2.21 is remarkably good at predicting c_0, and produces theoretical values that are within 15% of those found by experiment (Caro et al 1978).

A point of particular interest from eqn 2.21 is that the PWV is proportional to the square root of Young's modulus or elasticity. It is possible to measure PWV using a

Doppler technique, and thus to follow changes in E. Furthermore, E is affected by blood pressure and thus blood pressure changes could be monitored by observing changes in PWV.

2.5.2 Arterial pulsations

Radial arterial pulsations are of relevance to Doppler studies in several respects, and are in general sources of annoyance. The arterial wall/blood interface is a comparatively strong reflector of ultrasound and gives rise to high-amplitude low-frequency Doppler signals known as 'wall thump'. The changes in diameter during the cardiac cycle also introduce an additional source of error into the ultrasonic measurement of blood flow (Section 11.2.3), and make it difficult to examine the local blood flow close to an arterial wall without some type of tracking device.

The radial dilatation is mathematically difficult to predict exactly because it is affected by the exact characteristics of the vessel wall, but Womersley (1957) showed that, in an elastic tube with very stiff longitudinal tethering, the total excursion of the wall 2ξ can be written:

$$2\xi = R(1 - \sigma^2)(Pe^{i\omega t}/\rho c_0^2) \qquad 2.22$$

where R is the tube radius, σ Poisson's ratio (normally taken to be 0.5), $Pe^{i\omega t}$ the pressure wave, ρ the density of blood, and c_0 the Moens–Korteweg velocity. In this case the dilatation is exactly in phase with the pressure wave, but if the wall has other properties the wall movement may lead the pressure wave (for example if the constraint is less stiff) or lag behind it (if a highly constrained wall also has viscous properties).

2.6 THE EFFECTS OF GEOMETRIC CHANGES

Each curve, branch and constriction in the arterial system will modify both the velocity profile and the local pressure gradient.

2.6.1 Constrictions and projections

The effects of constrictions and projections on blood flow are particularly important because Doppler ultrasound is often used to identify or quantify one of these entities. The mechanism of pressure loss across arterial stenoses is complicated and beyond the scope of this chapter, and the reader is referred to a series of papers by D.F. Young and his colleagues (Young and Tsai 1973a, 1973b, Young et al 1975, Seeley and Young 1976, Young et al 1977) for an introduction to the problem. In general, however, the losses are made up of three terms and may be written:

$$\Delta P = \Delta P_s + \Delta P_c + \Delta P_e \qquad 2.23$$

where ΔP_s is the pressure drop caused by the resistance of the stenosis (and predicted by Poiseuille's law if the flow is laminar), ΔP_c is a result of energy lost due to the contraction of the stream, and ΔP_e is the result of the post-stenotic expansion. The relative sizes of these three terms depend heavily on the geometry of the stenosis, but one particularly simple case is that of an orifice where only the final term, ΔP_e, is significant. The equation for the pressure drop across an orifice is of particular relevance because it can be used to estimate the pressure drop across a stenosed heart valve from Doppler velocity measurements.

The starting point for the derivation of the pressure drop is Bernoulli's equation for steady flow, which is an expression of conservation of hydraulic energy (energy per unit volume) in frictionless incompressible fluids, and which may be written:

$$\underset{\substack{\text{pressure} \\ \text{energy}}}{P} + \underset{\substack{\text{kinetic} \\ \text{energy}}}{\tfrac{1}{2}\rho v^2} + \underset{\substack{\text{gravitational} \\ \text{potential} \\ \text{energy}}}{\rho g z} = \text{constant} \qquad 2.24$$

where P is the pressure in the fluid, ρ its density, v the velocity of flow, g the acceleration due to gravity, and z the height relative to a given reference point. Applying this equation to the flow through an orifice we may write:

$$P_1 + \tfrac{1}{2}\rho v_1^2 = P_2 + \tfrac{1}{2}\rho v_2^2 \qquad 2.25$$

where the subscripts 1 and 2 refer to conditions before and at the orifice. The potential energy term has been omitted because z_1 and z_2 are assumed to be nearly equal. Collecting terms and writing the pressure drop as ΔP leads to:

$$\Delta P = \tfrac{1}{2}\rho(v_2^2 - v_1^2). \qquad 2.26$$

Finally if the stenosis is tight then $v_2^2 \gg v_1^2$ and the second term in the bracket may be ignored, and eqn 2.26 reduces to:

$$\Delta P \simeq 4v_2^2 \qquad 2.27$$

where ΔP is in mmHg and v_2 is in m s^{-1}.

Strictly, P_2 was the pressure within the orifice. However, because of the flow disturbances that occur downstream of a sudden expansion, very little of the kinetic energy gained during the contraction of the stream is converted back to pressure energy, but is dissipated as heat. Thus eqn 2.27 may be used to estimate the pressure drop across stenosed heart valves (Section 16.3.2.1).

For non-steady flow the Bernoulli equation has an additional term due to the energy that is required to accelerate the flow, and eqn 2.25 should be rewritten:

$$P_1 + \tfrac{1}{2}\rho v_1^2 = P_2 + \tfrac{1}{2}\rho v_2^2 + \rho \int_1^2 \frac{dv}{dt}\, ds \qquad 2.28$$

where s is the position along a streamline. The extra term only becomes important for large values of dv/dt, specifically during heart valve opening and closure. In practice this means there is a lag between velocity and pressure drop and that eqn 2.27 is not applicable to early systole; despite this it gives a clinically acceptable estimate of the pressure drop in both mitral and aortic stenoses (Hatle and Angelsen 1982).

It can be seen from eqn 2.27 that the relationship between flow and pressure drop across a stenosis where exit effects dominate is non-linear. This would not be the case for a stenosis where the pressure drop resulted purely from a Poiseuille type of loss, but in general the relationship between flow and pressure drop is non-linear and may be written in the form:

$$\Delta P = AQ + BQ^2 \qquad\qquad 2.29$$

where A and B are constants.

The effects of stenosis on the volumetric flow depend not only on the stenosis itself, but on the characteristics of the vascular bed proximal to and distal to the stenosis, but there is a tendency for tight stenoses to reduce both the mean flow and the pulsatility of the flow waveform. It is a common observation that the arterial lumen needs to be reduced by as much as 80% before the mean resting flow rate is affected (Mann et al 1938, Shipley and Gregg 1944, May et al 1963, Kindt and Youmans 1969, Eklof and Schwartz 1970), and therefore considerable effort has been directed towards the detection of subtle changes in the Doppler waveform shape caused by lesser degrees of narrowing (Chapter 10).

Doppler ultrasound may also be used to detect the flow disturbances caused by the encroachment of atheroma into the arterial lumen. If the obstacle projecting into the stream is a small one, small eddies will form but will soon die away. If the obstacle is larger, the eddies will grow in size and depending on the Reynolds number may cause a complete breakdown of laminar flow.

For a sharp-edged projection into the lumen of height h, laminar flow will only be maintained if:

$$h < 4r/Re \qquad\qquad 2.30$$

where r is the radius of the tube and Re the Reynolds number (Goldstein 1938, cited in McDonald 1974b).

Examples of the Doppler signal found distal to atheromatous plaques are to be found in Figs 10.2b and 10.2c. Flow disturbances caused by a stenosis may propagate for some distance but cannot usually be detected beyond 12 to 16 unstenosed diameters downstream (Clark 1980, Evans et al 1982).

2.6.2   Curves

When flow enters a curved section of a tube it experiences a centrifugal force, which is proportional to the velocity squared and inversely proportional to the radius of curvature. The resulting change in flow pattern depends on the velocity profile existing before the curve was entered. If the profile is parabolic, the fluid at the centre of the tube has the highest velocity and will experience the greatest force, and will therefore tend to move towards the outside wall of the tube. This means that the velocity profile becomes skewed towards this wall, and also that secondary flows are set up such that individual particles tend to follow helical paths (see Fig. 2.9). If the velocity profile is flat when the curve is entered, the velocity profile becomes skewed in the opposite direction (Fig. 2.10). This is because the pressure at the outside of the bend is larger than that at the inside (otherwise the fluid would not be forced around the bend), and therefore, because the total energy of a streamline cannot change, the kinetic energy (and hence the velocity) of the particles on the inside of the bend must be greater than those of particles on the outside.

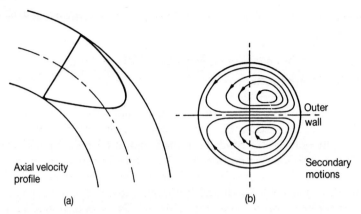

Figure 2.9   (a) Distortion of a parabolic velocity profile resulting from tube curvature. (b) Cross-section of the vessel showing the form of the secondary flow. (Reprinted with permission from Caro et al (1978), Oxford University Press.)

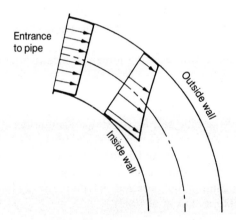

Figure 2.10   Distortion of a flat velocity profile resulting from tube curvature. (Reprinted with permission from Caro et al (1980), Oxford University Press.)

It is important to realize that these effects are non-linear and therefore the behaviour of a complex pulsatile flow waveform cannot be predicted by calculating the effects of the pulsatile components of flow separately and superimposing the answers thus obtained on steady flow profiles.

2.6.3   Branches

The arterial system branches repeatedly, and at each junction new velocity profiles must develop, and there may be changes in the mean velocity, Reynolds number and pressure gradient.

McDonald divided arterial junctions into three main classes: bifurcation junctions, fusion junctions and side branches. Fusion junctions are almost exclusively confined to the venous system, the only arterial example being the fusion of the two vertebral arteries into the basilar artery, but the other two types and various hybrids of the two are found throughout the body.

The effects on flow profile will depend on the exact geometry of the junction. At a pure symmetric bifurcation the flow is split into two and the largest velocities, which were in the centre of the stream of the parent vessel, will then be very close to the inside walls of the daughter vessels, leading to skewed velocity profiles.

At the same time, because each of the two streams must turn a bend, secondary motions of the type described in Section 2.6.2 are set up. These secondary motions sweep some of the fast-moving fluid on the inside of the daughter tubes round to the top and bottom walls, leading to an M-shaped velocity profile in the plane perpendicular to the junction (Caro et al 1978).

Figure 2.11 illustrates the types of profile found two diameters downstream of a symmetric bifurcation. These disturbed profiles gradually return to a more normal shape as viscous effects exert themselves, and a steady-state profile is reached at a distance comparable with the usual entrance length (eqn 2.6).

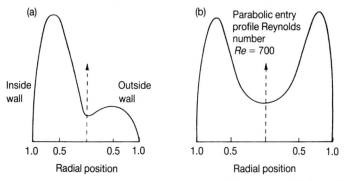

Figure 2.11   Velocity profiles two diameters downstream of a symmetric bifurcation, at a parent tube Reynolds number of 700. (a) Profile in the plane of the junction; (b) profile in the perpendicular plane. (Reprinted with permission from Schroter and Sudlow (1969) Resp Physiol 7, 341–355, Elsevier Science Publishers BV.)

There are so many configurations for side branches leaving a parent vessel that it is difficult to generalize on the effects of this type of junction. The type of flow found in a junction when a daughter tube branches off a parent tube at right angles is illustrated in Fig. 2.12. There is flow separation both at the inner wall of the daughter tube and on the wall of the main tube opposite and just downstream of the branch. Strong secondary motions are set up in the daughter tube due mainly to the acute bend, and some distance is required before a steady-state profile is established.

In general the total cross-sectional area of branched vessels exceeds that of the parent vessel and therefore both the mean velocity of flow and the Reynolds number reduce as vessels branch, and this leads to a greater stability in flow.

2.6.4 Tapers

All blood vessels, with the exception of capillaries, show some degree of both geometrical

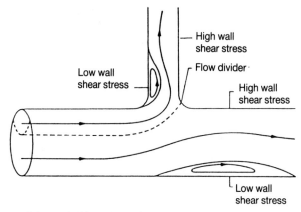

Figure 2.12   Flow at right angled junction. The dashed line is the surface dividing the fluid which flows down the side branch from that continuing in the main tube; the solid lines are stream lines. Note the closed eddies in the two regions of separated flow. Regions of high and low shear are also indicated. (Reprinted with permission from Caro et al (1978), Oxford University Press.)

and elastic taper (that is, the elastic modulus of the vessel changes progressively with distance). The major effects of the convergence found in the arterial system is to stabilize the laminar flow, raise the critical Reynolds number, and slightly flatten the velocity profile (Whitmore 1968).

2.7   VELOCITY PROFILES IN HUMAN ARTERIES

Experimental studies of velocity profiles in arteries have been marred by the difficulties of making point measurements in anything but the largest of arteries, and the majority of published results relate to the profiles found in canine aortas. These results may, to some extent, be extrapolated to man, but for other smaller arteries the profiles that develop must be inferred from the theories outlined earlier in this chapter. Hopefully this is a situation that will be rectified as pulsed Doppler techniques become more sophisticated.

Figure 2.13 illustrates the time-averaged profiles (normalized to the centreline time-averaged velocity) at various positions in the canine aorta. As predicted by the theory discussed in Section 2.3.2, the profile starts off very blunt and slowly becomes more rounded as flow progresses down the aorta. There is also a slight asymmetry across the vessel diameter which is much more marked for the velocity profiles at peak systole (Fig. 2.14). This asymmetry is thought to be a result of both the tight curve of the aortic arch and the consequent lateral pressure gradient (Section 2.6.2), and to be due to some extent to the local distortion caused by the major branches feeding the head and arms.

For the human aorta both the diameter and the mean Reynolds number are greater than those found in the dog, and therefore the steady-state profile will take longer to develop (eqn 2.6), and even though the human aorta is longer the profiles will be less developed at corresponding anatomical points. Substituting approximate values for diameter (~ 3.1 cm) and mean Reynolds number (~ 1500) into eqn 2.6 gives an entrance length of approximately 1.4 m, which suggests that the whole of the human aorta must be regarded as an entrance region. As the arterial system branches both the diameter and the Reynolds number fall, and each new junction results in new inlet phenomena. In some

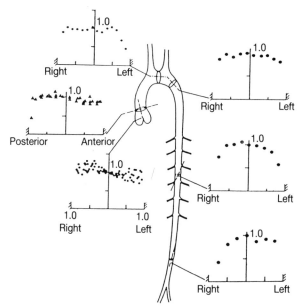

Figure 2.13   Time-averaged velocity profiles (normalized to the centreline time-averaged velocity) measured at various positions in the canine aorta. Note how the profiles become progressively more rounded as the measurement site moves more distal. (Reprinted with permission from Schultz (1972) Pressure and flow in large arteries. In: Cardiovascular fluid dynamics, Vol 1 (Ed. DH Bergel), Academic Press.)

arteries (such as the renal arteries) a steady-state mean profile can never be established, but in others (for example the superficial femoral artery) entrance effects are rapidly dispelled. Some entrance lengths for various human arteries, estimated from arterial dimensions and blood flow values quoted by a number of authors, are given in Table 2.1. In children the smaller values of Re and arterial diameter will result in a much more rapid achievement of steady-state conditions.

Because of the role of flow disturbances in atherogenesis, special interest has recently been focused on both the coronary (Batten and Nerem 1982) and carotid (Ku et al 1985) circulations, and it is to be hoped that these and similar studies will lead to a thorough understanding of flow in human arteries.

2.8   SUMMARY

Blood flow in arteries is complex but, despite this, useful approximations can be made using fairly simple models. The flow in the normal circulation is for the most part both laminar and pulsatile, and may be considered the sum of a steady flow component and a series of oscillatory components. Once established, the mean flow component results in a parabolic profile, whilst the pulsatile components cause oscillating profiles which vary between parabolic and blunt (but with phase differences across the vessel diameter) depending on the value of Womersley's parameter α. At each branch and bifurcation the velocity profiles are disturbed and the return to steady-state 'established' flow requires a finite distance. Each branch and curve also gives rise to secondary flows which also persist

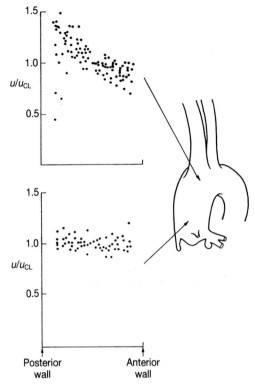

Figure 2.14   Peak systolic velocities across two diameters in the ascending aorta of the dog. Note that the asymmetry across the vessel diameter is much more marked than for the time-averaged profiles (Fig. 2.13). (Reprinted with permission from Caro et al (1978), Oxford University Press.)

for a finite distance. Constrictions of, and projections into, the arterial lumen cause energy losses over and above simple Poiseuille losses and may cause disturbed or even turbulent flow conditions.

In this chapter it has only been possible to consider those aspects of blood flow that are most relevant to the recording of a Doppler signal. Much more detailed information on these and other facets of blood flow are to be found in the books recommended in Section 2.11.

2.9   NOTATION

c_0	Moens–Korteweg velocity
D	Diameter of tube or artery
E	Young's modulus of elasticity
f	Frequency
$F(t)$	A periodic function in time
F_0	The mean value of $F(t)$
F_p	The amplitude of the pth Fourier component of $F(t)$
g	Acceleration due to gravity
h	Wall thickness

L	Length of a tube
M'_{10}	Amplitude of a special function arising from Womersley's theory (see the appendix)
P	Pressure
$Pe^{i\omega t}$	Pressure wave
Q	Volumetric flow
r	Distance from the centre of the tube
R	Tube radius
Re	Reynold's number
s	Position along a streamline
t	Time
U	Free stream velocity
$v(r)$	Velocity of lamina at radius r
$v(y)$	Velocity of lamina at radial coordinate r/R
v_{max}	Velocity of central lamina
$\bar{v}$	Spatial mean velocity
v_{core}	Core velocity
$V(t)$	Periodic mean velocity signal
V_0	Mean component of $V(t)$
V_p	Amplitude of pth harmonic of $V(t)$
x	Distance from the tube entrance
X	Entrance length
y	Non-dimensional radial coordinate, r/R
z	Height above a reference point
α	Womersley's parameter characterizing kinematic behaviour
δ	Boundary layer thickness
ΔP	Pressure drop across length of tube
ε'_{10}	Phase of a special function arising from Womersley's theory (see the appendix)
μ	Absolute viscosity
ν	Kinematic viscosity
ξ	Wall movement
ρ	Density
σ	Poisson's ratio
ϕ_p	The phase of the pth harmonic of a Fourier series
χ	Phase of a special function relating velocity to flow
$\lvert\psi\rvert$	Amplitude of a special function relating velocity to flow
ω	Angular frequency

2.10   REFERENCES

Batten JR, Nerem RM (1982) Model study of flow in curved and planar arterial bifurcations. Cardiovasc Res 16, 178–186.

Callum KG, Thomas, ML, Browse NL (1983) A definition of arteriomegaly and the size of arteries supplying the lower limbs. Br J Surg 70, 524–529.

Caro CG, Pedley TJ, Schroter RC, Seed WA (1978) The mechanics of the circulation, Oxford University Press, Oxford.

Clark C (1980) The propagation of turbulence produced by a stenosis. J Biomech 13, 591–604.

Eklof B, Schwartz SI (1970) Critical stenosis of the carotid artery in the dog. Scand J Clin Lab Invest 25, 349–353.

Evans DH (1982) Some aspects of the relationship between instantaneous volumetric blood flow and continuous wave Doppler ultrasound recordings. III. Ultrasound Med Biol 8, 617–623.

Evans DH, Macpherson DS, Asher MJ, Bentley S, Bell PRF (1982) Changes in Doppler ultrasound sonograms at varying distances from stenoses. Cardiovasc Res 16, 631–636.

Ganong WF (1971) Review of medical physiology. Lange Medical Publications, Los Altos, California.

Goldstein S (1938) Modern developments in fluid dynamics, Chapter 7, pp 142, Clarendon Press, Oxford.

Hatle L, Angelsen B (1982) Doppler ultrasound in cardiology – Physical principles and clinical applications, pp 24–26, Lea and Febiger, Philadelphia.

Kindt GW, Youmans JR (1969) The effect of stricture length on critical arterial stenosis. Surg Gynecol Obstet 128, 729–734.

Ku DN, Giddens DP, Phillips DJ, Strandness DE (1985) Hemodynamics of the normal human carotid bifurcation: in vitro and in vivo studies. Ultrasound Med Biol 11, 13–26.

Mann FC, Herrick JF, Essex HE, Baldes EJ (1938) The effect on the blood flow of decreasing the lumen of a blood vessel. Surgery 4, 249–252.

May AG, Van de Berg L, DeWeese JA, Rob CG (1963) Critical arterial stenosis. Surgery 54, 250–259.

McDonald DA (1974a) The viscous properties of blood. In: Blood flow in arteries, Chapter 3, pp 55–70, Edward Arnold, London.

McDonald DA (1974b) Blood flow in arteries, p 95, Edward Arnold, London.

Milnor WR (1982) Hemodynamics, Chapter 6, pp 135–156, Williams and Wilkins, Baltimore.

Payen DM, Levy BI, Menegalli DJ, Lajat YI, Levenson JA, Nicolas FM (1982) Evaluation of hemispheric blood flow based on non-invasive carotid blood flow measurements using the range-gated Doppler technique. Stroke 13, 392–398.

Seeley BD, Young DF (1976) Effect of geometry on pressure losses across models of arterial stenoses. J Biomech 9, 439–448.

Shipley RE, Gregg DE (1944) The effect of external constriction of a blood vessel on blood flow. Am J Physiol 141, 289–296.

Strandness DE, Sumner DS (1975) Hemodynamics for surgeons, Chapter 9, pp 209–289, Grune and Stratton, New York.

Whitmore RL (1968) Rheology of the circulation, Pergamon Press, Oxford.

Womersley JR (1955) Oscillatory motion of a viscous liquid in a thin-walled elastic tube. I: The linear approximation for long waves. Phil Mag 46, 199–221.

Womersley JR (1957) Oscillatory flow in arteries: the constrained elastic tube as a model of arterial flow and pulse transmission. Phys Med Biol 2, 178–187.

Young DF, Tsai FY (1973a) Flow characteristics in models of arterial stenoses. I. Steady flow. J Biomech 6, 395–410.

Young DF, Tsai FY (1973b) Flow characteristics in models of arterial stenoses. II. Unsteady flow. J Biomech 6, 547–559.

Young DF, Cholvin NR, Roth AC (1975) Pressure drop across artificially induced stenoses in the femoral arteries of dogs. Circ Res 36, 735–743.

Young DF, Cholvin NR, Kirkeeide RL, Roth AC (1977) Hemodynamics of arterial stenoses at elevated flow rates. Circ Res 41, 99–107.

2.11   RECOMMENDED READING

Caro CG, Pedley TJ, Schroter RC, Seed WA (1978) The mechanics of the circulation, Oxford University Press, Oxford.

McDonald DA (1974) Blood flow in arteries, Edward Arnold, London.

Milnor WR (1982) Hemodynamics, Williams and Wilkins, Baltimore.

Whitmore RL (1968) Rheology of the circulation, Pergamon Press, Oxford.

Chapter 3

SUMMARY OF THE BASIC PHYSICS OF DOPPLER ULTRASOUND

3.1 INTRODUCTION

Medical ultrasound and its propagation in tissue have been discussed in several texts (McDicken 1981, Kremkau 1984, Hill 1986, Wells 1977, Woodcock 1977). For an in-depth study of the physics of ultrasound the reader is referred to these publications. In this chapter ultrasound physics is considered specifically from the point of view of Doppler techniques.

There is more interest in quantifying Doppler signals than there is in quantifying ultrasound images, and an understanding of wave propagation phenomena assists in the measurement of parameters such as velocity, flow and pressure gradients, and in an evaluation of potential sources of error.

3.2 ULTRASONIC WAVES

Ultrasonic waves, like audible sound waves, are compressional waves produced by the push–pull action of the source on the propagating medium. These waves are sometimes called longitudinal waves since the oscillatory motion of the particles of the medium is parallel to the direction of propagation. The source is normally a transducer in which the vibrating element is a piece of piezoelectric ceramic or plastic driven by an appropriate voltage signal. In most Doppler techniques the frequencies employed are in the range 1–10 MHz. This range is slightly lower than that in pulse echo imaging where 2–20 MHz is used. The Doppler range is lower since, with pulsed Doppler, aliasing problems are more troublesome at high frequencies (Section 8.4.1.1). Corresponding to the 1–10 MHz range, the wavelengths are on average 0.7–0.07 mm in soft tissue.

Doppler instruments generate either continuous wave (CW) or pulsed wave (PW) ultrasound. In the latter case the pulses vary in length from 2 to 10 cycles depending on the design of the instrument. Typically the intensities (I_{spta}) generated by CW and PW devices are 100 and 400 mW cm^{-2} respectively. The pressure amplitudes for CW and PW units are around 0.05 and 3.0 MPa respectively. It should be noted that there is a wide spread of outputs around these values. Since these output values are of direct concern to the question of safety they are examined in detail in Chapter 13.

Other types of ultrasonic wave such as shear (transverse) or surface waves have not been applied in medical ultrasonics. Shear waves are strongly attenuated in soft tissue.

From the point of view of Doppler techniques the parameters that describe a wave, i.e. amplitude, frequency and phase, are important as in pulse echo imaging. Indeed frequency and phase are more important for Doppler methods since the velocity of blood is obtained from the shift in frequency, and directional information from the phase of the scattered waves (Chapter 6). The phenomena, discussed below, that affect ultrasound waves are also important. This may not be immediately obvious to the investigator unless careful consideration is given to possible distortions of the ultrasound beam in tissue.

3.2.1 Speed of sound

The speed of sound, c, in soft tissue depends on its bulk modulus K, and density ρ, and may be written:

$$c \simeq (K/\rho)^{\frac{1}{2}}. \qquad 3.1$$

(The approximation sign is used here because, rigorously, K should be replaced by K', a function of the *adiabatic* bulk modulus and the shear modulus – see Wells 1969, pp. 3–4.)

The bulk modulus K is given by applied pressure, p_a, divided by the fractional change in volume, $-\Delta V/V$, i.e.:

$$K = -p_a V/\Delta V. \qquad 3.2$$

In general the less compressible a material, the higher the speed of sound (Goss et al 1978, Wells 1977). This can be confirmed for materials of interest in medical ultrasound (Table 3.1). From this table it should be noted that the values for soft tissue are closely clustered around the average of 1540 m s^{-1}. The speed of ultrasound in blood is reported to be in the range 1540–1600 m s^{-1}. Range calibration of Doppler instruments for a mixture of soft tissues is therefore simply implemented and beam shapes are not grossly distorted by phenomena such as refraction and scattering at tissue boundaries. However, the high velocity in bone does cause problems. To obtain diagnostic information with either imaging or Doppler beams, it is usually necessary to avoid bone unless it is thin.

In the frequency range 1–20 MHz the speed of ultrasound is effectively independent of frequency, i.e. dispersion effects need not be taken into account. The effect of small temperature changes can also be ignored.

The speed of sound, c, appears in the denominator of the basic Doppler equation (eqn 1.1). This is the speed of sound in blood. It is usually taken to be 1570 or 1580 m s^{-1}.

The speed of ultrasound is not of prime importance in Doppler techniques. However, it should be remembered that the sample volume in a duplex system (Chapter 4) may not be positioned exactly at the range indicated in the image if the average velocity for the Doppler beam path is significantly different from 1540 m s^{-1}. For example, an average velocity of 1480 m s^{-1}, perhaps due to fat in the beam path, would result in a sample volume at a depth of 5 cm being displaced 2 mm from the location indicated on the display. This can result in an erroneous signal or a complete loss of signal.

3.2.2 Intensity and power

As for any form of radiation, the intensity at a point in a transmitted field is the rate of flow

Table 3.1 Speed of ultrasound, and acoustic impedance in some common materials. Data from Wells (1969), Goss et al (1978) and Bamber (1986). The acoustic impedance cannot be calculated where the density of the material is not known

Material	Speed $(\mathrm{m\ s}^{-1})$	Acoustic impedance $(\mathrm{kg\ m}^{-2}\ \mathrm{s}^{-1})$
Air (NTP)	330	0.0004×10^6
Amniotic fluid	1510	—
Aqueous humour	1500	1.50×10^6
Blood	1570	1.61×10^6
Bone	3500	7.80×10^6
Brain	1540	1.58×10^6
Cartilage	1660	—
Castor oil	1500	1.43×10^6
CSF	1510	—
Fat	1450	1.38×10^6
Kidney	1560	1.62×10^6
Lens of eye	1620	1.84×10^6
Liver	1550	1.65×10^6
Muscle	1580	1.70×10^6
Perspex	2680	3.20×10^6
Polythene	2000	1.84×10^6
Skin	1600	—
Soft tissue average	1540	1.63×10^6
Tendon	1750	—
Tooth	3600	—
Vitreous humour	1520	1.52×10^6
Water (20 °C)	1480	1.48×10^6

of energy through unit area at that point. The intensity is related to the pressure amplitude, p, of the wave by:

$$I = \tfrac{1}{2}(p^2/\rho c) \qquad 3.3$$

and to the particle displacement amplitude, x_0, and the particle velocity amplitude, u_0, by the following equations:

$$I = \tfrac{1}{2}\rho c (2\pi f)^2 x_0^2 = \tfrac{1}{2}\rho c u_0^2. \qquad 3.4$$

The points selected for measurement of intensity are commonly the focus of the beam or within 1 or 2 cm of the transducer face. For a continuous wave beam, the intensity may be measured at the spatial peak, to give I_{sp}, or it may be averaged across the beam to give the spatial average, I_{sa}. With pulsed ultrasound, temporal averaging as well as spatial averaging may be carried out. For example, the spatial peak intensity may be averaged over several pulses to give I_{spta} (intensity – spatial peak, temporal average) or over the pulse length for I_{sppa} (intensity – spatial peak, pulse average). Other quantities of interest are I_{sptp} (intensity – spatial peak, temporal peak) and I_{sata} (intensity – spatial average, temporal average). The relevance of these quantities to biological effects is discussed in Chapter 13. From the machine user's point of view, it is probably sufficient to ascertain I_{sp} or I_{spta} and, if possible, I_{sptp}. The latter is in fact a difficult quantity to measure accurately.

Knowledge of these quantities allows the outputs of different machines to be compared and their safe usage considered. A well designed Doppler unit will have an I_{sp} or I_{spta} value of less than 100 mW cm^{-2}, although for adult cardiology and adult brain blood flow it is normally considered acceptable to use higher values. It will be seen in Chapter 13 that there is also interest in pressure amplitude when safety aspects are being considered.

The power of an ultrasonic beam is the rate of flow of energy through the cross-sectional area of the beam. This is a popular quantity to quote since it is relatively easy to measure and is not dependent on identifying the spatial peak.

The transmitted intensity or power of a Doppler unit is often fixed and sensitivity is altered by manipulating the receiver gain. From the point of view of safety it is actually better to fix the gain at the maximum level and manipulate the power.

Power and intensity controls on ultrasonic units are often labelled using the decibel notation. For example, taking the maximum intensity position of the control as I_0, other values, I, are calibrated relative to it as

$$\text{Output intensity in decibels} = 10 \log(I/I_0). \qquad 3.5$$

This method of labelling power and intensity controls is used since it is fairly difficult to measure absolute values in mW or mW cm^{-2} and a separate calibration is required for each transducer.

Ultrasonic intensity is normally measured with a hydrophone which takes the form of a small probe with a piezoelectric element in it (Chapter 5).

Ultrasonic power is normally measured with a radiation pressure balance. Associated with the flow of energy in a beam there is a flow of momentum. When this flow of momentum is intercepted by an object, a force is experienced by the object which is directly related to the power of the beam if the whole beam is intercepted (Livett et al 1981). For a beam of power P incident normally on a perfectly reflecting flat surface:

$$\text{Radiation force} = 2P/c. \qquad 3.6$$

The force experienced is 0.135 mg mW^{-1}. Power levels down to 0.1 mW can be measured with such devices. If the object totally absorbs the beam, the force is half that of the reflection case. To measure these small forces, the object is often placed under water on the pan of a sensitive chemical balance (Hill 1970, Kossoff 1965, Rooney 1973). With a small object that only intercepts part of a beam it is possible to measure intensity at a point but this approach has largely been replaced by the hydrophone one. Portable radiation balances have been designed for easy transport to the ultrasound machines (Farmery and Whittingham 1978) and are commercially available (Fig. 13.1).

In Doppler applications the main interest in intensity and power is in considerations of safety. At present several pulsed Doppler devices are capable of generating unacceptably high levels of intensity and care must be taken in their application (Chapter 13).

3.2.3   Diffraction and interference

Diffraction is the term used to describe the spreading out of a wave from its source as it passes through a medium. The way in which a wave spreads out is highly dependent on the shape and size of the source relative to the wavelength of the sound. For instance the field pattern of a disc-shaped crystal, i.e. its diffraction pattern, may be reasonably cylindrical for a short distance, after which it starts to diverge. The diffraction pattern from a small element of an array transducer is divergent from close to the transducer. Within a diffraction pattern there may be fluctuations in intensity, particularly close to the transducer. Diffraction also occurs beyond an obstacle such as a slit aperture or an array of slits which partially blocks a wavefront. When a beam interacts with a less geometrically regular object, the term 'scattering' is usually applied to describe spreading out of the beam. The narrowness of a beam or the sharpness of a focus is limited by diffraction, indeed they are said to be 'diffraction limited'.

Interference occurs when two or more waves overlap in the same medium. The resultant pressure at any point is obtained by adding algebraically the pressures from each wave at the point. This principle of superposition is widely used to predict ultrasound field shapes using mathematical models. For example, the shape of a transducer field is derived by considering the crystal face to be subdivided into many small parts. The diffraction pattern for each small part is calculated and the effect of them overlapping and interfering gives the resultant ultrasound field shape. The shapes of ultrasonic fields and beams as used in Doppler techniques are discussed in Chapter 5.

Interference and diffraction are rarely considered in detail since, as techniques stand at present, ultrasound beams and fields are rarely well defined.

3.2.4 Standing waves and resonance

Standing waves are formed when two waves of the same frequency travelling in opposite directions interfere. They are most readily seen if the waves have similar amplitude. Examination of the pressure variations shows alternate stationary regions of high and low pressure amplitude, i.e. a pattern of nodes and antinodes. When standing waves are present in tissue, the flow of red cells in capillaries can be arrested, i.e. blood stasis occurs (Dyson et al 1974). The cells gather in bands separated by half a wavelength (Fig. 3.1). Further analysis of this phenomenon has been carried out by ter Haar and Wyard (1978). Blood stasis is unlikely to occur with diagnostic techniques because standing waves are unlikely to be set up.

Standing wave patterns can result from a sound wave being reflected back and forth between two flat parallel surfaces. If the separation of the surfaces is a whole number of half wavelengths a marked increase in the pressure amplitudes is observed. This is due to the waves adding constructively to give a resonance. A marked resonance, the fundamental, occurs when the separation of the surfaces is half a wavelength. Ceramic transducer elements are made equal in thickness to one-half of the wavelength corresponding to their operating frequency to give efficient generation and detection of ultrasound. Continuous wave Doppler transducers have little or no damping and resonate at their operating frequency. Pulsed wave Doppler transducers have some damping. In general, when the size of a structure has a special numerical relationship to the wavelength, a resonant vibration will occur. One interesting possibility is that small bubbles in blood could be driven by ultrasound to resonate at a frequency which is related to blood pressure (Fairbank and Scully 1977).

Standing waves and interference are of interest to the designer of transducers. They are two of the phenomena that contribute to the overall performance of a transducer and provide some insight into its complexity of operation. Given this complexity and scope for variation it is always worth thoroughly assessing the performance of each transducer in clinical use.

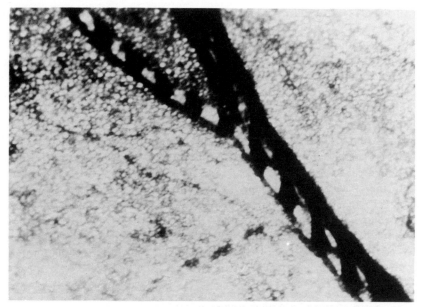

Figure 3.1   Blood stasis due to standing waves in a vein of a chick embryo. Standing waves were generated by 3 MHz continuous wave ultrasound. (By courtesy of Dyson M)

3.2.5   Non-linear propagation

The discussion so far has assumed that the waveforms, pulsed or continuous, retain the same shape as they pass through material. This is only true for very low amplitude waves in a non-attenuating medium such as water. For waves of larger amplitude, the speed of sound is higher in the portions of tissue influenced at any instant by the increased pressure than it is in the portions at reduced pressure. The speed is changed because the density changes with pressure. The positive half-cycles of the waveform therefore catch up on the negative half-cycles resulting in distortion of the waveform (Fig. 3.2). At increasing depth in the propagating medium the distortion increases until the waveform becomes a sawtooth shape. When sharp discontinuities appear in the waveform it is called a shock wave.

The speed of sound, c, and the particle velocity, u, at any point are related by:

$$c \simeq c_0 + (1 + B/2A)u \qquad\qquad 3.7$$

where c_0 is the velocity of sound for low-amplitude waves and B/A is the non-linearity parameter of the medium.

This variation of the speed within the waveform leads to the distortion mentioned above. The distortion increases as the amplitude of the wave, and hence the particle velocity, u, increases.

The parameter B/A is a property of the medium. For liquids and soft tissues it takes values in the range 2–13 (Bamber 1986). It does not appear to be simply related to physical properties, being influenced by factors such as concentration of macromolecules, solute–solvent interactions and tissue structure. Temperature and pressure do not strongly influence values of B/A. Table 3.2 shows some typical values for B/A.

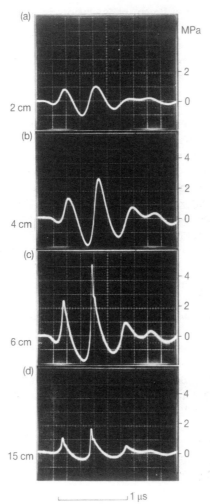

Figure 3.2 Distortion of a pulse of 2 MHz ultrasound due to non-linear propagation in water. Pulse shapes are recorded at 2, 4, 6 and 15 cm ranges. (By courtesy of Duck FA, Starrit (1984) Br J Radiol 57, 231–240)

Table 3.2 Values of the non-linearity parameter. Data from Bamber (1986)

Material	B/A
Water	5.0
Blood	6.3
Liver	7.8
Fat	11.1

The significance of non-linear propagation in medical ultrasonics has been appreciated in recent years (Cartensen et al 1980, Muir and Cartensen 1980, Law et al 1985, Bacon 1984, Duck and Starritt 1984). From the theory it is evident that non-linear propagation is more readily observed for high-amplitude waves and also for high-frequency waves since each pressure cycle contributes to the total effect. Shock wave formation occurs readily in water since the high-frequency components introduced to the waveform are not attenuated (see Section 3.3.3). It is therefore of particular relevance in techniques that involve waterbaths or any liquid path along which the beam passes. The presence of shock wave formation at diagnostic intensity levels in absorbing tissue (liver and calf muscle) has also been demonstrated (Duck et al 1986). In this work it was shown that for a 2.5 MHz pulse of initial amplitude 0.58 MPa, the second harmonic increased by about 10 dB in the first 5 cm of transmission. Non-linear distortion is therefore influential in the conditions pertaining in diagnostic ultrasound. In the case of absorbing media, the high-frequency components generated by this distortion are preferentially absorbed and eventually a smoother waveform of reduced amplitude results.

The following points can be made in relation to non-linear propagation:

1. From the limited data available, soft tissue, blood and other liquids appear to have similar properties.
2. Due to non-linear propagation the frequency components of the continuous or pulsed carrier wave vary with depth. A complete analysis of the significance of this does not appear to have been carried out. However, in the light of the extensive work with Doppler methods in test phantoms, errors due to non-linear distortion are probably relatively small.
3. Non-linear distortion is more severe with PW devices than with CW devices since larger amplitude waves are employed in the former case.
4. Increased pressure amplitudes in the form of shock waves need to be considered from a hazard standpoint (Chapter 13).

3.3   ULTRASONIC PHENOMENA

3.3.1   Scattering

When an ultrasonic wave travelling through a medium strikes a discontinuity of dimensions similar to or less than a wavelength, some of the energy of the wave is scattered in many directions. Scattering is the process of central importance in diagnostic ultrasonics since it provides most of the signals for both echo imaging and Doppler techniques. The discontinuities may be changes in density or in compressibility or both. Detailed information is not available on the nature of the discontinuities in soft tissue but they are usually considered to be a random distribution of closely packed scattering centres. These centres are modelled as discrete centres or as fluctuations in a continuum (Chivers 1977). Computer models of a wave interacting with such structures can predict a scattered wave with properties like those observed in practice. For example, statistical fluctuations in simulated echo signals are similar to those that give rise to the speckle pattern in images (Burkhardt 1978, Abbot and Thurstone 1979). The red cells in blood, singly or in groups, act as scattering centres which produce the signals used in Doppler techniques.

> The ratio of the total power, S, scattered by a target (or by unit volume of material) to the incident intensity, I, is called the total scattering cross-section, σ_t, of the target or the volume of material:
>
> $$\sigma_t = S/I. \tag{3.8}$$
>
> This ratio is used to compare the scattering powers of different structures.
>
> The differential scattering cross-section, $\sigma(\phi)$, is the ratio of the power scattered into unit solid angle located in a particular direction in space, $S(\phi)$, to the incident intensity, I.
>
> $$\sigma(\phi) = S(\phi)/I. \tag{3.9}$$
>
> Of particular interest is the differential cross-section at 180° to the direction of the incident ultrasound. This is known as the backscatter cross-section and determines the size of the signal sent back to the transducer in imaging and Doppler techniques.
>
> Theoretical treatment of scattering in a material is possible if sufficient simplifications are made. Dickinson (1986) considers scattering of a plane wave at centres in an inhomogeneous medium. The medium is considered to exhibit no absorption, small fluctuations in density and compressibility, and no multiple reflection. This results in an expression for the intensity of the scattered wave at a distance s which may be written in the form:
>
> $$I(s, \phi) \propto \frac{k^4}{s^2} [g_1(k, \beta) + g_2(k, \rho)\cos \phi]^2 \tag{3.10}$$
>
> where $k = 2\pi/\lambda$ and ϕ is the angle between the incident and scattered waves. $g_1(k, \beta)$ and $g_2(k, \rho)$ are functions that depend on ultrasound frequency, and on the distribution of compressibility and density (respectively) in the target volume.

The following conclusions can be drawn from this idealized case:

1. The scattered wave is spherical with modifications due to the terms in brackets which depend on the distribution of compressibility and density.
2. If the fluctuations in compressibility and density are small and $g_1(k, \beta)$ and $g_2(k, \rho)$ are constant over a significant frequency range, the scattered intensity has a fourth power dependence on frequency due to k^4, i.e. there is Rayleigh scattering.
3. The scattered intensity is dependent on the angle of observation, ϕ.

This analysis is reassuring in that it confirms that our understanding of scattering from limited experimental measurements in soft tissue and blood is reasonably correct. However, attempts to apply theory to scattering emphasize the complexity of the situation. Usually the model has to be simplified to such an extent that the results are little more than confirmation of intuitive reasoning. The development of techniques involving scattering proceed more by experimental investigation than by theoretical modelling.

Measurements of scattering cross-sections in the literature are very scarce. This is due in part to the difficulties in making such measurements. For example, there are problems in separating the forward scattered ultrasound from the original beam and the physical size of the beam or side lobes restricts the angular range of measurement. Estimates of the contribution of scattering to the total attenuation in tissue range up to 40%. The contribution is obviously dependent on the tissue and the frequency of the sound, but most investigators report the scattering contribution to be less than 10%. Scattering increases with the amount of fat and collagen but decreases with the water content of tissue. One exception to the paucity of scattering data is for blood which has been investigated quite thoroughly.

Shung et al (1976) give values for the scattering cross-section at 5, 8.5 and 15 MHz (Table 3.3) and show that the values are in agreement with theoretical ones calculated with a formula by Morse and Ingard (1968). The angular distribution of scattering in blood has been shown to be asymmetric, the back-scattered cross-section being about 6 dB greater than that at forward angles (Shung et al 1977). Scattering from blood is further considered in Section 8.2.1.

Table 3.3 Approximate scattering cross-section values for blood with a haematocrit of 40% over a range of frequency. Data extracted from Shung et al (1976)

Frequency (MHz)	Scattering cross-section (cm^2)
5.0	2×10^{-14}
8.5	1.4×10^{-13}
15	1.3×10^{-12}

From the point of view of Doppler techniques the study of scattering is important since it improves our understanding of CW and PW instruments. In day-to-day usage of instruments, however, the operator need not be concerned with scattering except to note that the signal from blood is very much weaker than that from soft tissue. The sample volume for soft tissue is therefore much larger than that for blood (Chapter 5). Wall thump filters are normally included in Doppler devices to reduce low-frequency signals from moving tissue.

3.3.2   Reflection and refraction

Reflection is a special case of scattering which occurs at smooth surfaces on which the irregularities are very much smaller than a wavelength. The physical property of tissue which is of importance in the reflection process is the acoustic impedance, z, of the material. This is the ratio of the wave pressure over the particle velocity, i.e. p/u. For a plane wave in a weakly absorbing medium:

$$\text{Acoustic impedance, } z = \rho c. \qquad\qquad 3.11$$

In the SI system of units, acoustic impedance has the units of $kg\ m^{-2}\ s^{-1}$ or Rayl. For soft tissue, acoustic impedances are in the range 1.3–1.9×10^6 Rayl. The acoustic impedance of blood is 1.6×10^6 Rayl. The higher the density or stiffness of a material, the higher is its acoustic impedance. Table 3.1 lists some commonly encountered values of the acoustic impedance.

Consideration of reflection at perpendicular incidence to a flat boundary between two media of impedances $\rho_1 c_1$ and $\rho_2 c_2$ provides some appreciation of the magnitude of echo amplitudes (Table 3.4). These values were calculated using the formula:

$$\text{Reflected amplitude} = \text{Incident amplitude} \times \frac{\rho_1 c_1 - \rho_2 c_2}{\rho_1 c_1 + \rho_2 c_2}. \qquad 3.12$$

Substantial errors may exist in these values since there is a spread in reported values for the velocity of sound in tissues.

Table 3.4   The ratio of the reflected wave to the incident wave amplitude and the percentage energy reflected for perpendicular incidence. Data from McDicken (1981)

Reflecting interface	Ratio of reflected to incident wave amplitude	Percentage energy reflected
Fat/muscle	0.10	1.08
Fat/kidney	0.08	0.64
Muscle/blood	0.03	0.07
Bone/fat	0.69	48.91
Bone/muscle	0.64	41.23
Lens/aqueous humour	0.10	1.04
Soft tissue/water	0.05	0.23
Soft tissue/air	0.9995	99.9
Soft tissue/PZT crystal	0.89	80
Soft tissue/polyvinylidene difluoride	0.47	0.22
Soft tissue/castor oil	0.06	0.43

When a wavefront strikes a smooth surface at oblique incidence it is reflected at an equal and opposite angle. In this case the amplitude of the echo is given by a more complex formula (Kinsler and Frey 1962).

Refraction is the deviation of a beam when it crosses a boundary between two media in which the speeds of sound are different. The resultant angle of propagation is given by the familiar Snell's Law

$$\sin i / \sin r = c_1 / c_2 = \mu \qquad 3.13$$

where μ is the refractive index.

Table 3.5 shows the deviation experienced by a beam for increasing angles of incidence at a muscle/blood interface. Although deviations of a few degrees can occur at soft tissue boundaries, experience in ultrasonic imaging indicates that blood vessels can often be clearly depicted, so refraction is not a severe problem.

From the point of view of Doppler techniques, reflection is primarily of interest in the visualization of blood vessel walls and cardiac structures when duplex and flow mapping systems are being used. Since reflected pulses have a finite length, the accuracy with which structures such as vessel diameter can be measured is limited. This in turn significantly

Table 3.5 Deviation experienced by an ultrasound beam passing through a muscle/blood interface at different angles of incidence.

Angle of incidence (degrees)	Deviation (degrees)
0	0
10	0.07
20	0.13
30	0.21
40	0.31
50	0.43
60	0.62
70	0.97
80	1.87
90	6.44

affects the accuracy of measurement of blood flow by several techniques. Refraction distorts an ultrasonic beam as it crosses the muscle/blood interface of a vessel. Greater deviations are to be expected when a wall is more rigid due to the presence of plaque. Distortion of ultrasonic beams at curved surfaces has been modelled and shown to be significant (LaFollette and Ziskin 1986). However, the case of refraction at a blood vessel has not been treated in detail. Distortion of a beam at a vessel wall can be expected to alter the intensity pattern in the vessel and will add to the error in measurement of mean velocity.

3.3.3 Absorption and attenuation

Absorption is the process of conversion of wave motion energy into heat. For ultrasonic propagation at low megahertz frequencies through biological materials the largest contribution to absorption is that due to relaxation mechanisms (Wells 1975). In these mechanisms the stress imposed by the wave on the medium is relaxed by the flow of energy to various energy states of the tissue. For example, thermal relaxation is the flow of wave energy to internal molecular energy and structural relaxation is flow to changed structural states of different energy. When the stress is released some of the energy flows back to the wave energy but it is out of phase with the original wave and the resultant amplitude is reduced. Each relaxation mechanism is most effective at one particular frequency, the relaxation frequency. As might be expected in tissue, many such relaxation processes are present and the effect of individual ones cannot be observed (Bamber 1986). Isolated processes can only be studied in simple liquids. Absorption increases rapidly with frequency in tissue over the frequency range used in diagnostic techniques.

In practice, absorption on its own is rarely of interest since other processes are simultaneously contributing to the total attenuation of the wave. These processes are scattering, reflection, refraction, non-linear propagation and beam divergence. Since many of these factors are strongly frequency dependent it is not surprising that attenuation rises rapidly over the frequency range 1–20 MHz. Table 3.6 shows the attenuation coefficients for a variety of tissues. It is common practice to quote attenuation coefficients in dB cm^{-1} when echo imaging or Doppler techniques are being considered.

$$\text{Intensity (or power) attenuation coefficient} = \frac{10}{x} \log_{10}\left(\frac{I}{I_0}\right). \qquad 3.14$$

$$\text{Amplitude attenuation coefficient} = \frac{20}{x} \log_{10}\left(\frac{p}{p_0}\right) \qquad 3.15$$

where x is the thickness of the tissue layer. These coefficients have the same numerical value.

Table 3.6 Attenuation coefficients for common tissues in dB cm^{-1} at 1 MHz. The value at a higher frequency may be obtained approximately by multiplying by the frequency in MHz

Tissue	Attenuation coefficient (dB cm^{-1})
Blood	0.2
Muscle	1.5
Liver	0.7
Brain (adult)	0.8
Brain (infant)	0.3
Bone	10.0
Fat	0.6
Water	0.002
Soft tissue (average)	0.7
Castor oil	1.0

From Table 3.6 it can be seen that the attenuation in blood is low. The decrease in the scattered signal from blood across a 1 cm vessel is 0.8 dB at 2 MHz and 2.0 dB at 5 MHz. This source of error is usually neglected in calculations of mean velocity. It can be allowed for by computation or, in the case of PW Doppler, by attenuation compensation (TGC) in the receiving amplifier. No data exist for attenuation in plaque in its various forms. However, solid plaque attenuates ultrasound strongly owing to high reflection and refraction and therefore casts shadows in the region behind it.

The frequency dependence of attenuation results in the high-frequency components of an ultrasonic pulse being preferentially reduced in amplitude relative to the low-frequency components. The resulting distortion of the ultrasonic pulse is rarely taken into account in present-day practice. It has been evaluated and shown to reduce the central frequency of the pulse bandwidth by amounts which can be significant (Holland et al 1984).

Absorption and attenuation are very important in Doppler techniques, and their influence on any examination must be considered. A frequency of ultrasound is selected which will provide the required penetration. Typically Doppler devices operating at around 7 MHz are used for superficial vessels while 2 MHz is employed for deep ones. The high attenuation of bone and gas severely limits access to the adult brain and thorax. When Doppler signals are to be detected at a series of ranges, as in colour flow mapping, attenuation compensation is supplied by electronic circuitry which applies increasing gain as signals are received from increasing depth (TGC).

3.4   THE DOPPLER EFFECT

The Doppler effect is the change in the observed frequency of a wave due to motion. This motion may be of the source or the observer (Fig. 3.3). When the observer moves towards the source, the increased frequency, f_r, due to passing more wave cycles per second is given by

$$f_r = f_t \frac{c+v}{c}$$ 3.16

where f_t is the transmitted frequency and v is the velocity of the observer.

If the velocity of the observer is at an angle θ to the direction of the wave propagation, v is replaced by the component of the velocity v in the wave direction, $v \cos \theta$:

$$f_r = f_t \frac{c+v \cos \theta}{c} .$$ 3.17

If the observer is at rest and the source moves with velocity v in the direction of wave travel, the wavelengths are compressed. The resulting observed frequency is:

$$f_r = f_t \frac{c}{c-v} .$$ 3.18

Taking angle into account:

$$f_r = f_t \frac{c}{c-v \cos \theta} .$$ 3.19

Both of these motions which give rise to changes in the observed frequency are in fact slightly different effects since in the first the wave is not altered and in the second it is compressed.

In medical ultrasonic applications, an ultrasonic beam is backscattered from moving blood cells. Both of the above effects combine to give the resultant Doppler shift in frequency. The observed frequency is then given by:

$$f_r = f_t \frac{c+v \cos \theta}{c} \cdot \frac{c}{c-v \cos \theta}$$ 3.20

$$= f_t \frac{c+v \cos \theta}{c-v \cos \theta} .$$ 3.21

The Doppler shift,

$$f_d = f_r - f_t$$

$$= f_t \frac{c+v \cos \theta}{c-v \cos \theta} - f_t .$$ 3.22

Since $c \gg v$:

$$f_d = \frac{2 f_t v \cos \theta}{c} .$$ 3.23

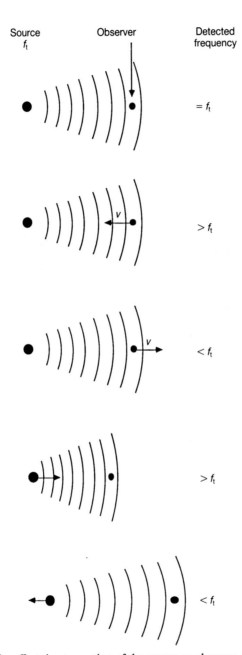

Source
f_t

Observer

Detected
frequency

$= f_t$

$> f_t$

$< f_t$

$> f_t$

$< f_t$

Figure 3.3   The Doppler effect due to motion of the source or observer. The detected frequency is increased or decreased depending on the direction of motion

It is instructive to consider some numerical evaluations of this formula for values encountered in medical ultrasonics as presented in Table 3.7. The Doppler shift is seen to be around one part in 1000. The Doppler shift frequencies are also in the audible range.

Table 3.7 Doppler shifts for the range of velocities and ultrasonic frequencies encountered in practice. An angle of insonation of 45° has been assumed

Speed (cm s^{-1})	Ultrasonic frequency (MHz)			
	2	3	5	10
1	18 Hz	27 Hz	46 Hz	92 Hz
10	183 Hz	275 Hz	459 Hz	918 Hz
50	915 Hz	1.37 kHz	2.29 kHz	4.59 kHz
100	1.83 kHz	2.75 kHz	4.59 KHz	9.18 kHz
500	9.15 kHz	13.75 kHz	22.95 kHz	45.90 kHz

3.5   SUMMARY

In this chapter some basic physics encountered in the use of ultrasonic techniques has been briefly described. Further detail can be obtained in other texts. Points of particular relevance to Doppler techniques have been emphasized. The properties of ultrasonic waves have been discussed with particular attention being paid to intensity and power. The ultrasonic phenomena which occur in tissue such as scattering and attenuation have been considered in some detail since they strongly influence the magnitude of signals received from tissue. Finally the Doppler effect has been explained.

3.6   NOTATION

B/A	Non-linearity parameter
c	Velocity of sound in tissue
c_0	Velocity of low-amplitude sound waves
f	Frequency
f_r	Received or observed frequency
f_t	Transmitted or source frequency
$g_1(k, \beta)$ $g_2(k, \rho)$	Functions appearing in the scattering equation
i	Incident angle
I	Intensity
k	Wave number ($=2\pi/\lambda$)
K	Bulk modulus
p_a	Applied pressure
p	Pressure
P	Power
r	Reflected or refracted angle
s	Distance

S	Total scattered power
$S(\phi)$	Power scattered into a unit solid angle
u	Particle velocity
v	Velocity of a source, observer or target
x	Thickness of a tissue layer
x_0	Displacement amplitude
z	Acoustic impedance
β	Adiabatic compressibility
$\Delta V/V$	Fractional change in volume
θ	Angle between a velocity and direction of wave propagation
λ	Wavelength
μ	Refractive index
ρ	Density
σ_t	Total scattering cross-section
σ_θ	Differential scattering cross-section
ϕ	Scattering angle

3.7   REFERENCES

Abbott JG, Thurstone FL (1979) Acoustic speckle: theory and experimental analysis. Ultrasonic Imaging 1, 303–324.

Bacon DR (1984) Finite amplitude distortion of the pulsed fields used in diagnostic ultrasound. Ultrasound Med Biol 10, 189–195.

Bamber JC (1986) Attenuation and absorption. In: Physical principles of medical ultrasonics (Ed. CR Hill), Chapter 5, Ellis Horwood, Chichester.

Burckhardt CB (1978) Speckle in ultrasound B-mode scans. IEEE Trans Sonics Ultrasonics SU-25, 1–6.

Cartensen EL, Law WK, McKay ND, Muir TG (1980) Demonstration of nonlinear acoustical effects at biomedical frequencies and intensities. Ultrasound Med Biol 6, 359–368.

Chivers RC (1977) The scattering of ultrasound by human tissues – some theoretical models. Ultrasound Med Biol 3, 1–13.

Dickinson RJ (1986) Reflection and scattering. In: Physical principles of medical ultrasonics (Ed. CR Hill), Chapter 6, Ellis Horwood, Chichester.

Duck FA, Starritt HC (1984) Acoustic shock generation by ultrasonic imaging equipment. Br J Radiol 57, 231–249.

Duck FA, Starritt HC, Hawkins AJ (1986) Observations of finite-amplitude distortion in tissue. Proc Inst Acoustics 18, 71–77.

Dyson M, Pond JB, Woodward, B, Broadbent J (1974) The production of blood cell stasis and endothelial damage in blood vessels of chick embryos treated with ultrasound in a stationary wave. Ultrasound Med Biol 1, 133–148.

Fairbank WM, Scully MO (1977) A new noninvasive technique for cardiac pressure measurement: resonant scattering of ultrasound from bubbles. IEEE Trans Biomed Eng BME-24, 107–110.

Farmery MJ, Whittingham TA (1978) A portable radiation-force balance for use with diagnostic ultrasonic equipment. Ultrasound Med Biol 3, 373–379.

Goss SA, Johnston RL, Dunn F (1978) Comprehensive compilation of empirical ultrasonic properties of mammalian tissues. J Acoust Soc Am 64, 423–457.

Hill CR (1970) Calibration of ultrasonic beams for biomedical applications. Phys Med Biol 15, 241–248.

Hill CR (Ed.) (1986) Physical principles of medical ultrasonics, Ellis Horwood, Chichester.

Holland SK, Orphanoudakis SC, Jaffe CC (1984) Frequency-dependent attenuation effects in pulsed Doppler ultrasound: experimental results. IEEE Trans Biomed Eng BME-31, 626–631.

Kinsler LE, Frey AR (1962) Fundamentals of acoustics, Wiley, New York.

Kossoff G (1965) Balance technique for the measurement of very low ultrasonic power outputs. J Acoust Soc Am 38, 880–881.

Kremkau FW (1984) Diagnostic ultrasound: principles, instrumentation and exercises, 2nd edn, Grune and Stratton, New York.

LaFollette PS, Ziskin MC (1986) Geometric and intensity distortion in echography. Ultrasound Med Biol 12, 953–963.

Law WK, Frizzell LA, Dunn D (1985) Determination of the nonlinearity parameter B/A of biological media. Ultrasound Med Biol 11, 307–318.

Livett A J, Emery EW, Leeman SJ (1981) Acoustic radiation pressure. J Acoust Soc Am 76, 1–11.

McDicken WN (1981) Diagnostic ultrasonics: principles and use of instruments, 2nd edn, Wiley, New York.

Morse PM, Ingard, KU (1968) Theoretical acoustics, McGraw-Hill, New York.

Muir TG, Cartensen EL (1980) Prediction of nonlinear acoustic effects at biomedical frequencies and intensities. Ultrasound Med Biol 6, 345–357.

Rooney JA (1973) Determination of acoustic power outputs in the microwatt–milliwatt range. Ultrasound Med Biol 1, 13–16.

Shung KK, Sigelmann RA, Reid JM (1976) Scattering of ultrasound by blood. IEEE Trans Biomed Eng BME-23, 460–467.

Shung KK, Sigelmann RA, Reid JM (1977) Angular dependence of scattering of ultrasound from blood. IEEE Trans Biomed Eng BME-24, 325–331.

ter Haar G, Wyard SJ (1978) Blood cell banding in ultrasonic standing wave fields: a physical analysis. Ultrasound Med Biol 4, 111–123.

Wells PNT (1969) Physical principles of ultrasonic diagnosis, Academic Press, London.

Wells PNT (1975) Absorption and dispersion of ultrasound in biological tissue. Ultrasound Med Biol 1, 369–376.

Wells PNT (1977) Biomedical ultrasonics, Academic Press, London.

Woodcock JP (1979) Ultrasonics, Adam Hilger, Bristol.

Chapter 4

DOPPLER SYSTEMS: A GENERAL OVERVIEW

4.1 INTRODUCTION

The last decade has seen a rapid increase in the use of Doppler ultrasound devices for monitoring the cardiovascular system. Developments in Doppler technology have led to a vast increase in the number of non-invasive blood velocity investigations carried out in many areas of medicine. As with many rapidly expanding technologies there have been a considerable number of types of instrument developed and used in their institutions of origin, whereas only a few are in widespread clinical use.

The majority of Doppler devices presently in wide use may be classified into one (or sometimes more) of the following groups:

1. Velocity detecting systems
2. Duplex systems
3. Profile detecting systems
4. Velocity imaging systems

Some devices that do not easily fall into these categories are discussed in Chapter 18. The four types of system listed above are discussed in this chapter. The explanations are based on the systems approach, and the reader will be referred to the relevant sections of other chapters for more details of the particular processes involved. Because of the relatively recent appearance of colour flow mapping systems a short section on the practical aspects of using these devices is included.

4.2 VELOCITY DETECTING SYSTEMS

The simplest Doppler units are stand-alone systems that produce an output signal related to the velocity of the targets in a single sample volume, be it large or small. The transducers for such systems are not connected to any type of position-sensing gantry, and are often hand held. Such systems may be very basic and produce a non-directional audio output, or may be quite sophisticated producing directional signals sampled from predetermined depths in the tissue; they may also derive various types of information from the Doppler signal and output one or more Doppler envelope signals.

Because these systems are not combined with an imaging facility the angle of insonation, θ, between the ultrasound beam and the direction of blood flow is unknown except at some favourable anatomical locations, and therefore their outputs cannot be calibrated in absolute terms. They can, however, be used for detecting the presence or absence of flow (and perhaps its direction), monitoring *changes* in flow, and for recording the shape of the flow waveform, which may contain considerable information about the integrity of the cardiovascular system. They may also be used to make velocity measurements where the angle θ can be reliably estimated (particularly if θ is small, because the cos θ term in eqn 1.1 is then close to unity and changes very slowly). Examples of suitable sites for velocity measurement are the aortic arch insonated through the suprasternal notch, and the middle cerebral artery insonated through the temporal bone.

4.2.1 Continuous wave velocity detecting systems

All Doppler systems may be broadly categorized as either continuous wave (CW) or pulsed wave (PW). CW systems are the simpler of the two, but have some advantages over PW systems and are therefore still widely used even in sophisticated instruments. CW Doppler units both transmit and receive ultrasound continuously, and because of this they usually have no range resolution except in the sense that signals from a large distance from the transducer are much more attenuated than those from short distances. Also because the transmission is continuous it is necessary to use separate transmitting and receiving crystals, although these are usually housed in the same probe (see Fig. 5.3).

A block circuit diagram of a simple non-directional CW Doppler unit is shown in Fig. 4.1. The master oscillator usually produces a frequency of between 2 and 10 MHz, but higher and lower frequencies have been used. The frequency chosen depends on the depth of interest since ultrasound attenuation is highly dependent on frequency (Section 3.3.3); for superficial vessels frequencies of around 8 MHz are used, and for deep vessels frequencies as low as 2 MHz. The oscillations are amplified by the transmitting amplifier and the output used to drive the transmitting crystal. The crystal converts the electrical energy into acoustic energy (Section 5.2), which is then propagated as a longitudinal or compression wave into the body. The ultrasound energy is reflected and scattered by both moving and stationary particles within the ultrasound beam, and a small portion finds its way back to the receiving crystal which re-converts the acoustic energy into electrical energy. This small signal is then amplified by a radio frequency amplifier and mixed with a reference signal from the master oscillator. The process of mixing produces both the sum of the transmitted and received frequencies, and the required difference frequency or Doppler shift frequency. The low pass filter removes all

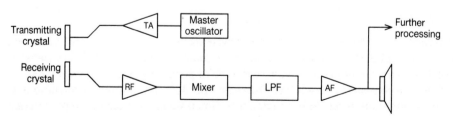

Figure 4.1 Block diagram of a simple non-directional continuous wave Doppler system. TA, transmitting amplifier; RF, radio-frequency amplifier; LPF, low pass filter; AF, audio frequency amplifier

signals outside the audio range, and this leaves only the Doppler difference frequency which is amplified and used to drive a loudspeaker or headphones, or sent for further processing. The process of recovering the Doppler audio signal from the Doppler shifted ultrasound (radio frequency) signal is known as demodulation.

Simple non-directional Doppler units are of value for detecting the movement of blood, for example to check whether an artery or graft is patent, or for measuring blood pressure in conjunction with a sphygmomanometer, but otherwise their value is limited.

Simple units of the type described above are unable to distinguish between motion towards and away from the transducer, and most modern instruments have additional circuitry that allows this distinction to be made. There are a number of ways of realizing this (Sections 6.2.5 and 6.4), but the most widespread technique is that of quadrature phase detection, whereby the received signal is mixed with two quadrature reference signals (signals separated by a 90° phase shift) from the master oscillator (see Fig. 6.8). This results in two audio frequency signals, both containing the Doppler information, but shifted by $\pm 90°$ to each other, depending on whether flow is towards or away from the probe. Further electronic processing is still required to unscramble the directional information into two channels. The technique most commonly used for this is one of phase domain processing (Section 6.4.2) which results in separate forward and reverse flow signals which may be listened to through stereo headphones, recorded on audio tape (Section 7.2) or sent for further processing.

Directional information is important clinically so that arterial and venous flows may be separated, because the direction of flow in an artery may change completely (for example when steal occurs), and because in peripheral arteries the direction of flow may reverse once or more during each cardiac cycle. It is also of great value in studies of the heart where the flow patterns are so complex.

Although with experience an operator may derive considerable information by listening to the audio signal, more objective methods of analysis are desirable. Further processing of the Doppler signal allows the Doppler shift frequencies to be presented in terms of a time-varying trace so that written records may be made and the time-varying changes in the blood velocity patterns analysed. The earliest attempts at producing such traces were based on the zero-crossing technique (Franklin et al 1961). The zero-crossing detector is a crude form of frequency processor, which functions by counting the number of times a signal crosses its own mean value in a given time. Under 'ideal' conditions this produces an output that is proportional to the root mean square frequency of the input signal. In practice the zero-crossing detector suffers from a variety of problems (Section 9.5) and is highly susceptible to noise. Its performance is particularly poor when a wide range of Doppler shift frequencies are present, as is the case with CW studies where both slow-moving blood near the arterial wall and fast-moving blood in the vessel centre are insonated simultaneously. Despite these drawbacks zero-crossing detectors are still incorporated in some low-cost Doppler instruments to provide a chart recorder output. These outputs are best avoided.

There are now a range of methods available for obtaining a pictorial record of the Doppler shift signal, of which the best and most commonly used is real-time spectral analysis (Section 9.2). The output of spectral analysers is usually represented as a sonogram (Fig. 4.2). In this type of display the horizontal axis (t) represents time, the vertical axis frequency (f), and the intensity at coordinates (t, f) the power of the signal at that frequency and at that point in time. Both the time-varying maximum frequency and

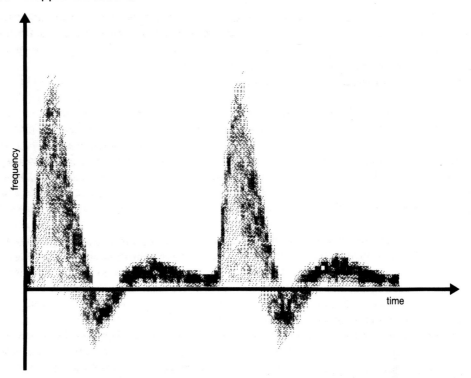

Figure 4.2 Sonogram of the Doppler signal recorded from a common femoral artery. The horizontal axis represents time, the vertical axis the Doppler shift frequency, and the intensity of each pixel the power of the signal at the corresponding frequency and time. Flow towards the transducer appears above the time axis, flow away below it

mean frequency envelopes may also be extracted from the output of the analyser. The maximum frequency envelope, or outline of the Doppler spectrum versus time, is the most commonly used parameter for Doppler waveform analysis (Chapter 10), whilst the intensity-weighted mean frequency envelope is most commonly used for computing blood flow velocity and volumetric flow (Chapter 11). Furthermore, the combination of the mean and maximum frequency gives the user some information about the instantaneous flow profiles in the vessel (for instance, for parabolic flow, the maximum frequency or velocity should be approximately twice that of the mean frequency, whereas for plug flow the two should be approximately the same).

The use of a spectral analyser with simple CW Doppler units significantly increases their cost, and therefore less expensive analysis techniques were developed specifically to extract the time-varying maximum frequency (Section 9.4) and the time-varying mean frequency envelopes (Section 9.3) from the Doppler signal.

4.2.2   Pulsed wave velocity detecting systems

Because CW Dopplers continuously transmit ultrasound they usually provide no information about the range at which movement is occurring. Whilst this may not be a problem when high-frequency ultrasound is used to study superficial vessels (particularly since the maximum range is limited anyway by rapid attenuation), it can cause considerable problems in studying deep structures, particularly the heart and vascular organs such as the brain. Even for superficial vessels it is sometimes difficult to separate the signals from arteries and veins (for example in the popliteal fossa) with CW Doppler. Pulsed Doppler systems overcome these problems by transmitting short bursts of ultrasound at regular intervals, and receiving only for a short period of time following an operator adjustable delay. The length of the delay determines approximately the range from which signals are gathered (see Section 8.4.1) (Wells 1969, Peronneau and Leger 1969, Baker 1970).

PW Doppler units differ from CW units in respect of both their electronics and their transducers. Additional circuitry is necessary to gate the transmitted and received signals at appropriate times, and to sample and hold the demodulated signals, but the demodulation process itself is identical. PW transducers contain only a single piezo-electric element which is used both to transmit and later to receive ultrasound, and contain acoustic damping so that short pulses may be transmitted and received. Further details of PW electronics and PW transducers may be found in Sections 6.3 and 5.2.1 respectively.

PW systems have one major drawback; they are only able to detect velocities unambiguously up to a finite maximum which is related to the depth of examination (Section 8.4.1.1). The maximum Doppler shift frequency a pulsed Doppler unit is able to detect is half the pulse repetition frequency (PRF). As the depth of the region of interest increases, the PRF must be decreased to allow the pulses sufficient time for the round journey, and thus for vessels deep within the body only lower velocities may be detected. The maximum velocity problem is of particular annoyance when examining high-velocity jets in the heart, and many cardiac instruments are equipped with both a PW and a CW option.

Further details of the theory of PW systems can be found in Section 8.4.1.

4.2.3 Adaptive pulsed Doppler systems

Some PW systems that do not incorporate a CW facility attempt to overcome the problems of measuring high velocities by increasing their PRF beyond the maximum required to prevent range ambiguity. Doubling the PRF will cause the device to monitor not only velocities at the depth of interest, but also velocities at a depth which is in between the probe and the depth of interest (Section 8.4.1.2). Continuing to increase the PRF will add more equally spaced sample volumes between the probe and initial depth of interest, but this does not usually present a problem as the higher velocities usually only occur at one point in the beam.

4.3 DUPLEX SYSTEMS

Duplex scanners are devices that combine a pulse-echo B-scan system (McDicken 1981a) and a Doppler system so that the Doppler shift signal can be recorded from known anatomical locations (Barber et al 1974, Phillips et al 1980). There are a number of ways of combining the two modalities, but they all share certain characteristics; the permissible directions for obtaining Doppler information all lie within the scan plane of the pulse-echo imager, and the direction of the Doppler beam at any instant is indicated by a cursor superimposed on the image. If the Doppler is PW, the position and usually the extent of the sample gate are also indicated (Fig. 11.1).

Many duplex systems combine mechanical sector scanners for imaging with a separate Doppler transducer which may, or may not, be offset from the imaging transducer (Fig. 4.3). The operator images the blood vessel or portion of the heart of interest and places the Doppler sample volume at the required location using the cursor as a guide. The image is then frozen and the Doppler probe activated. The operator may then, if the scanner configuration allows, make adjustments to the position of the sample volume using either the frozen B-scan image or a regularly updated image for guidance. Such systems do not permit the simultaneous operation of both the imaging and Doppler facilities, but with phased array systems (Section 5.2.5) where the ultrasound beam is steered electronically it is possible to share the ultrasonic information between the imaging and Doppler components and provide the user with a real-time ultrasound picture and real-time Doppler information simultaneously.

(a)   (b)

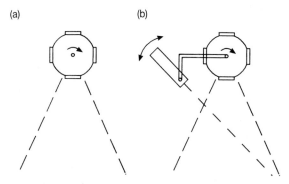

Figure 4.3   Schematic diagram of two types of duplex transducer. (a) Spinning wheel with four separate crystals, one of which is a PW Doppler crystal. When switched into Doppler mode the wheel stops with the Doppler crystal pointing in the selected direction. Once this has occurred only the range and not the direction of the sample volume can be changed. (b) Spinning wheel imaging transducer with offset Doppler probe. With this type of system the wheel usually stops spinning during Doppler recordings, but both the angle and range of the Doppler sample volume may be changed at any time

4.3.1   Velocity measurement with duplex scanners

The measurement of absolute velocity using the Doppler technique depends on knowing the angle of insonation of the Doppler beam relative to the axis of flow (eqn 1.1). Using a duplex system it is possible to image the vessel walls and hence estimate the flow axis. Duplex scanners incorporate a special angle-measuring cursor which can be rotated about the Doppler beam cursor, and which produces a direct readout of the appropriate angle (Fig. 11.1). It is also usually possible to use this angle information to automatically convert the system output from Doppler shift frequency to velocity.

4.3.2   Flow measurement with duplex scanners

The volumetric flow through a vessel is equal to the product of the average velocity of flow over the cardiac cycle and the vessel cross-sectional area. Both these quantities may be measured using a duplex scanner. The former is calculated from the instantaneous mean velocities, obtained as described above, averaged over the cardiac cycle. The vessel area is usually calculated by measuring the vessel diameter from the B-scan image, and assuming the vessel cross-section to be circular. Details of both the method and potential sources of error in duplex flow measurements are given in Section 11.2.

4.4 PROFILE DETECTING SYSTEMS

In every vessel there is a spatial variation of velocity across the vessel which varies with time (Section 2.7), and is referred to as the velocity profile. The shape of this velocity profile can be useful in calculating volume flow, and can also be used as an indicator for the presence of arterial disease. For instance, if a plaque reduces the lumen of the vessel, the velocity profile will be distorted. To date the use of velocity profile detecting systems has been largely restricted to research laboratories. However, as we will see later (Section 4.5.3), by colour coding the velocities in the vessel an image may be built up which provides the user with a much simpler display that can be more easily interpreted in a clinical situation.

Using a simple pulsed Doppler system it is possible to obtain a measure of the velocity profile in a vessel by sequentially moving the sample volume from one side of the vessel to the other, and noting the velocity information at each point of measurement (Peronneau et al 1972, Histand et al 1973). This technique is time consuming and clumsy and does not allow the simultaneous measurement of velocities across the vessel.

An alternative to using a single-gate PW Doppler is to have a multichannel Doppler system where the information from a number of range gates is processed in parallel. Velocity profiles may then be reconstructed by combining the signals from each of the range gates (Baker 1970, Keller et al 1976). The major disadvantage of this approach is that each individual range gate requires its own separate circuitry, and therefore the electronics required can become both bulky and costly. Furthermore, each of the channels must be accurately tuned and matched to prevent artefactual results and this can be both time consuming and tedious.

A much more elegant solution is to use a single serial processing system that behaves like a multigate system, and a number of instruments capable of this have been described in the literature. Analogue 'infinite gate' systems based on delay line cancellation and phase detection (Grandchamp 1975, Brandestini 1975, Nowicki and Reid 1981) showed early promise, but have been superseded by serial digital processing systems (Brandestini 1978, Hoeks et al 1981, Reneman et al 1986) which comprise a standard pulsed Doppler 'front end' followed by analogue-to-digital converters and digital circuitry. These latter systems are very versatile and their outputs may be processed to display either the velocity versus time traces for preselected gates (Fig. 4.4a), or alternatively the velocity profiles in the vessel at preselected times (Fig. 4.4b). These systems are described further in Section 6.6.1.

4.5 VELOCITY IMAGING SYSTEMS

If the sample volume of a Doppler system is moved around inside the body there will be some positions, corresponding to the locations of blood vessels, from which strong Doppler signals will be received. At other points, where large vessels are absent, there will be an absence of Doppler signals. Therefore if the position of the Doppler sample volume

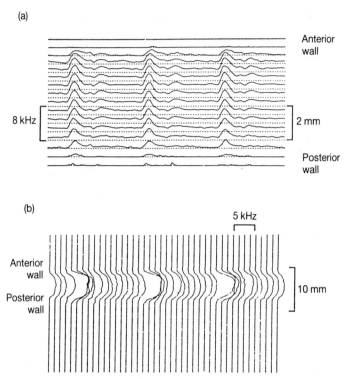

Figure 4.4   (a) The instantaneous velocity waveforms as recorded simultaneously from various gates along the ultrasound beam in the common carotid artery of a healthy subject using the system described by Hoeks et al (1981). (b) The axial velocity profiles at discrete time intervals during the cardiac cycle synthesized from the velocity waveforms in (a). (Reprinted with permission from Reneman et al (1986), © Pergamon Journals Ltd)

is mapped onto a two-dimensional display which is intensity modulated in response to either the amplitude or the frequency of the Doppler shift signal, blood flow images may be produced. This is exactly analogous to ultrasonic pulse echo imaging except that the Doppler signal rather than the size of the reflected ultrasound pulse is used to build the image.

4.5.1   Continuous wave Doppler imaging

The simplest type of Doppler imaging system consists of a focused CW transducer connected to a position sensing arm in such a way that the probe can be moved in two dimensions over the skin surface (Reid and Spencer 1972, Spencer et al 1974). The image on the screen is built up by multiple manual sweeps of the transducer over the vessel of interest. The position of the 'active' pixel of the image mimics the motion of the transducer and is intensified if a sufficiently large Doppler signal is present. This results in a plan view of the vessels (Fig. 4.5), which is the same projection as is obtained during conventional arteriography. The relatively rapid attenuation of ultrasound by tissue limits the depth from which information may be obtained to a few centimetres. In pulse echo terms this type of projection is equivalent to C-scanning.

CW imaging systems may be improved considerably by colour coding the Doppler shift

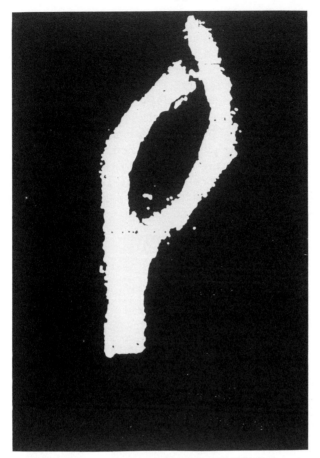

Figure 4.5 Anteroposterior projection of a carotid bifurcation built up using a raster scanning motion over the skin surface

information (Curry and White, 1978) (Plate 4.1). Most usually a temperature scale is used which progresses from dull red for low forward velocities to white for high forward velocities (Chan and Pizer 1976). Flow in the opposite direction may also be imaged using a blue temperature scale. The colour chosen for each pixel is that corresponding to the peak frequency obtained from the sample during the cardiac cycle, and this of course means that the probe must dwell at each measurement point for at least one complete heart beat. Colour coded systems are particularly useful because a reduction in vessel lumen shows up both as a reduction in image size and as a local increase in blood velocity. Because the ultrasound beam is of a finite size, the increase in blood velocity through relatively mild stenoses may be much easier to detect than the change in the diameter of the flow channel.

The long scanning time limitation of CW Doppler systems may be overcome by using a linear array transducer to sweep rapidly across the vessel and to sample each scan line several times during the cardiac cycle. Arenson et al (1982) have taken this one step further with an ingenious real-time two-dimensional 'long pulse' CW imaging system which combines a stepped linear array of Doppler crystals with a rotating mirror.

4.5.2 Pulsed wave Doppler imaging

Because they have no depth resolution, CW imaging systems can only provide a single view of the artery under examination, i.e. its projection on the skin surface. If, however, a PW Doppler system is used, it is possible to produce an image from any selected plane (Hokanson et al 1971, Mozersky et al 1971, Fish 1972), (Fig. 4.6). In order to produce a scan whose plane is perpendicular to the body surface a pulsed Doppler probe is moved in just one dimension (x-axis) across the skin; the other dimension of scan (z-axis) being produced by altering the range of the Doppler sample gate. With the probe in its initial position the delay between the pulse transmission and the sample gate opening is increased periodically (and usually automatically) in small steps corresponding to perhaps 0.5 mm of tissue. The probe is then moved in the x-scan direction in small increments and the vertical stepping of the z-axis is repeated for each value of x until the entire target vessel has been explored and mapped.

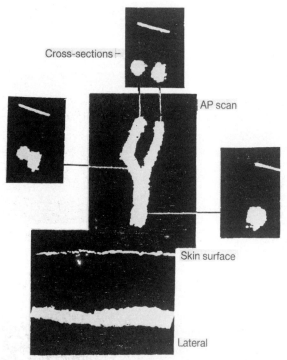

Figure 4.6   Lateral, AP, and cross-sectional projections of a carotid bifurcation, produced by a PW Doppler imaging system

Because the sample volume must remain at each point for at least one cardiac cycle this procedure can be extremely time consuming, and a complete scan may take of the order of 15 minutes. This is both wasteful of time and practically difficult because the part of the body being scanned must be kept completely immobile for this period. To overcome these difficulties Fish (1975) devised a multichannel instrument which allowed the flow to be simultaneously interrogated at 30 sites along the ultrasound beam. Further developments of the same system allow the user to display velocity profiles at various points in the cardiac cycle (Section 4.4) and to measure volumetric flow (Section 11.3) (Fish 1977,

1981). A further increase in scanning speed was achieved by Pourcelot (1979) who used a linear array transducer to sweep the pulsed ultrasound beam from side to side, but the images produced by this prototype system exhibited very poor resolution.

4.5.3   Real-time two-dimensional colour flow mapping

To a large extent the images of vessels produced by Doppler flow mapping systems and ultrasound pulse echo systems are complementary. The former provides an image of the flow channel, but not of the vascular wall; the latter an image of the healthy vessel wall, but not necessarily of the flow channel as many non-calcified lesions mimic blood with their low acoustic impedance. Thus the combination of these two types of image is potentially of great value. The first step towards producing such images was to combine a multigate Doppler imaging system with a pulse echo B-scan system (Eyer et al 1981). The pulse echo information was used to produce a monochrome image, and this was then supplemented with flow information encoded by means of a colour scale (Plate 4.2). As with previous multigate systems, the slow spatial sweep of the Doppler beam over the region of interest took 20–30 seconds, since it was necessary to dwell for a complete cardiac cycle with the Doppler transducer in each new sampling direction. (This particular device used a sector rather than a linear scanning action.) Nevertheless the system seemed capable of producing clinically useful images.

Although the system described above could not produce real-time two-dimensional Doppler images, like other multigate systems it could be used to produce real-time one-dimensional images. Eyer and his colleagues made use of this facility to colour code time–motion pulse echo scans of arteries and veins, and so to elucidate changes of flow in one section of the artery with time (Plate 4.3). (Time–motion or M-mode scans are commonly used in cardiac studies; a pulse echo transducer is fixed in a given direction and the ranges of significant interfaces are plotted on a chart as a function of time – McDicken 1981b.) One area where colour coding of the M-mode scan is of particular value is for examining the motion of blood through the heart. Flow can be coded using a red or blue scale depending on its direction, and this makes it easy to recognize abnormalities such as retrograde flow in regurgitant valvular disease (see for example Plate 16.3).

A relatively recent and exciting development in velocity imaging is the introduction of real-time colour flow mapping (CFM) systems which combine conventional real-time pulse echo B-scanning with *real-time* two-dimensional Doppler imaging (Namekawa et al 1982, Omoto et al 1984, Kasai et al 1985). As with the earlier slower systems the pulse echo information is used to produce an ordinary gray-scale image, whilst the Doppler shift information from all or part of the scan area is superimposed as a colour image. Different shades of blue and red are used to represent different velocities away from and towards the probe, and the presence of turbulence may be represented either by a mosaic pattern or by adding green to the display. In this latter case, as turbulence increases, flow coded as red on account of its direction approaches yellow, and flow coded as blue approaches cyan.

Details of a method of extracting velocity information from Doppler shift signals from many points on a line are described in Section 6.6.2. From the user's point of view, however, it is sufficient to realize that such methods provide a measure of mean velocity and its direction within each sample volume. A number of examples of images made using commercially available CFM systems can be found in Plates 4.4–4.8.

CFM systems have been found particularly useful for examining the heart where not only anatomical changes can be observed, but also the time-varying velocities that occur in valves and chambers (see Chapter 16). CFM systems have also been applied to many other sites where they can be used to make images of even relatively small vessels and to observe the flow distributions over the cardiac cycle in real time (see Chapters 14, 15 and 17).

Naturally even such sophisticated systems are limited by the same maximum velocity constraints as simple PW systems (Section 4.2.2) and care is necessary to ensure that high-velocity flow patterns are correctly interpreted.

4.6   COLOUR FLOW MAPPING SYSTEMS – PRACTICAL CONSIDERATIONS

Flow mapping is usually performed simultaneously with real-time echo imaging of the surrounding tissues. The reflected ultrasound, generated by each transmission pulse, is processed in two parallel paths to produce the pulse echo and flow map images which are then superimposed. However, it is also possible to perform separate pulse echo imaging and flow mapping. This eliminates any compromises that are necessary for simultaneous imaging. Shorter pulses can be used when only pulse echo real-time imaging is performed and velocity may be determined more accurately if all of the available time is given to the Doppler mode. Simultaneous imaging is normally preferred since flow and related tissue structures can be studied.

4.6.1   Transducers for CFM

All of the transducer types used for pulse echo imaging can be used in flow mapping, i.e. mechanical scanners, linear arrays and phased arrays (see Chapter 5). The ultrasonic frequencies employed to date are in the range 2–10 MHz. Since many blood vessels of interest run parallel to the skin surface, an angular wedge of tissue-mimicking material is often placed along the face of a linear array to give nonperpendicular insonation of the flowing blood. Alternatively the beam from the linear arrays may be electronically steered to emerge from the array at an angle to its front surface. Other transducers may also give rise to difficulties with regard to the optimum angle of insonation. As for duplex systems, the transducer type should be carefully selected to suit the application.

Transducers designed for invasive endosonography are also applied in flow mapping. The advantages and disadvantages of endosonographic probes as encountered in real-time pulse echo imaging are also present in flow mapping. In other words, the inconvenience may be more than balanced by the use of high-frequency ultrasound to obtain high-resolution images and strong Doppler signals.

Although it is not ideal to use the same transducer for pulse echo imaging and Doppler, since the latter requires longer pulses, the convenience of a single probe for accessing the tissues of interest is usually considered to justify the compromise.

4.6.2   Flow map features

The number of lines of velocity information which can be produced per second is restricted by the fact that at least three pulses must be transmitted in each beam direction (and their associated echoes received) for an estimate of the mean velocity at each point to be made. In practice around ten pulses are usually used to improve this estimate. To maintain a reasonable line density in each image frame, the field of view width, the frame rate, or the depth of penetration must be limited. The relationship between line density and these

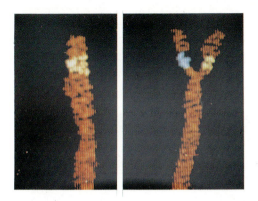

Plate 4.1 Colour-coded images of diseased carotid bifurcations. The left-hand image is from a patient with a completely occluded right internal carotid which is therefore not displayed. There was also an atheromatous plaque in the external carotid causing a stenosis, displayed as a moderately increased velocity (yellow) on the Doppler scan. the right-hand image is from a patient with a tight stenosis of the left external carotid artery displayed on the Doppler scan as markedly increased velocity (blue), and two moderate stenoses of the internal carotid, displayed as moderately increased velocity (yellow). (Reprinted with permission from Curry and White (1978), © Pergamon Journals Ltd.)

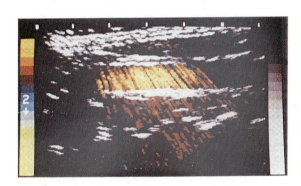

Plate 4.2 Colour-coded B-mode flow map of the common carotid artery obtained from the mid-neck region. (Reprinted with permission from Eyer et al (1981), © Pergamon Journals Ltd.)

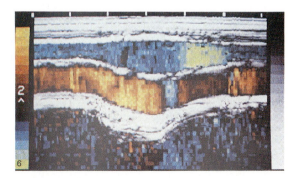

Plate 4.3 Colour-coded M-Mode image of the jugular vein and common carotid artery during a valsalva manoeuvre. Increasing time is defined to be from left to right with the entire horizontal axis covering 3 s. (Reprinted with permission from Eyer et al (1981), © Pergamon Journals Ltd.)

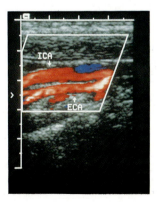

Plate 4.4 Doppler colour flow map of a normal carotid bifurcation with external carotid artery branches. (Courtesy of Acuson.)

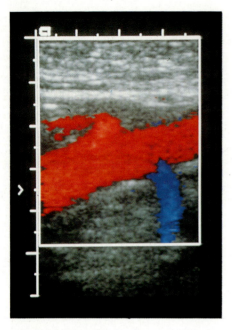

Plate 4.5 Doppler colour flow map of abdominal aorta and renal arteries. (Courtesy of Acuson.)

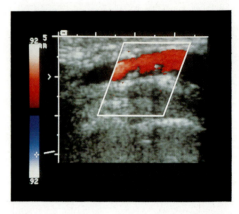

Plate 4.6 Doppler colour flow map of a femoral artery with an irregular plaque. (Courtesy of Acuson.)

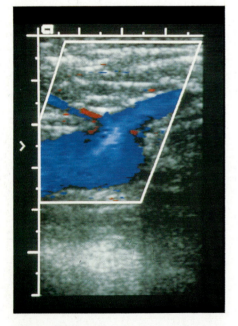

Plate 4.7 Doppler colour flow map of femorosaphenous junction with branch. (Courtesy of Acuson.)

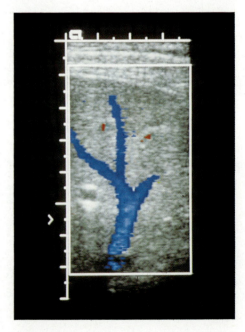

Plate 4.8 Doppler colour flow map of normal hepatic veins. (Courtesy of Acuson.)

image parameters is the same as for real-time echo imaging, i.e.:

$$\text{Line density} \propto \text{PRF}/W \cdot \text{FR} \cdot Z_{max} \cdot N_p \qquad 4.1$$

where PRF is the pulse repetition frequency, W the field of view width, FR the frame rate, Z_{max} the depth of penetration, and N_p the number of pulses transmitted along each line. Flow mapping frame rates can typically be varied from 1 to 20 per second and the depth of penetration from 2 to 20 cm. The field of view width may often be varied from the whole area visualized by the real-time pulse echo image to a small portion of it. Operating with a small width allows a high frame rate and hence more detailed representation of flow in the selected region. Blood flow may alter rapidly and may be difficult to comprehend in real time, therefore a cine-review loop facility is very valuable. An example of such a facility is one in which 60 image frames are stored in electronic memory for replay in slow motion or frame by frame.

Manufacturers often state in their literature that a large number of colour shades, e.g. 128, are employed to indicate the magnitude of velocity of blood, its direction and the presence of turbulence; for example, 64 shades of red for flow towards the transducer and 64 shades of blue for flow away, with a mosaic signifying the presence of turbulence. In practice, the operator only appreciates four or five shades in each colour, though the quality of the image is helped by having more shades present than can be immediately perceived. Most machines exhibit several options by which velocity can be colour coded and the operator can select that most suited to each application.

4.6.3   Resolution in colour flow mapping

The lateral resolution in a flow map image is determined by the ultrasound beam width and the line density. The beam width dependence of lateral resolution is as in real-time pulse echo imaging. The axial, or range, resolution in a flow map image is inferior to that of a pulse echo image since it depends on the sample volume length which in turn depends on the transmitted pulse length and the gated range length (Section 8.4.1). When the pulse echo and the flow map image are generated by processing the same echo signals, the axial resolution is usually poorer than that obtained by a purely pulse echo system, since the addition of flow mapping requires longer pulses to be transmitted. One recently introduced system avoids this compromise by using different pulses for pulse echo and Doppler imaging. From an image resolution point of view there appears to be little difference between flow maps generated by different types of transducer. This is probably due to the fact that the low line density of the flow image is a dominant factor which reduces resolution in all scanners. Typical flow map resolution figures are quoted in Table 4.1.

Table 4.1 Typical axial and lateral resolution in colour flow map systems of different ultrasonic frequency

Resolution (mm)	Frequency (MHz)		
	3	5	7
Axial	1.5	1	0.7
Lateral	3.5	2	1.5

Different shades of colour indicate different speeds of flow, e.g. dull red for low speed through increasing lightness of red to white for higher speeds. A colour bar pattern, calibrated in speed, is presented at the side of the display. Most systems will allow a particular colour or speed selected on the bar pattern to be clearly indicated on the displayed image. This can be done, for example, by altering that colour to be a completely different one, e.g. green, in both the colour bar and the image. Velocity resolution by direct observation of a colour-coded image is quite crude and is likely to be little better than 20%. The use of the calibration marker improves this situation, typically giving around 10% resolution.

Temporal resolution is directly related to the image frame rate so at best events separated by one-twentieth of a second can be resolved. At low frame rates, the temporal resolution is sacrificed with a view to establishing only the presence and direction of flow.

4.6.4   CFM system controls

Flow mapping introduces a small number of additional controls over and above those found in ultrasonic imaging and duplex systems. Duplex systems can often be updated to include flow mapping. A typical set of controls will contain about six of the following:

1. *Gain* to alter the sensitivity of the flow mapping independently of the rest of the system. TGC (time gain compensation) is not normally supplied independently of the pulse echo systems.
2. *Filters* (high and low pass) to exclude wall thump and extraneous high frequencies respectively.
3. *Threshold* to exclude the weakest signals which may be largely noise.
4. *Colour code selection* to allow alteration in the way velocity is presented in colour.
5. *Frame averaging* to provide a complete picture by ensuring that all parts of the field of view are scanned during the systolic motion of blood and to reduce the noise level in the image.
6. *ECG triggering* to generate a flow map related to a particular phase in the cardiac cycle (see Plate 16.4).
7. *Spectral analysis* to produce a sonogram corresponding to a selected site in the flow map. High-quality recording is usually possible for later spectral analysis.
8. *M-scan/flow mode* for one selected beam direction; an M-scan and flow velocities at points along that line can be presented simultaneously (see Plates 16.1, 16.3 and 16.8).
9. *Mixed displays*, for example, flow map and sonogram at a selected site (see Plate 14.6, 16.5 and 17.4).
10. *Video recording* of the data in the flow map image in an analogue or digital form for replay or later analysis.
11. *Image zoom* magnification of limited region of flow map for more detailed study.

The controls associated with colour flow mapping should confront the operator with few problems.

Colour flow mapping introduces a new dimension to Doppler studies. It is instructive to compare its attributes with those of CW and PW Doppler techniques (Table 4.2).

Table 4.2   Comparison of CW Doppler, PW Doppler and colour flow mapping features. Values quoted are averages; a large spread may be found

Mode	CW Doppler	PW Doppler	Colour flow mapping
Site examined	One large sample volume	One or line of small sample volumes	Two-dimensional array of small sample volumes
Transducer type	Multiple element	Single or multiple element	Single or multiple element
Duplex imaging	Yes	Yes	Yes
Display mode	Sonogram	Sonogram	Two-dimensional colour image and sonogram
Directional sensitivity	Yes	Yes	Yes
Turbulence detection	Spectral broadening	Spectral broadening	Mixed colour pattern
Transmitted pulses for one velocity estimate	Not applicable	50	At least three, typically ten
Aliasing limitation	No	Yes	Yes
Range ambiguity	Yes	Possible	Possible
Angle dependence	Yes	Yes	Yes
Tissue motion rejection	Yes	Yes	Yes
Velocity resolution	2%	2%	10%
Temporal resolution	10 ms	10 ms	100 ms
Output intensity (I_{spta})	50 mW cm^{-2}	500 mW cm^{-2}	100 mW cm^{-2}
Quantitative flow measurement	Possible	Possible	Possible

4.7   SUMMARY

Doppler systems have advanced from simple flow detectors to their present state of sophistication in a matter of only 30 years. There are now a wide range of devices available, each suited to a particular task and all useful in their own way. The hearts of all these instruments are very similar, and all are subject to the same physical constraints – even the most sophisticated colour flow mapping systems are subject to aliasing and the angle dependence of the Doppler shift frequency.

The purpose of this chapter has been to give the reader a general overview of the Doppler systems in common use. Some practical aspects of the use of colour flow mapping systems have been discussed. More detailed explanations of the physics, instrumentation and clinical applications of Doppler ultrasound follow throughout the rest of this book.

4.8   REFERENCES

Arenson JW, Cobbold RSC, Johnston KW (1982) Real-time two-dimensional blood flow imaging using a Doppler ultrasound array. In: Acoustical Imaging, Vol 12 (Eds EA Ash, CR Hill), pp 529–538, Plenum, New York.

Baker DW (1970) Pulsed ultrasonic Doppler blood-flow sensing. IEEE Trans Sonics Ultrasonics SU-17, 170–185.

Barber FE, Baker DW, Nation AWC, Strandness DE, Reid JM (1974) Ultrasonic duplex echo-Doppler scanner. IEEE Trans Biomed Eng BME-21, 109–113.

Brandestini M (1975) Application of the phase detection principle to a transcutaneous velocity profile meter. Proc 2nd Eur Congr Ultrasonics Med (Eds E Kazner, M de Vlieger, HR Muller, VR McCready), pp 133–144, Excerpta Medica, Amsterdam.

Brandestini M (1978) Topoflow – a digital full range Doppler velocity meter. IEEE Trans Sonics Ultrasonic SU-25, 287–293.

Chan FH, Pizer SM (1976) An ultrasonogram display system using a natural colour scale. J Clin Ultrasound 4, 335–338.

Curry GR, White DN (1978) Color coded ultrasonic differential velocity arterial scanner (Echoflow). Ultrasound Med Biol 4, 27–35.

Eyer MK, Brandestini MA, Phillips DJ, Baker DW (1981) Color digital echo/Doppler image presentation. Ultrasound Med Biol 7, 21–31.

Fish PJ (1972) Visualizing blood vessels by ultrasound. In: Blood flow measurement (Ed. VC Roberts), Chapter 3, Sector Publishing, London.

Fish PJ (1975) Multichannel, direction-resolving Doppler angiography. In: Ultrasonics in medicine: proceedings of the Second European Congress (Eds E Kazner, M de Vlieger, HR Muller, VR McCready), pp 153–159, Excerpta Medica, Amsterdam.

Fish PJ (1977) Recent progress in the field of Doppler devices. In: Recent advances in ultrasound diagnosis (Ed. A Kurjak), pp 54–63, Excerpta Medica, Amsterdam.

Fish PJ (1981) A method of transcutaneous blood flow measurement – accuracy considerations. In: Recent advances in ultrasonic diagnosis 3 (Eds A Kurjak, A Kratochwil), pp 110–115, Excerpta Medica, Amsterdam.

Franklin DL, Schlegel W, Rushmer RF (1961) Blood flow measured by Doppler frequency shift of back-scattered ultrasound. Science 134, 564–565.

Grandchamp PA (1975) A novel pulsed directional Doppler velocimeter. Proc 2nd Eur Congr Ultrasonics Med (Eds E Kazner, M de Vlieger, HR Muller, VR McCready), pp 122–132, Excerpta Medica, Amsterdam.

Histand MB, Miller CW, McLeod FD (1973) Transcutaneous measurement of blood velocity profiles and flow. Cardiovasc Res 7, 703–712.

Hoeks APG, Reneman RS, Peronneau PA (1981) A multigate pulsed Doppler system with serial data processing. IEEE Trans Sonics Ultrasonics SU-28, 242–247.

Hokanson DE, Mozersky DJ, Sumner DS, Strandness DE (1971) Ultrasonic arteriography: a new approach to arterial visualization. Biomed Eng 6, 420.

Kasai C, Namekawa K, Koyano A, Omoto R (1985) Real-time two-dimensional blood flow imaging using an autocorrelation technique. IEEE Trans Sonics Ultrasonic SU-32, 458–464.

Keller HM, Meier WE, Anliker M, Kumpe DA (1976) Non-invasive measurement of velocity profiles and blood flow in the common carotid artery by pulsed Doppler ultrasound. Stroke 7, 370–377.

McDicken WN (1981a) Diagnostic ultrasonics – principles and use of instruments, Chapter 10, B-scan instruments, pp 142–153, John Wiley, Chichester.

McDicken WN (1981b) Diagnostic ultrasonics—principles and use of instruments, Chapter 16, Time–motion scan instruments, pp 227–233, John Wiley, Chichester.

Mozersky DJ, Hokanson DE, Baker DW, Sumner DS, Strandness DE (1971) Ultrasonic arteriography. Arch Surg 103, 663–667.

Namekawa K, Kasai C, Tsukamoto M, Koyano A (1982) Real-time bloodflow imaging system utilizing autocorrelation techniques. In: Ultrasound '82 (Eds RA Lerski, P Morley), pp 203–208, Pergamon Press, New York.

Nowicki A, Reid JM (1981) An infinite gate pulse Doppler. Ultrasound Med Biol 7, 41–50.

Omoto R, Yokote Y, Takamoto S, Kyo S, Ueda K, Asano H, Namekawa K, Kasai C, Kondo Y, Koyano A (1984) The development of real-time two-dimensional Doppler echocardiography and its clinical significance in acquired valvular heart disease. Jap Heart J 25, 325–340.

Peronneau PA, Leger F (1969) Doppler ultrasonic pulsed blood flowmeter. Proc 8th Int Conf Med Biol Eng, pp 10–11.

Peronneau P, Xhaard M, Nowicki A, Pellet M. Delouche P, Hinglais J (1972) Pulsed Doppler ultrasonic flowmeter and flow pattern analysis. In: Blood flow measurement (Ed. VC Roberts), Chapter 2, pp 24–28, Sector, London.

Phillips DJ, Powers JE, Eyer MK, Blackshear WM, Bodily KC, Strandness DE, Baker DW (1980) Detection of peripheral vascular disease using the duplex scanner III. Ultrasound Med Biol 6, 205–218.

Pourcelot LG (1979) Real-time blood flow imaging. In: Echocardiology (Ed. CT Lancée), pp 421–429, Martinus Nijhoff, The Hague.

Reid JM, Spencer MP (1972) Ultrasonic Doppler technique for imaging blood vessels. Science 176, 1235–1236.

Reneman RS, Van Merode T, Hick P, Hoeks APG (1986) Cardiovascular applications of multi-gate pulsed Doppler systems. Ultrasound Med Biol 12, 357–370.

Spencer MP, Reid JM, Davis DL, Paulson PS (1974) Cervical carotid imaging with a continuous wave Doppler flowmeter. Stroke 5, 145–154.

Wells PNT (1969) A range-gated ultrasonic Doppler system. Med Biol Eng 7, 641–652.

Chapter 5

ULTRASONIC TRANSDUCERS, FIELDS AND BEAMS

5.1 INTRODUCTION

Several general discussions of transducers for medical imaging are available (Hill 1986, Wells 1977, Silk 1984, Mortimer 1982). In this chapter, transducers will be discussed from the point of view of Doppler techniques. As with ultrasonic imaging, the performance of the ultrasonic transducer is of great importance in a Doppler system. The significance of a poor transducer may not be as immediately obvious in Doppler techniques as in imaging where inferior quality images are produced. In the Doppler case poor transducers may result in additional flow signals from neighbouring vessels or noise in the Doppler signal and sonogram due to low sensitivity. Transducers designed for imaging are highly damped to generate short pulses and focused to provide a narrow beam; transducers for Doppler measurements have little or no damping since they are designed to generate long pulses or continuous wave ultrasound, and they are not necessarily focused. The best Doppler transducers are therefore more sensitive than those for imaging. Instrumentation which employs the same transducer for imaging and Doppler compromises the performance of one or other of the modes of operation. This compromise may have been deliberate on the part of the designer in order to make the instrument suited to a particular clinical application. For example, in heart studies there is only room for one small transducer at rib spaces on the chest.

The Doppler beam is the region in front of the transducer in which blood flow may be detected. The characteristics of the beam are of importance in the operation of any Doppler system. The shape of the beam is determined by a combination of the shape of the zone into which ultrasound is transmitted (the ultrasonic field) and the reception zone from which ultrasonic signals may be received. For a single element transducer used both for the transmission and reception of pulses, the ultrasonic field and the reception zone have the same shape and are orientated in the same direction (Fig. 5.1a). With a dual-element transducer the shapes of the field and reception zone are often the same but are orientated to converge in front of the transducer (Fig. 5.1b). The ultrasonic beam is then the region of overlap. In the case of multi-element arrays which employ electronic focusing, the transmission field and the reception zone have different shapes and again the beam is the region of overlap (Fig. 5.1c).

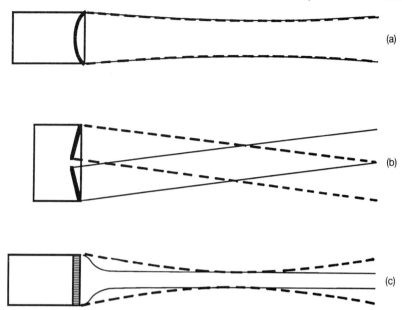

Figure 5.1 The transmission fields (dashed lines) and reception zones (solid lines) of single-element, dual-element and multi-element transducers

In pulsed Doppler techniques where the blood flow signal is obtained from a selected depth by electronic range gating (Chapter 6), the region in which blood flow is detected is called the sample volume. In techniques where the blood flow is detected over a longer range in front of the transducer, the sensitive region is usually referred to as the beam rather than the sample volume. The beam shape may result from the overlap of the transmission zone and reception zone as noted above for a dual crystal transducer or from a series of consecutive sample volumes along the axis of the transducer. The latter type of beam is employed in Doppler flow map imaging.

Distinctions between the transmission field, the reception zone, the beam and the sample volume are not always specified in the scientific literature. However, such distinctions are worth emphasizing since they increase the user's appreciation of the range over which signals are obtained with each type of instrument.

5.2 TRANSDUCERS

An ultrasonic transducer converts electrical energy into acoustic energy during transmission when its active element is excited by a voltage signal. Conversely, during reception, the acoustic energy of the returned ultrasound is converted into an electrical signal. The active elements of transducers usually depend on the piezoelectric effect for their operation. Initially transducer elements were made from naturally occurring materials such as quartz. Later, piezoelectric ceramics, for example lead zirconate titanate, were introduced and are at present still the most commonly used material. Piezoelectric plastic elements are now slowly being introduced, e.g. polyvinylidene difluoride (PVDF) (Chen et al 1978, Woodward 1977, Lancée et al 1985, Hunt et al 1983). These have acoustical properties closer to tissue than the solid ceramics, therefore the passage of ultrasound across the transducer/tissue interface is more efficient. Transducers based on a combination of ceramics and plastic are also found, for instance with a ceramic transmitting and a plastic receiving element. The design of ultrasonic transducers has been described in the scientific literature (Bainton and Silk 1980, McKeighen 1983, Hunt et al 1983), however, design details are often not completely available as they are of commercial value. The user or purchaser of transducers is therefore advised to assess them thoroughly in clinical tests. Nominally identical transducers do not always have identical performance since small variations in materials or structure can have a large influence on the final performance. In the following subsections the types of transducer encountered in equipment will be discussed with particular emphasis being put on their influence on Doppler techniques. Table 5.1 lists the types of transducer employed for each Doppler application.

Table 5.1   A guide to the type of transducer employed in different medical applications

Application	Scan type	Transducer
Cardiac	Sector	Phased array Mechanical scanner PW or CW single transducer
Superficial structures	Linear	Linear array Mechanical waterbath scanner
Paediatrics	Sector or linear	Phased array Curved array Mechanical scanner Mechanical waterbath scanner
Abdomen	Sector or linear	Phased array Curved array Mechanical scanner
Obstetrics	Sector or linear	Phased array Linear array Curved array Mechanical scanner PW or CW single transducer
Internal	Sector (90 °, 360 °) or linear	Phased array Linear array Mechanical scanner

5.2.1   Single-element transducers

A single-element transducer, as found in PW Doppler devices, is shown schematically in Fig. 5.2. Each face of the active element is coated with a thin metallic layer which acts as an electrode. Wire connections to these electrodes allow the excitation signal to be applied and the echo signal to be passed to the receiver amplifier. Focusing at a fixed range is achieved by using a lens (Tarnoczy 1965) or a concave piezoelectric element. In the latter case, to provide a flat face on the transducer and hence easier coupling to the skin a weaker convex lens is placed in front of the concave element. The combined effect of the concave element and the convex lens is designed to give focusing at the desired range. When a ceramic element is used, a single or multilayered waveplate is attached to the front face of

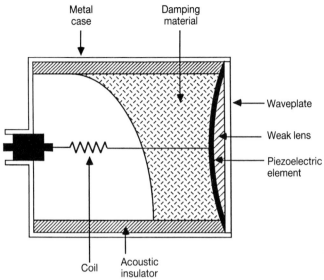

Figure 5.2   A single-element damped transducer for the generation and detection of pulsed wave ultrasound

the transducer to reduce the mismatch of acoustic impedance at the transducer/tissue interface and hence give an increased transfer of acoustic energy across the interface. Waveplate matching is most effective for CW Doppler transducers since the thickness of the plate can be made equal to one-quarter of the wavelength of the ultrasound. In the case of PW Doppler, the pulse has a frequency bandwidth and hence a spread of wavelengths which the waveplate structure cannot accommodate exactly. However, waveplating is still a worthwhile feature of pulsed transducers (Kossoff 1966). To reduce the vibrations of the active element after the excitation pulse has ceased some damping material, e.g. rubber, is bonded to its back surface. Ideally less damping is employed than in transducers designed for imaging, but it will be seen that it is fairly common practice to use the same transducer for both imaging and Doppler functions. The piezoelectric element and its attachments are supported on the end of a tube of acoustic insulator which is housed in a metal cylinder. The metal cylinder is earthed to act as a screen to prevent pick-up of electromagnetic signals. This is important in the light of the weak echo signals to be detected by the transducer. The front electrode is also earthed to complete the screened case round the transducer element. The back electrode is linked to the transmitter and receiver of the instrument. To improve the electrical matching between the element and the electronics an induction coil is sometimes placed between them. Finally the whole assembly may be inserted into a plastic tube and sealed at each end.

In the pulsed Doppler mode of operation, the transducer is excited with an electrical signal of frequency and duration corresponding to that of the required ultrasonic pulse. Typically the pulse is of length 5–10 cycles. In a few systems a short spike voltage is applied and the length of the ultrasonic pulse is determined by the ringing of the active element. This is the same technique as utilized in pulse echo imaging. It is not the best method for generating low-noise Doppler signals or of minimizing the ultrasonic exposure of the patient. The central frequency of pulsed Doppler transducers varies from 1 MHz for the examination of deep vessels to 10 MHz for superficial ones.

5.2.2   Double-element transducers

A double-element transducer, as used in CW Doppler devices, is shown schematically in Fig. 5.3. Its structure is similar to that of the PW unit except that no damping material is bonded to the back surface of the piezoelectric element. The two elements are fixed with their faces at an angle to each other. In operation, one element generates a transmission field which overlaps the reception zone of the other element. When the elements are placed side by side in this manner, they are usually D-shaped or rectangular. Another variation is one central disc element surrounded by an annular element. The operating frequencies of CW Doppler transducers cover the range from 2 MHz for fetal hearts and deep vessels to 10 MHz for vessels near the skin.

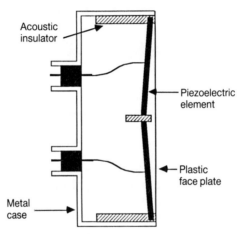

Figure 5.3   A dual-element undamped transducer for the generation and detection of continuous wave ultrasound

In Doppler fetal monitoring systems, similar transducers find application in the generation of ultrasound to detect fetal heart activity. To provide a wide beam which accommodates some fetal movement, several elements are used in the transmission and reception modes.

5.2.3   Linear array transducers

Linear array transducers are widely used for imaging (Vogel et al 1979, Kino and DeSilets 1979). In some duplex and Doppler flow map systems the linear array performs both the echo imaging and the Doppler function (Fig. 5.4a). In other array duplex systems, an independent Doppler transducer is linked to the end of the array and directs the Doppler beam into the field of view of the imager (Fig. 5.4b). When the array elements are used to generate the Doppler beam, it can be focused or defocused at the selected blood vessel. By changing the group of elements the point of origin of the beam can be moved along the array and it may also be possible to alter the angle of the beam to the array by employing beam steering techniques (Somer 1968). These two manipulations introduce a little flexibility into the application of linear arrays for Doppler, but the need to maintain contact between the flat surface of the transducer and the skin can be quite restricting. A separate Doppler transducer on a linear array is often advantageous from the points of

LINEAR ARRAYS

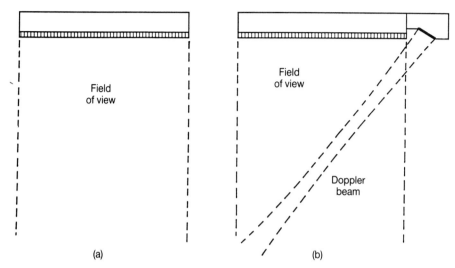

Figure 5.4   (a) A linear array in which the same transducer is employed for imaging and Doppler modes. The Doppler mode beam may be emitted at an angle across the field of view using electronic beam steering. (b) A linear array with a separate Doppler transducer

view of sensitivity and angle of vessel insonation, particularly in obstetrics and superficial vessel examination.

Linear arrays typically contain up to 200 piezoelectric elements whose operation is under computer control. There is great flexibility in the selection of the elements in use at any one time, the number ranging from 8 to 128. When 128 elements are employed sharp focusing may be obtained deep within the body due to the large size of the source. In the Doppler mode, fewer elements are normally employed since it is often desired to steer the beam at an angle to the blood vessel being studied. Since there are several elements available at any time, both CW and PW modes can be catered for with this type of transducer.

5.2.4   Curved array transducers

Introducing curvature to linear arrays has proven to be popular in imaging since the field of view is increased and contact with the skin surface is easier in some situations (Fig. 5.5a). Sharp curvature converts linear arrays into sector scanners. The smaller transducer size, or 'footprint', allows imaging and Doppler measurements to be made where access is limited (Fig. 5.5b) (Ishiyama et al 1985).

5.2.5   Phased array transducers

A modern phased array consists of 32–64 thin linear elements and has a length of 1–2 cm (Fig. 5.6) (von Ramm and Thurstone 1976, Somer 1968). The same transducer functions in both imaging and Doppler modes. As for linear and curved arrays, both CW and PW Doppler techniques can be accommodated by phased arrays. Occasionally a separate Doppler transducer is mounted next to the imaging phased array. The images produced

CURVED ARRAY

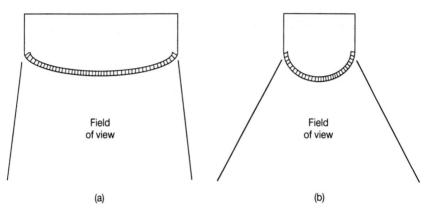

Figure 5.5   Curved arrays as used in imaging and Doppler modes

PHASED ARRAY

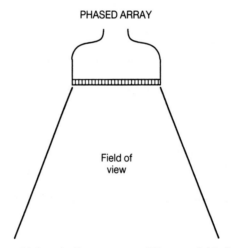

Figure 5.6   A phased array which typically generates a 90° sector field of view from a limited area of contact on the skin surface

by phased arrays have always been slightly inferior to those from linear array or single transducer scanners, since it is difficult to steer a beam without a significant amount of energy being transmitted in directions other than that of the main beam. The same problem occurs for Doppler beams from phased arrays so that moving structures well off the main beam axis may contribute spurious signals to the Doppler output from the site of interest.

Phased array transducers find widespread application in duplex and Doppler flow mapping instruments designed for cardiac examinations, since their small size and sector field of view permit access avoiding bone and lung.

5.2.6   Annular array transducers

The electronic focusing of both linear and phased arrays is only in the direction of the plane of scan, i.e. perpendicular to the length of the elements. To provide focusing in the out-of-plane direction two-dimensional arrays containing a large number of elements are required. One or two experimental arrays of this type have been developed. However, annular arrays using a smaller number of elements can achieve true electronic focusing about the axis of the transducer (Fig. 5.7). Theoretical modelling of the beam shape from an annular array indicates that at least five ring elements are necessary to generate a good quality focused beam (Melton and Thurstone 1978, Parks et al 1979, Arditi et al 1981, Dietz et al 1979). At present, commercial scanners are becoming available with eight annular elements. The flexibility of array technology allows the Doppler beam to be focused or defocused in the region of the blood vessel of interest. Annular arrays can be employed to perform both CW and PW Doppler functions.

ANNULAR ARRAY

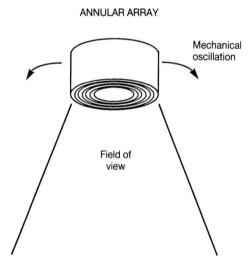

Figure 5.7   An annular array which can provide true electronic focusing about the central axis of the beam

In the quantitative measurement of blood flow it may be desirable to insonate the vessel of interest with a uniform ultrasonic beam (Chapter 11). It has been shown that a uniform beam can be generated with an annular array composed of just two elements (Evans et al 1986).

5.2.7   Internal transducers

Simple single and double-element transducers are used to examine exposed vessels during surgery or in the orifices of the body (Duck et al 1974). However, more information is obtained if the transducer is incorporated into a real-time scanner for internal application. As for external application, both duplex and flow mapping Doppler modes are valuable. At present, commercially available internal scanners are based on either rotating transducer or phased array principles of operation. It is also possible to envisage the use of linear arrays. The frequency of operation of internal Doppler units is normally in the range 5–10 MHz. When arrays or double-element transducers are used both CW and PW Doppler operations are possible.

5.3 ULTRASONIC FIELDS AND BEAMS

The wavelength of ultrasound in soft tissue for all commonly used frequencies is less than a millimetre. As a result of these short wavelengths, highly directional fields and beams can be generated using transducers with an active element of dimensions of around 1 cm. The small size of the transducers has significant consequences for the convenience and versatility of diagnostic ultrasonics. The high directionality of ultrasonic beams is of paramount importance since virtually all imaging and Doppler techniques are dependent on it. In current practice of Doppler ultrasound, these beam shapes and dimensions are rarely well known to the user. When the subject matures this uncertainty should be resolved. Improved knowledge of beam properties will benefit clinical application since the source of Doppler signals will then be identified more precisely.

5.3.1 Continuous wave fields and beams

The transmitted field from a flat disc element oscillating uniformly at a well defined frequency in a thickness mode of vibration is well documented since it is often used as an introduction to the ultrasound fields employed in medical imaging. Complete analytical solutions are available for the pressure amplitude distribution along the transducer axis and in the far field (Kinsler and Frey 1962). The pressure amplitude along the transducer axis is given by

$$p(z) = p_0 \, \sin\!\left(\frac{\pi}{\lambda}\left[(a^2 + z^2)^{\frac{1}{2}} - z\right]\right) \qquad\qquad 5.1$$

where p_0 is the maximum pressure amplitude, z is the distance along the axis, and a is the radius of the disc.

Figure 5.8a shows the plot of the intensity distribution along the transducer axis. It can be seen that there are rapid fluctuations out to a distance of a^2/λ from the transducer face. This region is known as the near or Fresnel field. The last axial maximum occurs at the end of the near field beyond which the intensity falls as the inverse square of distance. This region is known as the far or Fraunhofer field. Fluctuations also occur in the direction perpendicular to the axis (Fig. 5.8b). In the near field they are rapid whereas in the far field they are slow.

The reception zone is identical in shape to the transmission field since reception is the converse of transmission for a transducer in which there is no electronic focusing. The resultant ultrasonic beam which depends on the combination of the transmission field and the reception zone exhibits even greater fluctuations than these two individual components. It will be recalled that the beam sensitivity at each point is obtained by multiplying the transmission field and the reception sensitivity for that point.

The fluctuations in the beam are of significance in the clinical application of CW Doppler instruments. Vessels that fall within the near field are not uniformly insonated. The measured mean velocity will be in error and the maximum velocity may also be inaccurate if the high velocity coincides with a minimum in the beam sensitivity. One obvious way to ensure more uniform insonation of the moving cells is to arrange for the vessel to fall within the far field. Table 5.2 is a list of near field lengths for some typical piezoelectric element dimensions and operating ultrasonic frequencies.

The fluctuations in the beam will also increase the beam width spectral broadening effects discussed in Chapter 8.

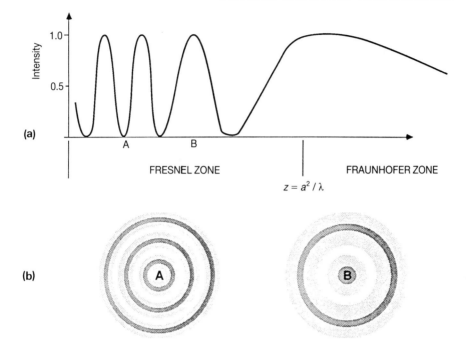

Figure 5.8   (a) Intensity variation along the axis of a continuous wave field. (b) Intensity variations, represented by the gray scale, of a continuous wave field perpendicular to the axis at points A and B

Table 5.2   Near field lengths for some typical transducer element diameters and related operating frequencies

Frequency (MHz)	Diameter (mm)	Near field length (mm)
1	20	65
2	15	73
5	8	52
10	5	42
15	3	22
20	2	16

5.3.2 Pulsed wave fields and beams

Theoretical calculation can also be made to determine the intensity distribution from transducers operating in the pulsed mode. The field may be found by summing the contributions from the frequency components of the pulse or by using the impulse response approach (Duck 1981). These calculations have been performed for circular, annular, rectangular and bowl-shaped sources.

Numerical calculations on a computer are commonly employed to predict the transmitted field from transducers. This approach is particularly appropriate in the case of transducers of complex shape, e.g. annular arrays. Good agreement is obtained between computed fields and those measured experimentally (Beaver 1974).

An example of a pulsed wave field calculated numerically is shown in Fig. 5.9. It is seen that the fluctuations in the field are much smaller than in the CW case. This is due to the minima and maxima of the pressure distribution for each frequency component of the pulse occurring at different locations and hence smoothing the field.

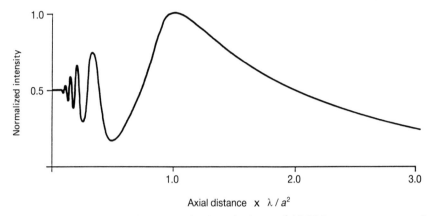

Figure 5.9   Intensity variation along the axis of a pulsed wave field. This pattern corresponds to a 3 cycle pulse. Fluctuations perpendicular to the beam are smaller than in the continuous wave case. (By courtesy of D Pye, University of Edinburgh)

5.4   DOPPLER SAMPLE VOLUMES

In the continuous wave case the sample volume is identical to the whole of the Doppler beam. Note that in a CW system the sample volume position is usually but not necessarily fixed. It can be moved by altering the positions of the transmitted field or reception zone. This type of device has been made using two single-element transducers (McHugh et al 1981) and commercially with a single-element transducer linked to a phased array.

For PW systems the length of sample volume depends on the sum of the transmitted pulse length and the length of the gated range (Section 8.4.1). The width of the sample volume is determined by the width of the beam at the position of the range gate. It is also worth noting that a change of the sensitivity of the Doppler unit will alter the effective length of the transmitted pulse and the effective beam width. The sample volume is therefore not a fixed quantity but depends to some extent on the settings of the sensitivity controls. Likewise the width of the beam in a Doppler flow mapping system also depends on the sensitivity settings.

5.4.1   CW sample volumes

Few detailed plots of the sample volume of CW beams have been published (Evans and Parton 1981, Wells 1970). Figure 5.10 illustrates some results of Evans and Parton. These distributions were obtained by using one crystal to transmit a long pulse with the other acting as an echo receiver, which is equivalent to CW operation from a beam plotting point of view. The reflecting target was a small ballbearing which was systematically moved in the region in front of the transducer. In this approach it is not the magnitude of a Doppler signal that is plotted but the echo amplitude from the ballbearing. The distributions are therefore similar but not identical to those that would be obtained using

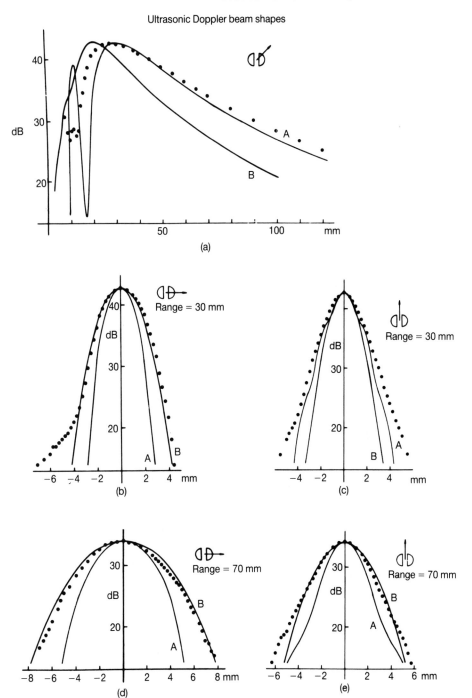

Figure 5.10 Echo amplitude plots from a dual-element Doppler transducer along the transducer axis and perpendicular to it at two different ranges (30 mm and 70 mm). The transducer operated at 3.9 MHz. The points represent experimental results, the lines the predictions of two theoretical models. (Reprinted with permission from Evans and Parton (1981), Pergamon Press Ltd)

moving blood cells as the target. These results give an appreciation of the size of a CW beam sample volume, in particular its extended range along the transducer axis.

The above plots were found to agree best with those of a model of transducer action in which the active element is considered to be restricted at its edge resulting in some apodization (i.e. a non-uniform weighting of contributions from different parts of the element). This is a realistic situation since the means of mounting a crystal usually restrict its action at the edge.

5.4.2   PW sample volumes

Pulsed wave sample volumes have also not been studied in extensive detail. Figures 5.11 and 5.12 present the results of Walker et al (1982) and Hoeks et al (1984).

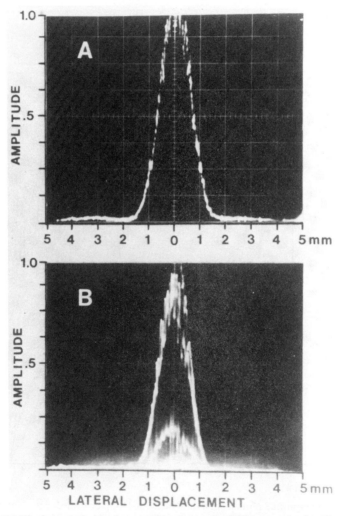

Figure 5.11   (A) The lateral shape of a sample volume of a PW beam obtained by examining the Doppler signal from a moving string. (B) The lateral shape of a sample volume of a PW beam obtained by examining the echoes from a stationary string. The transducer was a 5 MHz annular array. (Reproduced from Walker et al (1982) by permission of John Wiley & Sons, Inc.)

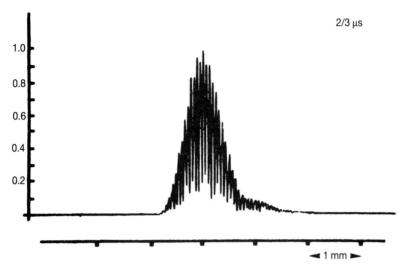

Figure 5.12   The axial sample volume shape of a PW beam at 6.1 MHz (emission duration 2/3 μs). The fine structure in the trace is due to the measurement technique. (Reproduced from Hoeks et al (1984), by permission of Pergamon Press Ltd)

The shape of the sample of a PW beam is often described as a 'teardrop' (Fig. 5.13). This arises from the combination, i.e. convolution, of the three-dimensional shape of the transmitted pulse and gated range.

The lack of results revealed by this discussion is due in part to the lack of development of convenient and accurate methods for measuring sample volumes. There is also some debate as to whether sample volume shape is best measured by a small moving point reflector or by a line of moving reflectors that mimic a stream of moving cells.

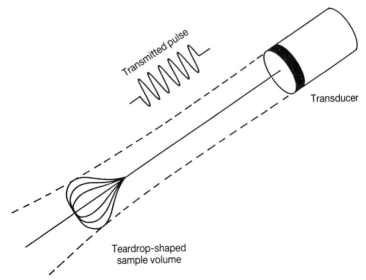

Figure 5.13   Teardrop shape of the sample volume of a PW Doppler beam

5.5 MEASUREMENT OF FIELDS, ZONES, BEAMS AND SAMPLE VOLUMES

Measurement of the ultrasonic field (transmission zone) of a transducer is undertaken by moving a small detector systematically in the region in front of the transmitting element. The transducer face and the detector (hydrophone) are normally immersed in water. The hydrophone motion is usually restricted to a plane containing the axis of the field and to a few planes perpendicular to the axis (Fig. 5.14). The latter scans by the hydrophone check on the symmetry of the field which cannot be assumed.

A typical hydrophone detecting element is a piece of piezoelectric material of diameter 0.5–1 mm, although devices have been made as small as 0.1 mm in diameter. For detailed plotting of an ultrasonic field of frequency in the range 1–10 MHz, i.e. of wavelength 1.5–0.15 mm, the latter are required. However, even with the larger devices, a useful plot of the field shape can be obtained. The development of plastic (PVDF) piezoelectric material has allowed hydrophones to be constructed in the form of a thin membrane which disturbs the field very little during the measurement (Preston et al 1983, Shotton et al 1980). For example the membrane may be 0.05 mm thick and 10 cm in diameter with a 1 mm active element at its centre (Fig. 5.15). For rapid beam plotting an array of active elements can be incorporated into the membrane (Preston et al 1985). PVDF hydrophones are stable, have a known frequency response and are easy to use, requiring only an amplifier and a fast oscilloscope. Piezoelectric elements are also mounted on needles to act as hydrophones (Lewin 1981). This approach lends itself to the manufacture of very small devices.

The signal from a hydrophone at any point is proportional to the wave pressure at that point. Values of the pressure amplitude are recorded and displayed numerically or as shades of gray in a two-dimensional contoured image of the section through the beam (Figs 5.16a and 5.16b) (Whittingham and Roberts 1986). The intensity at any point is usually calculated using the square of the pressure amplitude. This latter procedure is not strictly valid in the near field of the transducer where the pressure and particle velocity are

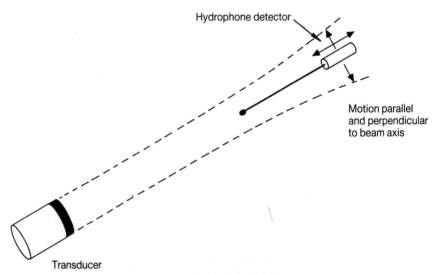

Figure 5.14 Schematic diagram of a hydrophone field plotting system

Figure 5.15   A membrane hydrophone. The active element is a small disc of diameter less than 1 mm at the centre. (By courtesy of R Preston)

not in phase. Current practice is therefore to characterize the beam in terms of pressure rather than intensity. It should also be remembered that the beam characteristics measured in water are different from those in tissue where dispersive absorption is present and non-linear propagation is less significant. Nevertheless, a great deal can be deduced about a beam in tissue from a plot obtained in water.

The plotting of a reception zone in isolation would require a point transmitter to be moved in a regular fashion in front of the transducer. Another approach would be to calculate the shape of the reception zone from the transmitted field shape and the beam shape. Reception zone shapes are rarely determined in practice.

An approximation to a Doppler beam shape may be obtained by operating the transducer in a pulse echo mode in which the reflecting target is a small ballbearing. As the target is systematically moved in front of the transducer the echo is detected after amplification in the receiver of the pulsed Doppler unit. A long ultrasonic pulse of 10 cycles or more simulates the operation of a CW unit. This approach is reasonable when the properties of the transducer are under study. However, if the influence of the whole system on the beam shape is to be known, a small moving target that generates a Doppler shift signal is required. Test objects for this purpose based on moving string, ballbearings and liquid jets are described in detail in Chapter 12.

An elegant way of visualizing an ultrasonic field is known as the 'schlieren' method (Willard 1947, Wells 1977, Follett 1986). It depends on the fact that pressure fluctuations in an ultrasonic field cause fluctuations in the refractive index of water. In a schlieren system, a wide beam of light is directed through a water tank and brought to a sharp focus using a lens or mirror. When no ultrasound is transmitted into the tank, a stop placed at the focus blocks the light beam. However, when ultrasound is present the light is refracted in the region where the refractive index is varying and hence bypasses the stop. The optical system then forms an image of the ultrasound field adjacent to the stop. This image can be

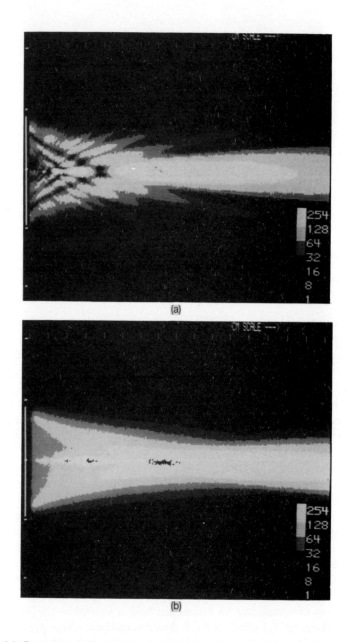

Figure 5.16 (a) Gray-tone hydrophone plot of the pressure amplitudes in the field of a CW transducer. (b) Gray-tone hydrophone plot of the pressure amplitudes in the field of a PW transducer. (Reproduced from Whittingham and Roberts (1986), by permission of The Institute of Physical Sciences in Medicine)

viewed directly, projected on to a screen or photographed. The technique functions for both continuous and pulsed wave ultrasound. In the pulsed case, the optical beam is supplied by a stroboscope which is synchronized to the ultrasonic pulses. At low and medium diagnostic intensity levels, the schlieren image may be qualitatively interpreted as exhibiting higher brightness at regions of higher intensity. Quantitative interpretation is much more difficult due to the complex three-dimensional structure of the ultrasonic field. The value of a schlieren system is that it can provide a quick appreciation of field characteristics, e.g. focal region, asymmetry or pulse dimensions (Fig. 5.17).

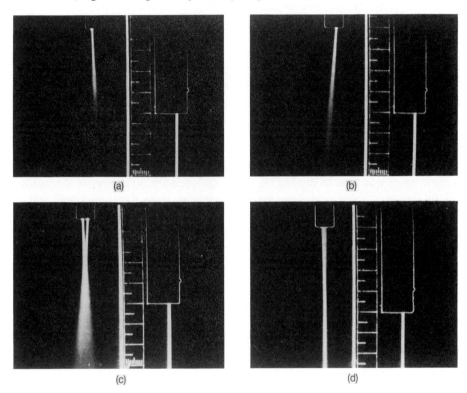

Figure 5.17 Schlieren photographs of ultrasound transmitted into water from the crystals of a continuous wave Doppler transducer. The transducer is a 10 MHz device with rectangular crystals of dimension 5 mm × 2 mm and separated by 0.75 mm. The scale shows 1 cm and 0.5 cm divisions. (a) One crystal transmitting. (b) The other crystal transmitting. (c) A double exposure obtained by driving each crystal in turn to demonstrate overlap of transmission and reception zones. If the crystals are driven simultaneously a strong inference pattern is obtained. (d) One crystal transmitting with the transducer rotated 90° about its axis relative to the position corresponding to the top pictures. The lack of circular symmetry in the field is illustrated. (By courtesy of DH Follett, Bristol, UK)

5.6 STERILIZATION OF TRANSDUCERS

Transducers are often very expensive items and care is required in their sterilization. Steam sterilization in an autoclave is most effective but will probably damage the transducer or the bonding of its case. The most practical approach is gas sterilization with ethylene oxide but this normally takes several days. A degree of sterilization can be achieved with antiseptic liquids. If there is doubt about sterility professional advice should be sought, but apart from autoclaving no guarantees of complete sterility are usually forthcoming.

5.7   DAMAGE TO TRANSDUCERS

Dropping transducers is the most common cause of damage. Smaller impacts can often damage the windows of mechanical transducers of duplex systems. Replacement of the windows is usually an expensive repair. The life of mechanical scanners can be greatly extended by ensuring that they are not moving when they are not in use. A few machines automatically switch off the drive when the transducer is replaced in its holder. Although electronic array transducers are more robust than mechanical scanning devices, when they are damaged it is not usually possible to repair them.

Care should be taken in selecting the cleaning liquid for use on a probe as some are known to attack plastic and rubber. It is probably best to use water.

5.8   SUMMARY

In this chapter the transmitted field was distinguished from the reception zone and the ultrasonic beam was seen to depend on their combined effect. The types of transducers found in simple, duplex and flow mapping systems were then described and their suitability for different applications indicated. The ultrasonic fields, beams and sample volumes associated with CW and PW equipment were discussed. Finally care of transducers was considered.

5.9   REFERENCES

Arditi M, Foster FS, Hunt JW (1981) Transient fields of concave annular arrays. Ultrasonic Imaging 3, 37–61.

Bainton KF, Silk MG (1980) Some factors which affect the performance of ultrasonic transducers. Br J NDT January, 15–20.

Beaver WL (1974) Sonic nearfields of a pulsed piston radiator. J Acoust Soc Am 56, 1043–1048.

Chen WH (1983) Analysis of the high-field losses of polyvinylidene fluoride transducers. IEEE Trans Sonics Ultrasonics SU-30, 238–249.

Chen WH, Shaw HJ, Weinstein DG, Zitelli LT (1978) PVF2 transducers for NDE. Proc IEEE Ultrasonics Symp, pp 780–783.

Dietz DR, Parks SI, Linzer M (1979) Expanding-aperature annular array. Ultrasonic Imaging 1, 56–75.

Duck FA (1981) The pulsed ultrasonic field. In: Physical aspects of medical imaging (Ed. BM Moores, RP Parker, BR Pullan), Wiley, Chichester.

Duck FA, Hodson CJ, Tomlin PJ (1974) Esophageal Doppler probe for aortic flow velocity monitoring. Ultrasound Med Biol 1, 233–241.

Evans DH, Parton L (1981) The directional characteristics of some ultrasonic Doppler blood-flow probes. Ultrasound Med Biol 7, 51–62.

Evans JM, Skidmore R, Wells PNT (1986) A new technique to measure blood flow using Doppler ultrasound. In: Report No. 47, Physics in medical ultrasound (Ed. JA Evans), Institute of Physical Sciences in Medicine, London.

Follett DH (1986) Light diffraction by ultrasound as evidence of finite amplitude distortion. Proc Inst Acoustics 8, 55–62.

Hill CR (1986) The generation and structure of acoustic fields. In: Physical principles of medical ultrasonics (Ed. CR Hill), Ellis Horwood, Chichester.

Hoeks APG, Ruissen CJ, Hick P, Reneman RS (1984) Methods to evaluate the sample volume of pulsed Doppler systems. Ultrasound Med Biol 10, 427–434.

Hunt JW, Arditi M, Foster FS (1983) Ultrasound transducers for pulse-echo medical imaging. IEEE Trans Biomed Eng BME-30, 453–481.

Ishiyama K, Yanagawa T, Sato T, Yano M, Yoshikawa N (1985) Development of small radius convex scanning system with view angle of 98 degrees. Proc 4th Mtg World Fed Ultrasound Med Biol, p 542.

Kino GS, DeSilets CS (1979) Design of slotted transducers with matched backing. Ultrasonic Imaging 1, 189–209.

Kinsler LE, Frey AR (1962) Fundamentals of acoustics, 2nd edn, Wiley, New York.

Kossoff G (1966) The effects of backing and matching on the performance of piezoelectric ceramic transducers. IEEE Trans Sonics Ultrasonics SU-13, 20–30.

Lancée CT, Souquet J, Ohigashi H, Bom N (1985) Ferro-electric ceramics versus polymer piezoelectric materials. Ultrasonics 23, 138–142.

Lewin PA (1981) Miniature piezoelectric polymer ultrasonic hydrophone probes. Ultrasonics 19, 213–216.

McHugh R, McDicken WN, Thompson P, Boddy K (1981) Blood flow detection by an intersecting zone ultrasonic Doppler unit. Ultrasound Med Biol 7, 371–375.

McKeighen RE (1983) Basic transducer physics and design. Seminars Ultrasound 4, 50–59.

Melton HE, Thurstone FL (1978) Annular array design and logarithmic processing for ultrasonic imaging. Ultrasound Med Biol 4, 1–12.

Mortimer AJ (1982) Physical characteristics of ultrasound. In: Essentials of medical ultrasound (Eds MH Repacholi, DA Benwell), Humana, Clifton New Jersey.

Parks SI, Linzer M, Shawker TH (1979) Further development and clinical evaluation of the expanding aperture annular array system. Ultrasonic Imaging 1, 378–383.

Preston RC, Bacon DR, Livett AJ, Rajendran K (1983) PVDF membrane hydrophone performance properties and their relevance to the measurement of the acoustic output of medical ultrasound equipment. J Phys E: Sci Instrum 16, 786–796.

Preston RC, Taylor GEM, Thompson RC, Zeqiri B, Livett AJ (1985) The performance of the ultrasound beam calibrator BECA2. Proc. 4th Mtg World Fed Ultrasound Med Biol, p 537.

Shotton KC, Bacon DR, Quilliam RM (1980) A PVDF membrane hydrophone for operation in the range 0.5 MHz to 15 MHz. Ultrasonics 18, 123–126.

Silk MG (1984) Ultrasonic transducers for non-destructive testing, Adam Hilger, Bristol.

Somer JC (1968) Electronic sector scanning for ultrasonic diagnosis. Ultrasonics 6, 153–159.

Tarnoczy T (1965) Sound focusing lenses and waveguides. Ultrasonics 3, 115–127.

Vogel J, Bom N, Ridder J, Lancée C (1979) Transducer design considerations in dynamic focusing. Ultrasound Med Biol 5, 187–193.

von Ramm OT, Thurstone FL (1976) Cardiac imaging using a phased array ultrasound system. 1. System design. Circulation 53, 258–262.

Walker AR, Phillips DJ, Powers JE (1982) Evaluating Doppler devices using a moving string test target. J Clin Ultrasound 10, 25–30.

Wells PNT (1970) The directivities of some ultrasonic Doppler probes. Med Biol Eng 8, 241–256.

Wells PNT (1977) Biomedical ultrasonics, Academic Press, London.

Whittingham TA, Roberts TJ (1986) Practical beamshapes as visualised by the Newcastle beam plotting system. In: Report No. 47, Physics in medical ultrasound (Ed. JA Evans), Institute of Physical Sciences in Medicine, London.

Willard GW (1947) Ultrasound waves made visible. Bell Labs Rec 25, 194–200.

Woodward B (1977) The stability of polyvinylidene fluoride as an underwater transducer material. Acoustica 38, 264–268.

Chapter 6

BASIC DOPPLER ELECTRONICS AND SIGNAL PROCESSING

6.1 INTRODUCTION

The physical principles of both continuous and pulsed wave Doppler systems have been outlined in Chapter 4. The practical implementation of these principles will be outlined in this chapter which is intended primarily for those who have an interest in understanding the detailed workings of such devices.

6.2 CONTINUOUS WAVE SYSTEMS

Figure 6.1 is a block diagram of a typical CW Doppler system. The master oscillator operates at a constant frequency and drives the transmitting crystal of the probe via a transmitting amplifier. The returning ultrasound signal, containing echoes from both stationary and moving targets, is fed to the radio frequency (RF) amplifier by the receiving crystal. This amplified signal is then demodulated and filtered to produce audio frequency signals (usually in phase quadrature) whose frequencies and amplitudes provide information about motion within the ultrasound beam.

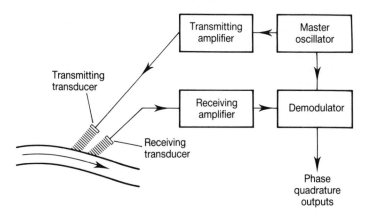

Figure 6.1 Block diagram of a continuous wave Doppler system

6.2.1 The master oscillator

The master oscillator is the heart of the Doppler system, and provides both the operating reference frequency, which is amplified by the transmitter circuit, and reference signals for the demodulation stage. It is important that the oscillator has high short-term stability so that artificial Doppler shift noise is not generated. Nowadays the necessary signals are provided by purpose-built integrated circuit crystal oscillator chips.

6.2.2 The transmitter

The function of the transmitter or transmitting amplifier is to drive the transmitting crystal with the correct time-varying voltage in order to produce the required acoustic signal. For continuous wave devices the voltage applied across the crystal is usually in the range of 2–5 V peak to peak. For a 50 Ω matched system the transmitter output stage has to be capable of delivering 100 mA. For single-frequency CW systems the transmitter is usually a transformer coupled output stage. A typical example of such a circuit is illustrated in Fig. 6.2. For multifrequency systems, where the frequencies range from 2 to 10 MHz, the output stage has to be wideband and is usually in the form illustrated in Fig. 6.3. These wideband transmitter systems do not excite the crystal with a sine wave, but with a square wave; the higher harmonics do not prove a problem as they are filtered by the narrow bandwidth of the piezoelectric crystal.

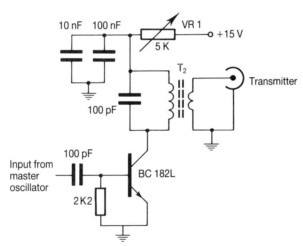

Figure 6.2 Transmitter stage for a single-frequency continuous wave Doppler velocimeter. The output transformer is tuned to the frequency of operation and the variable resistor VR1 provides adjustment for the overall output voltage

6.2.3 The probe

The CW Doppler probe consists of two crystals, one acting as a transmitter while the other acts as a receiver. The most usual configuration is known as a 'split D' and is produced by splitting a circular transducer. It is usual to angle each crystal so that a region of sensitivity is defined by the overlap of the transmission field of one

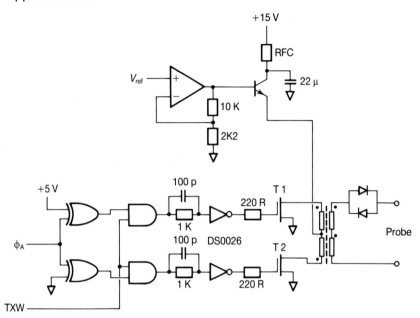

Figure 6.3   Circuit diagram of a wideband transmitter system for continuous and pulsed wave applications. ϕ_A, operating clock frequency; TXW, transmission burst width (PW operation only); V_{ref}, output control voltage

transducer and the reception field of the other. The piezoelectric crystals are attached to coaxial cables which in turn are connected to the transmitter and receiver. The crystals are housed and separated by an acoustic insulator which reduces the crosstalk between the transmitter and receiver. Excessive crosstalk will overload the RF amplifier and demodulator stages in the Doppler device leading to degradation of the directional separation of forward and reverse flow signals. Further details of Doppler transducers and the acoustic fields they produce may be found in Chapter 5.

6.2.3.1   Probe matching

Most continuous wave Doppler pencil probes do not contain electrical matching networks (Section 5.2.1). Consequently there can be variation of the acoustic output power from one probe to another. Furthermore, the variation in output impedance of the receiving transducer will cause variations in matching to the receiving amplifier which will in turn produce variations in signal-to-noise ratio. Changing probes can sometimes give the user the impression that one probe is more sensitive than another; although this may be true in certain circumstances, the apparent increase in sensitivity is mainly due to the increase in acoustic output power where there is a drop in the impedance of the transmitted crystal. Ideally the transmitting crystal and the receiving crystal should be matched to the connecting cable which is either 75 or 50 Ω resistive.

In order to obtain an acoustic match between the piezoelectric crystal and the body, a matching layer of epoxy resin of quarter wavelength thickness can be used. However, for the higher frequencies of operation this thickness becomes practically too small and sometimes three-quarter wave matching is used instead. In practice inexpensive CW probes forego this rigour and the thickness of araldite is not well controlled giving rise to considerable variation of sensitivity from one probe to another.

6.2.4 The receiver

The function of the receiver or receiving amplifier is to amplify the small voltages produced by the receiving crystal, and for Doppler shifted signals themselves this is of the order of 10–100 μV. However, leakage from the transmitter to receiver crystal constitutes a received signal approaching 10 mV. A typical receiver amplifier stage is illustrated in Fig. 6.4, where transformer matching has been employed in order to optimize signal-to-noise characteristics of the FET amplifier. The receiver must have

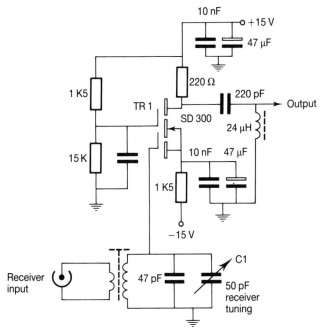

Figure 6.4 Circuit diagram of a continuous wave receiver. The variable capacitor C1 is tuned for optimum sensitivity at the frequency of operation

a sensitivity of better than 1 μV RMS, and a dynamic range capability in excess of 80 dB. For single-frequency CW devices the RF amplifier need only have a limited bandwidth in the region of 200 kHz; however, for wideband systems the receiver bandwidth needs to be of the order of the range of operating frequency, usually 2–10 MHz. A typical receiver circuit for a wideband system is illustrated in Fig. 6.5.

6.2.5 The demodulator

The purpose of the demodulator is to remove the carrier frequency and provide an output consisting of the Doppler sidebands. All but the simplest Doppler units now

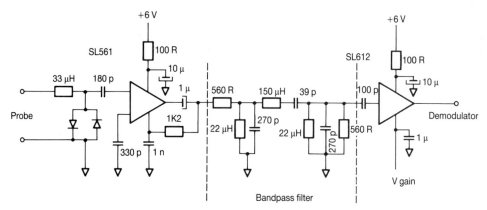

Figure 6.5   Circuit diagram of a wideband continuous or pulsed wave Doppler receiver. The 33 μH coil and the diode combination are protection from the transmission burst for pulsed Doppler applications. The bandpass filter illustrated is for a 2 MHz system. The final amplification stage is a voltage-controlled amplifier

use coherent demodulation techniques (i.e. the reference signal for demodulation is taken from the master oscillator rather than the clutter signal that arises from stationary or near stationary tissue). Straightforward coherent demodulation does not provide directional information because both upper and lower Doppler sidebands (corresponding to flow towards and away from the probe) are shifted into the same region of the baseband, and therefore alternative techniques must be used. Three such solutions that have been successfully employed are single sideband detection, heterodyne detection, and quadrature phase detection, and each of these has been discussed by Coghlan and Taylor (1976) and Atkinson and Woodcock (1982).

Single sideband (SSB) detection (see Fig. 6.6) is achieved by directly filtering the returning RF signal with a pair of low and high pass filters to remove the upper and lower sidebands respectively. The two channels are then independently coherently demodulated to produce separate audio signals, one composed of flow away from the probe, the other of flow towards the probe. A major problem with this method is that the filters must be exceedingly sharp (with a Q-factor approaching 10^6) if proper channel separation is to be obtained. The only practical method of achieving this is to use multistage crystal filters. A practical SSB detection system, which employs frequency conversion prior to the single sideband filter, has been described by deJong et al (1975).

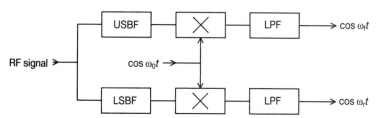

Figure 6.6   Single sideband detection system. USBF, upper sideband filter; LSBF, lower sideband filter; LPF, low pass filter

Heterodyne demodulation is achieved by coherently demodulating the returning ultrasound signal with a signal that is of a slightly lower frequency than that of the master oscillator, but which is derived from it by mixing its output with that of a heterodyne oscillator (Fig. 6.7).

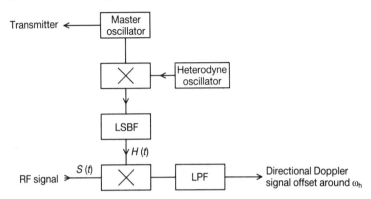

Figure 6.7   Heterodyne detection system. LSBF, lower side band filter

For the sake of simplicity consider an ultrasound signal that is composed of three distinct components: the carrier and two signals resulting from motion towards and away from the probe respectively. Such a signal may be written:

$$S(t) = A_0 \cos(\omega_0 t + \phi_0) + A_f \cos(\omega_0 t + \omega_f t + \phi_f) + A_r \cos(\omega_0 t - \omega_r t + \phi_r) \quad 6.1$$

where A, ω and ϕ refer to the amplitude, angular frequency and phase of each signal, and the subscripts 0, f and r to the carrier, forward and reverse signals. This composite signal is coherently demodulated with a difference signal, $H(t)$, derived by mixing the output of the master oscillator with that of the heterodyne oscillator, which may be written:

$$H(t) = \cos(\omega_0 t - \omega_h t) \quad 6.2$$

where ω_h is the angular frequency of the heterodyne oscillator. (Note that there is a second component of angular frequency, $\omega_0 + \omega_h$, which must be removed before demodulation takes place.)

The product of $S(t)$ and $H(t)$ is given by:

$$S(t) \cdot H(t) = \tfrac{1}{2}\{A_0 \cos(\omega_h t + \phi_0) + A_0 \cos(2\omega_0 t - \omega_h t + \phi_0)$$
$$+ A_f \cos(\omega_h t + \omega_f t + \phi_f) + A_f \cos(2\omega_0 t + \omega_f t - \omega_h t + \phi_f)$$
$$+ A_r \cos(\omega_h t - \omega_r t + \phi_r) + A_r \cos(2\omega_0 t - \omega_r t - \omega_h t + \phi_r)\} \quad 6.3$$

Eliminating all frequencies of the order of ω_0 by low pass filtering leaves:

$$\tfrac{1}{2}\{A_0 \cos(\omega_h t + \phi_0) + A_f \cos[(\omega_h + w_f)t + \phi_f] + A_r \cos[(\omega_h - \omega_r)t + \phi_r]\} \quad 6.4$$

Provided that the heterodyne frequency is higher than the highest Doppler difference frequency received from targets receding from the Doppler transducer, the reverse and forward Doppler components are completely separated and situated on either side of the heterodyne frequency, ω_h. The large clutter component that occurs at the heterodyne frequency (the first term in eqn 6.4) must be removed by a very sharp notch filter centred at that frequency. This type of output is particularly suited to spectral analysis since both the forward and reverse flow signals can be analysed using a single analysis channel. A practical heterodyne demodulation system has been described by Light (1970).

The third and most widely used type of directional demodulation is quadrature phase detection. The object of this type of demodulation is to preserve the real and imaginary Doppler difference components, and this is achieved by coherently demodulating the returning Doppler signal both with the master oscillator and with a signal derived from the master oscillator but shifted in phase by 90° (Fig. 6.8).

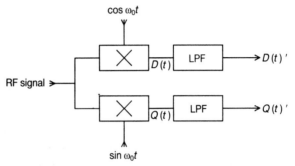

Figure 6.8   Quadrature phase detection system. LPF, low pass filter

Consider first the effect of multiplying the returning ultrasound signal given by eqn 6.1 by $\cos(\omega_0 t)$; the result is a 'direct' signal $D(t)$ given by:

$$D(t) = \tfrac{1}{2}A_0[\cos(\phi_0) + \cos(2\omega_0 t + \phi_0)]$$
$$+ \tfrac{1}{2}A_f[\cos(\omega_f t + \phi_f) + \cos(2\omega_0 t + \omega_f t + \phi_f)]$$
$$+ \tfrac{1}{2}A_r[\cos(\omega_r t - \phi_r) + \cos(2\omega_0 t + \omega_r t + \phi_r)] \qquad 6.5$$

Filtering out the DC component and terms of order $2\omega_0$ leads to:

$$D(t)' = \tfrac{1}{2}A_f \cos(\omega_f t + \phi_f) + \tfrac{1}{2}A_r \cos(\omega_r t - \phi_r) \qquad 6.6$$

Multiplying the signal given by eqn 6.1 by $\sin(\omega_0 t)$ leads in a similar way to a filtered 'quadrature' signal, $Q(t)'$, given by:

$$Q(t)' = -\tfrac{1}{2}A_f \sin(\omega_f t + \phi_f) + \tfrac{1}{2}A_r \sin(\omega_r t - \phi_r) \qquad 6.7$$

or

$$Q(t)' = \tfrac{1}{2}A_f \cos(\omega_f t + \phi_f + \pi/2) + \tfrac{1}{2}A_r \cos(\omega_r t - \phi_r - \pi/2) \qquad 6.8$$

It can be seen by comparing eqn 6.6 with eqn 6.8 that for a signal resulting from flow which is solely towards the probe the direct signal lags the quadrature signal by 90°, whilst for flow that is solely away from the probe the direct signal leads the quadrature signal by 90°. However, before forward and reverse flow can be completely separated, some type of further processing must occur (see Section 6.4).

Practically, demodulation is achieved using commercially available demodulator integrated circuits. Figure 6.9 illustrates such a circuit for phase quadrature demodulation. Because of the transmitter receiver breakthrough, the demodulator needs to have a large dynamic range so that it can detect low-level signals due to

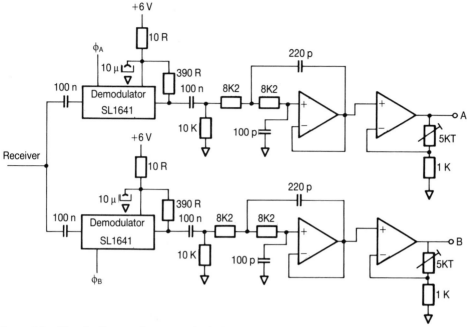

Figure 6.9   Circuit diagram for a practical phase quadrature demodulation system. ϕ_A, master clock; ϕ_B, master clock phase shifted by 90°; A, B, phase quadrature outputs

moving blood. Acoustic breakthrough is particularly severe at low frequencies, and with 2 MHz systems such as those used in cardiology it is usually necessary to filter the receiver signal with a crystal notch filter before demodulation can be successfully achieved.

It is important for phase quadrature demodulation that the two clock signals are truly shifted by 90°. This is best achieved using a logic technique such as that illustrated in Fig. 6.10.

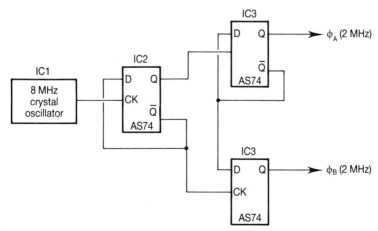

Figure 6.10   Logic system required for obtaining phase quadrature reference signals. Note that the master oscillator is set to four times the operating frequency of the Doppler system

6.2.6   Filters

The demodulator output signals are filtered so that the high-frequency noise is removed and the low-frequency signals due to wall motion are also removed. For systems that have fixed filtering, standard operational amplifier filtering techniques are used. However, for variable filtering, switch filter techniques are now usually employed. An example of a switch filtering circuit which employs custom-built integrated circuits is illustrated in Fig. 6.11.

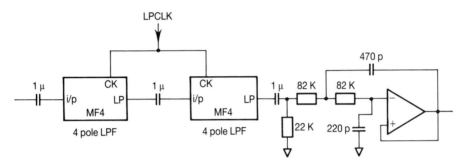

Figure 6.11   Example of a switched filter technique to provide a variable 8-pole low pass filter. The cutoff frequency is set by the LPCLK clock frequency. Similar configurations may be used for high pass filter applications

6.3   PULSED WAVE SYSTEMS

Pulsed Doppler systems are used to obtain Doppler information at a specific range from the face of the transducer. Figure 6.12 is a block diagram of a typical pulsed Doppler system. The main difference between a CW and a pulsed system is that the transducer is excited with bursts of pulses instead of being continuously excited. The burst of ultrasound travels into the body where it is Doppler shifted by moving structures along the sound path. Returning echoes from both stationary and moving targets are received by the same transducer. This process is then repeated for the next burst of ultrasound. The Doppler shift information is extracted by the demodulation process in exactly the same way as it is for the CW system. The demodulated signal contains the range-phase information, i.e. for each burst of ultrasound the difference in phase between the reference signal (provided by the master oscillator) and the received echo at the specified range. The output from the demodulator is sampled at a

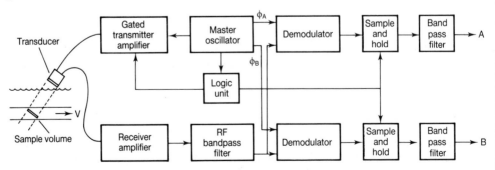

Figure 6.12   Block diagram of a pulsed Doppler system

specified point in time relative to the onset of the transmission pulse. The time of sampling defines the depth of interest and is chosen to correspond to the time it takes for sound to travel from the transducer to the depth of interest and back again. The sampling is achieved using a sample and hold amplifier which is updated after every transmission burst. The output from the sampler is filtered to remove the sampling frequency, thereby providing the Doppler shift frequency information at the range of interest. The physical principles of PW Doppler are discussed further in Section 8.4.1.

6.3.1   The transmitter

Pulsed Doppler probes are usually excited with peak-to-peak voltages of between 20 and 100 V and with a burst length ranging between 1 and 10 μs. The main constraint on the transmitter is that it has to have a sufficient bandwidth to cope with such a signal. Also, because a single element transducer is normally permanently connected to the receiver input, the output noise from the transmitter must be kept to an absolute minimum. An on–off ratio of greater than 150 dB is usually adequate. See Fig. 6.3.

6.3.2   The probe

As the length of the transmission burst is very small compared with the reception time, a single crystal transducer can be used which operates firstly as a transmitter of ultrasound and then as a receiver. As the reception path is identical to the transmission path, the problem of alignment is avoided. The probe is electrically matched to the input impedance of the receiver and is acoustically backed and matched so that it provides an ultrasound burst that matches the electrical excitation voltage produced by the transmitter, i.e. the bandwidth of the probe has to be as large as the bandwidth of the transmitted signal. See Chapter 5 for further details.

6.3.3   The receiver

The receiver amplifier receives signals from the probe. The signals from stationary targets and the small signals from moving blood are amplified equally. Therefore, the receiver requires a large dynamic range in order to avoid saturation from vessel wall signals. Furthermore, the receiver needs to be protected from the transmission burst as its input is directly connected to the transmitter. This is most commonly achieved using a protection network of the type shown in Fig. 6.5, which blocks the larger signals from the transmitter but allows the passage of small signals from the reflected and back-scattered ultrasound. The receiver must also have the capability to recover quickly from saturation and requires a bandwidth that is inversely proportional to the axial resolution of the system.

6.3.4   The demodulator

The phase quadrature demodulation system used by most CW systems may also be incorporated in the pulsed Doppler system where the requirements are similar. The required bandwidth is inversely proportional to the axial resolution of the system.

6.3.5   The sample-and-hold amplifier

The sample-and-hold circuit samples the output from the demodulator stage, and it is common practice to integrate over the sample period and then hold this integrated value until the next sampling period. A suitable circuit is illustrated in Fig. 6.13. The sampling duration together with the transmitted pulse length sets the range over which velocity information is gathered (see Section 8.4.1). For optimum performance, the length of the transmission burst is set to the same length as the sampling or receiver gate length (Peronneau et al 1974). It is desirable to reduce the transmitted output voltage when the transmitted pulse length is increased so that the transmitted ultrasound power may be kept constant. Most pulsed Doppler systems allow the user to vary the sample volume length, to match the vessel diameter. The required bandwidth of the system is inversely proportional to the gate length and so an improvement in signal to noise can be achieved by increasing gate length.

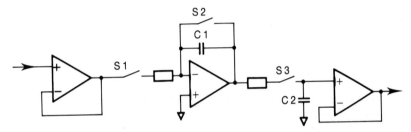

Figure 6.13   Pulsed Doppler integrating sample and hold system. Switch S1 samples the signal for a period defined by the gate length. During this time the signal is integrated. The final integrated voltage is then sampled and held via switch S3. The integrator is then reset by switch S2 and the sequence restarted on the next pulsed repetition cycle

6.3.6   Filters

The output from the sample-and-hold circuit contains not only the Doppler shift frequencies but also the sampling frequency which has to be removed. In order that the full range of Doppler shift frequencies can be utilized, i.e. up to half the pulse repetition frequency (PRF), a very sharp (at least 8 pole) low pass filter is used to eliminate the sampling frequency without degrading the Doppler signal. This filter must also be variable if a variable PRF system is used and this is best achieved using switched capacitor filters (see Fig. 6.11). As well as the low pass filter which removes the PRF, a high pass filter is also employed to remove the high-energy low-frequency signals that result from moving structures such as the heart and vessel walls. The high pass filter or 'wall' filter is usually variable over a range between 100 and 800 Hz, and is best implemented using switched capacitor filtering techniques.

6.3.7   Control logic

The control logic provides the necessary timing signals to implement the transmitter burst, demodulator phase quadrature reference signals, and to initiate the sample-and-hold amplifiers at the appropriate time relative to the transmission burst. Figure 6.14 shows a typical timing diagram for a pulsed Doppler system.

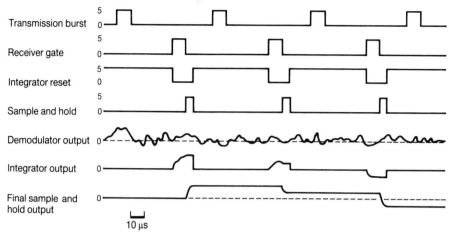

Figure 6.14   Timing diagram illustrating the sequence of events during the sampling of a Doppler signal in a PW system

6.4   DIRECTION DETECTION

If, as is now usual, the output from the Doppler system is in phase quadrature form, further processing is necessary to completely separate forward and reverse flow signals. Three methods of achieving this have been described: time domain processing, phase domain processing, and frequency domain processing (Coghlan and Taylor 1976).

6.4.1   Time domain processing

Time domain processing was developed by McLeod (1967) and is illustrated in Fig. 6.15. A logic unit is used to determine which of the two quadrature signals (described by eqns 6.6 and 6.8) is in the lead and which is lagging. Depending on this relationship the output of one of the quadrature channels (the one chosen is unimportant) is switched to either the forward flow or the reverse flow channel. This simple method works correctly only when the flow is unidirectional, because if simultaneous forward and reverse signals are present the phase relationship between the direct and quadrature signals is indeterminate. Furthermore, the method is susceptible to switching artefacts, particularly in the presence of high-amplitude low-frequency components such as vessel wall thump (Coghlan and Taylor 1976).

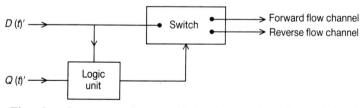

Figure 6.15   Time domain processor for use with quadrature phase detected signals

6.4.2   Phase domain processing

Equations 6.6 and 6.8 can be solved simultaneously to extract the forward and reverse components, and this is in essence the method by which phase domain processing is able to separate the two flow channels. The basis of the method by which this is achieved is illustrated in Fig. 6.16. Both the direct and quadrature channels are phase shifted by 90° and added to the other (unshifted) channel, and this results in two completely separate flow channels.

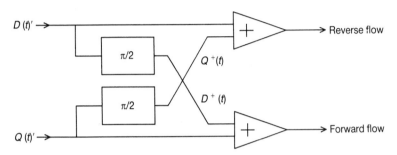

Figure 6.16   Phase domain processor for use with quadrature phase detected signals

Phase shifting the 'direct' signal given by eqn 6.6 leads to:

$$D^+(t) = \tfrac{1}{2}A_f \cos(\omega_f t + \phi_f + \pi/2) + \tfrac{1}{2}A_r \cos(\omega_r t - \phi_r + \pi/2) \qquad 6.9$$

Adding this signal to the unshifted 'quadrature' signal (eqn 6.8) eliminates the reverse flow component and gives:

$$F(t) = A_f \cos(\omega_f t + \phi_f + \pi/2) \qquad 6.10$$

Similarly shifting the 'quadrature' signal by 90° gives:

$$Q^+(t) = \tfrac{1}{2}A_f \cos(\omega_f t + \phi_f + \pi) + \tfrac{1}{2}A_r \cos(\omega_r t - \phi_r) \qquad 6.11$$

which when added to the unshifted direct channel (eqn 6.6) eliminates the forward component and yields:

$$R(t) = A_r \cos(\omega_r t - \phi_r) \qquad 6.12$$

There are a number of equivalent but more practical methods of achieving phase domain separation (Nippa et al 1975, Coghlan and Taylor 1976), and a modern practical circuit for this is shown in Fig. 6.17a. The phase shifting, which must be accurately 90° over a wide range of frequencies, is achieved by a series of networks such as those illustrated in Fig. 6.17b. The RC values for each stage may be calculated from tables (Bedrosian 1960).

6.4.3   Frequency domain processing

The third type of processing, frequency domain processing, is illustrated in Fig. 6.18. Both the direct and quadrature signals, $D'(t)$ and $Q'(t)$, are mixed with quadrature signals from a pilot oscillator. This results in the forward and reverse flow components being separated on either side of the pilot frequency (ω_p).

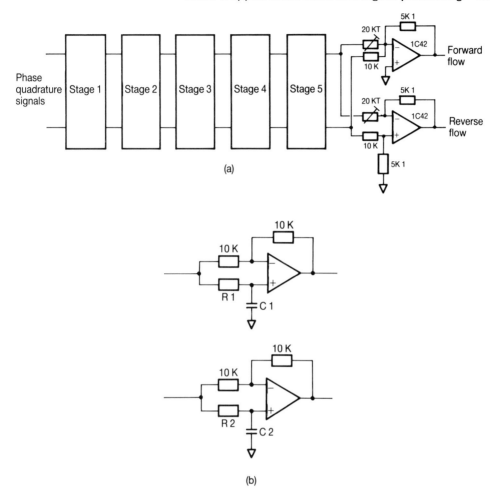

(a)

(b)

Figure 6.17   (a) Practical circuit for separating forward and reverse flow components using phase domain processing. Each stage consists of a pair of operational amplifiers as shown in (b). The RC values for each stage may be calculated from tables (Bedrosian 1960)

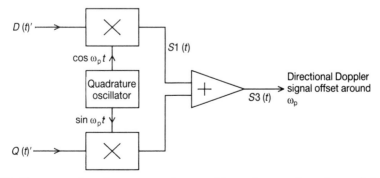

Figure 6.18   Frequency domain processor for use with quadrature phase detected signals

Multiplying eqn 6.6 by $A_p \cos(\omega_p t)$ where A_p and ω_p are the amplitude and angular frequency of the signals from the pilot oscillator gives:

$$S1(t) = \tfrac{1}{2}A_p\{A_f \cos(\omega_f t + \phi_f) \cos(\omega_p t) + A_r \cos(\omega_r t - \phi_r) \cos(\omega_p t)\} \qquad 6.13$$

which can be expanded to give:

$$S1(t) = \tfrac{1}{4}A_p\{A_f[\cos(\omega_p t - \omega_f t - \phi_f) + \cos(\omega_p t + \omega_f t + \phi_f)]$$
$$+ A_r[\cos(\omega_p t - \omega_r t + \phi_r) + \cos(\omega_p t + \omega_r t - \phi_r)]\} \qquad 6.14$$

Multiplying eqn 6.8 by $A_p \sin(\omega_p t)$ results in:

$$S2(t) = \tfrac{1}{4}A_p\{A_f[-\cos(\omega_p t - \omega_f t - \phi_f) + \cos(\omega_p t + \omega_f t + \phi_f)]$$
$$+ A_r[\cos(\omega_p t - \omega_r t + \phi_r) - \cos(\omega_p t + \omega_r t - \phi_r)]\} \qquad 6.15$$

Finally, adding eqns 6.14 and 6.15 results in:

$$S3(t) = S1(t) + S2(t) = A_p/2\{A_f \cos[(\omega_p + \omega_f)t + \phi_f] + A_r \cos[(\omega_p - \omega_r)t + \phi_r]\} \qquad 6.16$$

Practical aspects of implementation of this type of circuit have been discussed by Coghlan and Taylor (1976).

All three types of directional processing have been widely used, but for reasons mentioned earlier the time domain method is not recommended. Both phase and frequency domain processing give good channel separation if properly implemented; the former produces separate forward and reverse signals whilst the latter produces a single channel of output where the forward and reverse components are disposed on either side of a pilot frequency. Dual channel outputs may be useful for some types of analogue envelope extraction (see Section 6.5), whilst a single channel output is particularly suited to spectrum analysis since only one analysis channel is necessary to display flow in both directions (see Fig. 4.2).

6.5   ANALOGUE ENVELOPE DETECTORS

Many of the Doppler processing methods described later in this book (see Chapters 9, 10 and 11) rely on an envelope signal (such as the maximum or mean frequency waveform) derived from the Doppler audio signal. The best method of deriving such a signal is probably to Fourier transform the Doppler signal and digitally calculate the envelope signal, but analogue envelope detectors are still in widespread use and are therefore briefly discussed below.

6.5.1   Intensity weighted mean frequency processors

Under ideal conditions (see Chapter 9) the output of the intensity weighted mean frequency follower is proportional to volumetric flow, and has thus gained widespread popularity. Various methods have been proposed for deriving the mean frequency envelope, but the best and most popular are those of Arts and Roevros (Arts and Roevros 1972, Roevros 1974), and the so-called 'root f' followers of DeJong et al (1975) and Gerzberg and Meindl (1977, 1980). A phase-lock loop method has been described by Sainz et al (1976) but the details of the performance of this method with wideband signals are not clear. Block circuit diagrams of both the Arts and

Roevros and 'root f' followers are shown in Fig. 6.19, and details of the operation of the former are given below. This particular method has the advantage of operating directly on the phase quadrature signals.

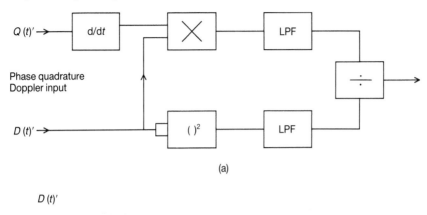

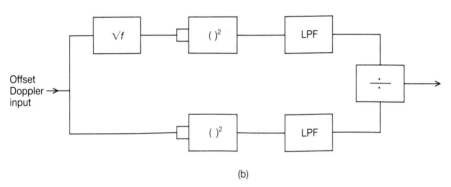

Figure 6.19   (a) The Arts and Roevros circuit for calculating the intensity weighted mean frequency from phase quadrature inputs. (b) The root f circuit for calculating the mean frequency from offset Doppler inputs. LPF, low pass filter

Consider two Doppler signals of frequency ω_1 and ω_2 having amplitudes of A and B respectively. Referring to eqns 6.6 and 6.8 shows that they will result in two quadrature signals given by:

$$D(t)' = A \cos(\omega_1 t + \phi_1) + B \cos(\omega_2 t + \phi_2) \qquad 6.17a$$

$$Q(t)' = -A \sin(\omega_1 t + \phi_1) - B \sin(\omega_2 t + \phi_2) \qquad 6.17b$$

Figure 6.19a shows the basic sequence of operations which leads to 'numerator' and 'denominator' signals which are then divided to form the mean. Consider first the numerator. Differentiating the quadrature signal $Q(t)'$ leads to:

$$dQ'/dt = -\omega_1 A \cos(\omega_1 t + \phi_1) - \omega_2 B \cos(\omega_2 t + \phi_2) \qquad 6.18$$

Multiplying this by the direct signal $D(t)'$ gives:

$$\text{Num}(t) = -\tfrac{1}{2}\omega_1 A^2 - \tfrac{1}{2}\omega_2 B^2 \tag{6.19a}$$

$$-\tfrac{1}{2}\omega_1 A^2 \cos(2\omega_1 t + 2\phi_1) - \tfrac{1}{2}\omega_2 B^2 \cos(2\omega_2 t + 2\phi_2) \tag{6.19b}$$

$$-\tfrac{1}{2}(\omega_1 + \omega_2)AB\{\cos[(\omega_1 - \omega_2)t + \phi_1 - \phi_2] + \cos[(\omega_1 + \omega_2)t + \phi_1 + \phi_2]\} \tag{6.19c}$$

This signal is then low pass filtered to leave only the DC and quasi-DC components. This removes the terms in eqn 6.19b and, provided ω_1 and ω_2 are not too similar, the terms in eqn 6.19c, to leave:

$$\text{Num}(t) = -\tfrac{1}{2}\omega_1 A^2 - \tfrac{1}{2}\omega_2 B^2 \tag{6.20}$$

In the case of a wideband signal from a Doppler unit, some difference components will leak through and these all contribute to the noise in the output signal. Fortunately this noise is not of sufficient size to invalidate the use of the processor.

The denominator signal is formed by squaring and low pass filtering the direct signal $D(t)'$. Squaring eqn 6.17a leads to:

$$\text{Denom}(t) = \tfrac{1}{2}A^2 + \tfrac{1}{2}B^2 \tag{6.21a}$$

$$+\tfrac{1}{2}A^2 \cos(2\omega_1 t + 2\phi_1) + \tfrac{1}{2}B^2 \cos(2\omega_2 t + 2\phi_2) \tag{6.21b}$$

$$+AB\{[\cos(\omega_1 - \omega_2)t + \phi_1 - \phi_2] + [\cos(\omega_1 + \omega_2)t + \phi_1 + \phi_2]\} \tag{6.21c}$$

Filtering as before leads to:

$$\text{Denom}(t) = \tfrac{1}{2}A^2 + \tfrac{1}{2}B^2 \tag{6.22}$$

Once again for a wideband signal some of the difference components will leak through and contribute to the noise. Finally, dividing the numerator by the denominator signal leads to:

$$\frac{\text{Num}(t)}{\text{Denom}(t)} = \frac{-(\omega_1 A^2 + \omega_2 B^2)}{A^2 + B^2} \tag{6.23}$$

which is the intensity weighted mean of the two original frequencies. A more complete proof that the circuit gives the true intensity weighted mean is given by Arts and Roevros (1972).

6.5.2 Other analogue frequency processors

A number of other types of analogue processor have been described, including the widespread but unsatisfactory zero-crossing detector (Franklin et al 1961, Flax et al 1970, Lunt 1975) which was at one time popular because of its simplicity and low cost, and the maximum frequency follower (Sainz et al 1976, Skidmore and Follett 1978) which is more usually implemented using digital techniques (Gibbons et al 1981). Theoretical aspects of the performance of these and a number of other 'signal location estimators' are considered in Chapter 9.

6.6 COMPLEX DOPPLER SYSTEMS

Several types of complex Doppler systems have been described in Chapter 4, including multigate profile detecting systems and real-time colour flow mapping systems. All of these systems (with the exception of infinite gate systems) share the same basic front-end design as simple PW systems, and it is only after quadrature phase detection that they differ in the way in which they handle the resulting range-phase signal. The detailed electronic arrangement of these complex systems is beyond the scope of this book, but the operation of two of the most important types of system are described in the following sections.

6.6.1   Multigate pulsed Doppler systems

Multigate pulsed Doppler systems may use either parallel or serial processing. If parallel processing is chosen, separate circuitry will be necessary for each gate and the resulting electronics will rapidly become both bulky and costly as the number of channels increases (Section 4.4). A much better solution is to use serial digital signal processing (Brandestini 1978, Hoeks et al 1981, Reneman et al 1986).

The arrangement of the serial processing system described by Hoeks et al is illustrated in Fig. 6.20. As far as the analogue-to-digital converters (ADCs), the system is the same as a standard single gate PW Doppler system employing quadrature phase detection (see Fig. 6.8); from this point on until the display the system is purely digital. Because of the large dynamic range of the quadrature signals (due in particular to the unwanted but large low-frequency signals from tissue interfaces) the ADCs are required to be 12-bit devices. The output of each of the ADCs is filtered by a pair of first-order high pass filters (HPFs) in order to reject the low-frequency clutter signal. The basic design of these filters is as illustrated in Fig. 6.21; the high pass output at a given time is the difference between the input and a running average of input signals from the same range. The number of memory locations in the filter determines the number of gates that can be used, and the multiplication factor K (which controls the proportion of the difference signal contributing to the running average) the filter cutoff frequency. The filtered quadrature signals are fed into a pair of zero-crossing detectors (ZCD) where each is compared with a threshold to produce a single bit of information (above or below threshold). The output from the comparators is stored in a memory, and if different from the previous processing cycle (indicating a zero-crossing) a signal of plus or minus one, depending on the detected direction of flow, is put on the input of the final low-pass filter (LPF). This filter is of the same design as those used to high pass filter the input (Fig. 6.21) with a cutoff frequency of either 6 or 12 Hz. During each processing cycle the instantaneous velocity distribution along the ultrasonic beam appears at the output of the low pass filter, and this digital signal may be further processed to display either the velocity versus time traces for preselected gates (Fig. 4.4a) or alternatively the velocity profile in the vessels at preselected times

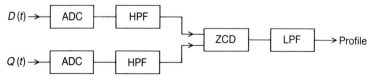

Figure 6.20   Block diagram of a multigate pulsed Doppler system with serial data processing. The input signals are taken from a circuit such as that illustrated in Fig. 6.8. All abbreviations are given in the text

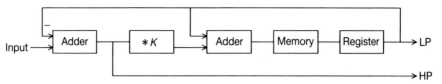

Figure 6.21   Block diagram of the first-order high pass/low pass filters used by Hoeks et al (1981)

(Fig. 4.4b). In addition most multigate systems, including the one described here, allow the user to select any channel for both audio presentation and full spectral analysis.

6.6.2   Real-time colour flow mapping systems

Real-time colour flow mapping systems (CFMs) combine conventional real-time grayscale imaging of anatomy with real-time colour depiction of blood flow (some examples of the type of images produced by this method are shown in Plates 4.4–4.8). A major technical difficulty in achieving real-time 'flow imaging' is one of spectral resolution, because the ultrasound beam can only be allowed to remain in any one orientation for a very short time if the image is to be regularly updated. If, for example, an image is composed of 50 raster lines, and is updated 20 times per second, the beam can only sample a given direction for 1 ms. Estimation of the mean frequency of a signal segment this short using fast Fourier analysis or the simple zero-crossing detector can only be very crude, so an alternative technique must be sought (the spectral resolution of a Fourier estimate is at very best the reciprocal of the data length, in this case 1/0.001 s or 1 kHz). The problem was first overcome by Namekawa et al (1982) using an autocorrelation technique to estimate both the mean frequency and, as a bonus, the variance of the Doppler spectrum. The theory of their method and its practical implementation are described below.

The mean angular frequency $\bar{\omega}$ of a Doppler power spectrum $P(\omega)$ may be defined as:

$$\bar{\omega} = \frac{\displaystyle\int_{-\infty}^{\infty} \omega\, P(\omega)\, \mathrm{d}\omega}{\displaystyle\int_{-\infty}^{\infty} P(\omega)\, \mathrm{d}\omega} \qquad\qquad 6.24$$

whilst its variance σ^2 is given by:

$$\sigma^2 = \frac{\displaystyle\int_{-\infty}^{\infty} (\omega - \bar{\omega})^2\, P(\omega)\, \mathrm{d}\omega}{\displaystyle\int_{-\infty}^{\infty} P(\omega)\, \mathrm{d}\omega} \qquad\qquad 6.25$$

$$= \overline{\omega^2} - (\bar{\omega})^2 \qquad\qquad 6.26$$

Furthermore, the autocorrelation function $R(\tau)$ is related to $P(\omega)$ by the Wiener–Khinchin theorem, i.e.:

$$R(\tau)=\int_{-\infty}^{\infty} P(\omega)\, e^{j\omega\tau}\, d\omega \qquad\qquad 6.27$$

Differentiating eqn 6.27 with respect to τ leads to two further relationships, namely:

$$\dot{R}(\tau)=j\int_{-\infty}^{\infty}\omega\, P(\omega)\, e^{j\omega\tau}\, d\omega \qquad\qquad 6.28$$

and

$$\ddot{R}(\tau)=-\int_{-\infty}^{\infty}\omega^{2}\, P(\omega)\, e^{j\omega\tau}\, d\omega \qquad\qquad 6.29$$

Equations 6.24 and 6.26 may now be written in terms of the autocorrelation functions for zero lag to give:

$$\bar{\omega}=-j\,\frac{\dot{R}(0)}{R(0)} \qquad\qquad 6.30$$

and

$$\sigma^{2}=\left(\frac{\dot{R}(0)}{R(0)}\right)^{2}-\frac{\ddot{R}(0)}{R(0)} \qquad\qquad 6.31$$

It is possible to evaluate eqns 6.30 and 6.31 directly but, as Kasai et al (1985) pointed out, this is rather time consuming, and they showed that if the autocorrelation function is treated as:

$$R(\tau)=\left|R(\tau)\right|\, e^{j\phi\tau} \qquad\qquad 6.32$$

the following approximations are valid:

$$\bar{\omega}=\dot{\phi}(0)\simeq\phi(T)/T \qquad\qquad 6.33$$

and

$$\sigma^{2}\simeq\frac{2}{T^{2}}\left(1-\frac{\left|R(T)\right|}{R(0)}\right) \qquad\qquad 6.34$$

where T is the time between subsequent ultrasonic pulses. The mean angular frequency and the variance of the Doppler signal may thus be calculated from its autocorrelation magnitudes and phases at $\tau=0$ and $\tau=T$.

The circuit of the complex autocorrelator used to derive the real and imaginary parts of the autocorrelation function is shown in Fig. 6.22. The two inputs are digital phase quadrature signals which have been filtered to remove signals from stationary and quasi-stationary tissue, and are represented by $x(t)$ and $y(t)$ (for a schematic of the

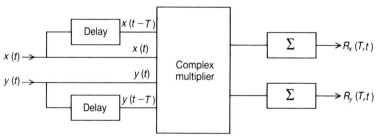

Figure 6.22   Complex autocorrelator described by Kasai et al (1985)

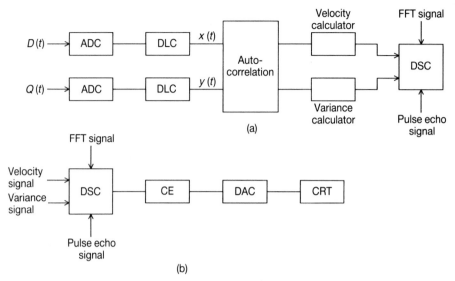

Figure 6.23   (a) The Doppler section of the real-time colour flow mapping system described by Kasai et al (1985). The input signal is as for Fig. 6.20. (b) Block diagram of the display module of a CFM system. All abbreviations are given in the text

entire Doppler component of the CFM system see Fig. 6.23). The complex multiplier performs the following computation:

$$Z(t) = [x(t) + jy(t)] [x(t-T) - jy(t-T)] \qquad 6.35$$

The real and imaginary parts of $Z(t)$ are then summed separately to give the real and imaginary parts of the autocorrelation denoted $R_x(T, t)$ and $R_y(T, t)$ respectively, i.e.

$$R_x(T, t) = \sum_{t-nT}^{t} \text{Re}[Z(t)] \qquad 6.36$$

$$R_y(T, t) = \sum_{t-nT}^{t} \text{Im}[Z(t)] \qquad 6.37$$

Finally eqns 6.33 and 6.34 are evaluated using the relationships:

$$\phi(T, t) = \tan^{-1} \frac{R_y(T, t)}{R_x(T, t)} \qquad 6.38$$

$$|R(T, t)| = [R_x^2(T, t) + R_y^2(T, t)]^{\frac{1}{2}} \qquad 6.39$$

and

$$R(0, t) = \sum_{t-nT}^{t} x^2(t) + y^2(t) \qquad 6.40$$

A schematic diagram of the Doppler component of the CFM system described by Namekawa et al (1982) and Kasai et al (1985) is shown in Fig. 6.23a. As in the case of the Hoeks system described in Section 6.6.1, the front end is similar to that of a standard single-gate PW system employing quadrature phase detection (although some degree of time-dependent gain may be necessary to compensate for increasing attenuation of signals from deeper within the body). The phase quadrature signals $D(t)$ and $Q(t)$ are digitized and then passed through delay line cancellers (DLCs) to eliminate the large signals from stationary and nearly stationary tissue. These signals are then fed to the complex autocorrelator which performs the operations described in the last paragraph, and thence to the velocity calculator and variance calculator. The results of both calculations are recorded in a digital scan converter (DSC), as are the B-mode or M-mode images that have been obtained using conventional pulse echo techniques, and the FFT analysed spectrum from any preselected sampling point.

The information in the DSC is further processed as illustrated in Fig. 6.23b. The colour encoder (CE) codes the Doppler shift information according to whether the phase of $\phi(T, t)$ falls in the first and second quadrant (i.e. $0 < \phi < \pi$) or in the third and fourth quadrants ($-\pi < \phi < 0$), as red or blue respectively; the faster the blood flow i.e. the greater $|\phi|$, the brighter the colour becomes. It should be noted that, as with a conventional pulsed Doppler system, if the pulse repetition frequency is not adequate, aliasing will occur and the colour will be wrongly encoded. The variance of the signal, which represents turbulence, is coded as shades of green and added to the velocity colours, so that as turbulence increases the red shades tend to yellow, and the blue shades to cyan. The B-mode image (or M-mode image) and the FFT-analysed data are converted to black and white signals in the conventional way. The output from the colour encoder is transformed to an analogue signal by the DAC and the composite image displayed on an RGB monitor in real time.

Table 6.1   Summary of demodulation techniques

Method	Type of output	Comments
Non-coherent	Single channel	No direction resolving capability; forward and reverse flow treated identically
Single sideband	Dual channel	Channel A contains forward flow, channel B contains reverse flow
Heterodyne	Single channel	Heterodyne frequency corresponds to zero velocity, lower frequencies to reverse flow and higher frequencies to forward
Quadrature detection + time domain processing	Dual channel	Only one channel switched in at any one time; confused by simultaneous forward and reverse flow
Quadrature detection + phase domain processing	Dual channel	Channel A contains forward flow, channel B contains reverse flow
Quadrature detection + frequency domain processing	Single channel	Heterodyne frequency corresponds to zero velocity, lower frequencies to reverse flow and higher frequencies to forward

6.7   SUMMARY

In this chapter the basic building blocks from which all pulsed and continuous wave Doppler units are constructed have been examined. A number of Doppler demodulation techniques have been described and their attributes are summarized in Table 6.1. Even the most advanced colour flow mapping systems share the same basic front-end design as simple PW units and suffer from the same limitations in terms of range and velocity ambiguity.

6.8   REFERENCES

Arts MGJ, Roevros JMJG (1972) On the instantaneous measurement of blood flow by ultrasonic means. Med Biol Eng 10, 23–34.

Atkinson P, Woodcock JP (1982) Doppler ultrasound and its use in clinical measurement, pp 54–74, Academic Press, London.

Bedrosian SD (1960) Normalised design of 90 degree phase-difference networks. IRE Trans Circuit Theory CT-7, 128–136.

Brandestini M (1978) Topoflow – a digital full range Doppler velocity meter. IEEE Trans Sonics Ultrasonics SU-25, 287–293.

Coghlan BA, Taylor MG (1976) Directional Doppler techniques for detection of blood velocities. Ultrasound Med Biol 2, 181–188.

DeJong DA, Megens PHA, DeVlieger M, Thon H, Holland WPJ (1975) A directional quantifying Doppler system for measurement of transport velocity of blood. Ultrasonics 13, 138–141.

Flax SW, Webster JG, Updike SJ (1970) Statistical evaluation of the Doppler ultrasonic blood flowmeter. Biomed Sci Instrum 7, 201–222.

Franklin DL, Schlegel W, Rushmer RF (1961) Blood flow measured by Doppler frequency shift of back-scattered ultrasound. Science 134, 564–565.

Gerzberg L, Meindl JD (1977) Mean frequency estimator with applications in ultrasonic Doppler flowmeters. In: Ultrasound in medicine 3B (Eds D White, RE Brown), pp 1173–1180, Plenum Press, New York.

Gerzberg L, Meindl JD (1980) The root f power-spectrum centroid detector: system considerations, implementation and performance. Ultrasonic Imaging 2, 262–289.

Gibbons DT, Evans DH, Barrie WW, Cosgriff PS (1981) Real-time calculation of pulsatility index. Med Biol Eng Comp 19, 28–34.

Hoeks APG, Reneman RS, Peronneau PA (1981) A multigate pulsed Doppler system with serial data processing. IEEE Trans Sonics Ultrasonics SU-28, 242–247.

Kasai C, Namekawa K, Koyano A, Omoto R (1985) Real-time two-dimensional blood flow imaging using an autocorrelation technique. IEEE Trans Sonics Ultrasonics SU-32, 458–464.

Light LH (1970) A recording spectrograph for analyzing Doppler blood velocity signals in real time. J Physiol 207, 42–44.

Lunt MJ (1975) Accuracy and limitations of the ultrasonic Doppler blood velocimeter and zero crossing detector. Ultrasound Med Biol 2, 1–10.

McLeod FD (1967) A directional Doppler flowmeter. Digest 7th Int Conf Med Biol Eng, p 213.

Namekawa K, Kasai C, Tsukamoto, M, Koyano A (1982) Real-time bloodflow imaging system utilizing auto-correlation techniques. In: Ultrasound '82 (Eds RA Lerski, P Morley), Pergamon, New York.

Nippa JH, Hokanson DH, Lee DR, Sumner DS, Strandness DE (1975) Phase rotation for separating forward and reverse blood velocity signals. IEEE Trans Sonics Ultrasonics SU-22, 340–346.

Peronneau PA, Bournat JP, Bugnon A, Barbet A, Xhaard M (1974) Theoretical and practical aspects of pulsed Doppler flowmetry: real-time application to the measurement of instantaneous velocity profiles *in vitro* and *in vivo*. In: Cardiovascular applications of ultrasound (Ed. RS Reneman), pp 66–84, North-Holland, Amsterdam.

Reneman RS, Van Merode T, Hick P, Hoeks PG (1986) Cardiovascular applications of multi-gate Doppler systems. Ultrasound Med Biol 12, 357–370.

Roevros JMJG (1974) Analogue processing of CW Doppler flowmeter signals to determine average frequency shift momentaneously without the use of a wave analyzer. In: Cardiovascular applications of ultrasound (Ed. RS Reneman), pp 43–54, North-Holland, Amsterdam.

Sainz A, Roberts VC, Pinardi G (1976) Phase-locked loop techniques applied to ultrasonic Doppler signal processing. Ultrasonics 14, 128–132.

Skidmore R, Follett DH (1978) Maximum frequency follower for the processing of ultrasonic Doppler shift signals. Ultrasound Med Biol 4, 145–147.

Chapter 7

RECORDING AND REPRODUCTION OF DOPPLER SIGNALS

7.1 INTRODUCTION

The recording and reproduction of Doppler signals can be undertaken reasonably easily since use can be made of equipment developed for the domestic market and imaging applications. The options available for these tasks are

1. Audio cassette recorder (analogue and digital)
2. Video cassette recorder
3. Photographic film
4. Photographic paper
5. Thermal video printer
6. Chart recorder

In selecting recording equipment many factors have to be taken into account. In the following paragraphs the most common options are considered, important specifications are outlined and typical values are given to aid in the selection of recording equipment and media. The treatment of these topics is far from exhaustive but it is adequate for setting up a system for accurate recording and reproduction of Doppler signals.

7.2 AUDIO CASSETTE RECORDERS

The analogue audio cassette recorder (AACR) is very convenient and relatively inexpensive. If a spectrum analyser is not available at the time of the clinical examination, recording of the audio Doppler signal allows sonograms to be obtained at a later date. In Fig. 7.1, a signal analysed directly is compared to one analysed after recording. Middle of the range domestic recorders are suitable for many purposes, but it will be evident from the following discussion that in one or two situations the replayed signal will be distorted to an extent that can produce errors. The important parameters in the specification of a recorder are listed in Table 7.1 along with values for typical units. AACRs are not suitable for recording quadrature signals since the phase information is not accurately reproduced (Smallwood 1985), and therefore the forward and reverse Doppler signals should be kept separate.

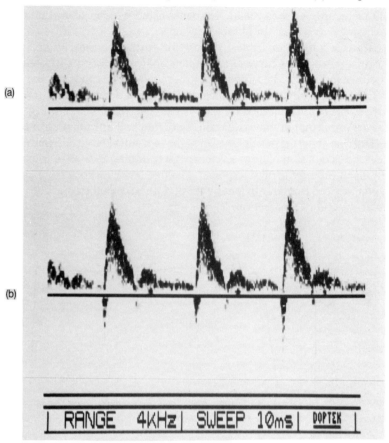

Figure 7.1   A Doppler signal analysed (a) directly from the Doppler unit, (b) after recording on tape. The difference between these sonograms is not significant for many purposes and is due mainly to different amplification of the signal in each technique

Table 7.1   Specification of typical analogue and digital tape recorders

	Analogue recorder	Digital recorder
1. Frequency response	30 Hz to 18 kHz ± 3 dB	2 Hz to 22 kHz ± 0.5 dB
2. Signal-to-noise ratio	58 dB	92 dB
3. Harmonic distortion	1%	< 0.0055%
4. Input signal range	20 mV to 7 V	250 mV to 10 V
5. Wow and flutter	0.2% (DIN)	Below measurable limits
6. Crosstalk	− 65 dB	Below measurable limits*

* ≤ 85 dB if only one ADC is used for both stereo channels.

Digital audio tape recorders (DAT) have now begun to appear on the market. These instruments have a much better specification than analogue devices (Table 7.1), and have additional facilities such as the ability to record 'sub-code' signals which may be used to identify the beginning of recordings and/or the date and time at which recordings are made. DAT recorders are as yet fairly expensive, but it seems likely that they will rapidly replace analogue recorders in Doppler systems.

The parameters listed in Table 7.1 are worth further consideration.

The *frequency response* curve of a complete analogue recording system (recorder plus tape) is shown in Fig. 7.2. The upper and lower limits, where the response falls 3 dB below the maximum, are obviously significant since they determine the range of Doppler frequencies that can be handled by the system. The small fluctuations in the frequency response are important in some situations since they will introduce error in the estimation of mean frequency, i.e. mean velocity, when the reproduced signal is analysed. This in turn will affect the calculation of flow. Even expensive domestic cassette recorders suffer from fluctuations in their frequency response. The frequency response of a complete cassette system is influenced to a large extent by the performance of the tape. The choice of tapes is covered at the end of this section.

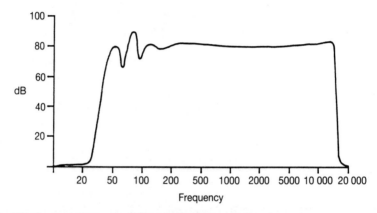

Figure 7.2   The frequency response of an audio tape recorder plus tape

DAT records have a wider frequency response and a much flatter response curve, and introduce minimal errors.

The *signal-to-noise ratio* of an analogue audio cassette system is determined largely by the intrinsic noise of the tape. This noise, together with the noise generated by the record/playback amplifiers, is usually not significant since it is much lower than the noise generated by the Doppler equipment (particularly by pulsed Doppler units). The enhanced SNR of DAT systems means that optimizing the recording levels is much less critical.

Harmonic distortion is a measure of the degradation undergone by a signal recorded and played back through the ACR. The distortion introduces additional frequency components (harmonics). The voltage level of the signal due to the harmonics is expressed as a percentage of an input test signal voltage. A typical value for analogue recorders is 1%, which is unlikely to be of significance in Doppler work.

The *input signal range* specifies the range within which the input should fall if it is not to be lost in noise or distorted. Careful attention should be paid to the meters on the recorder to ensure that a suitable signal is being supplied throughout the cardiac cycle.

Wow and flutter are due to variations in the rate at which the tape passes the record/play back head. Mechanical imperfections cause these variations which are measured against the desired speed and expressed in percentage terms. Flutter is the high-speed variation and wow the low-speed. Digital systems automatically remove speed variations.

Crosstalk is a measure of the degree of unwanted signal transfer between channels in stereo or multichannel equipment. Crosstalk occurs between electronic circuits as the signals are recorded and played back from tape. It is expressed in dB and represents the ratio of the desired signal in the channel to the transferred signal.

7.2.1   Miscellaneous points on cassette recorders

1. Front loaders enable equipment to be stacked.
2. Not all ACRs have microphone inputs. Of those that do, not all provide an input selection switch, without which the microphone must be plugged in for audio dubbing and unplugged to select Doppler inputs, which is hardly convenient. A voice commentary can be invaluable for identifying recordings.
3. Noise reduction systems (Dolby B, Dolby C, ANRS) are designed to reduce tape noise (hiss) but, since tape noise is lower than the background noise from Doppler systems, they have little to offer.
4. The convenience of the method of controlling the recorder should be considered at the time of purchase. A remote control unit is particularly useful.
5. If several recorders are being purchased it is advisable to obtain the same type since tapes may be transferred between them, and the numerical display of tape position is still appropriate when a tape is replayed.
6. Manufacturers recommend that the record/replay heads should be cleaned after every 10 hours of play to maintain performance. This is probably more frequent than typical practice.
7. Check that the input signal level is reasonably large to minimize the significance of noise in the recording or sonogram.

7.3   AUDIO CASSETTE TAPES

It was mentioned earlier that the quality of the tape is important in determining the performance of a system. Analogue tapes can be divided into four types, the main characteristics of which are listed in Table 7.2.

When selecting a tape type, the choice must be between chrome and metal, with metal providing the best performance but at almost twice the price of chrome.

Table 7.2   The specification of audio tapes

Type	Performance	Frequency range
1. Normal (LH) (Ferrite)	Average	30 Hz to 12 kHz
2. Chrome (CrO_2)	Good	30 Hz to 14 kHz
3. Ferrichrome (FeCr)	Compromise of (1) and (2)	30 Hz to 15 kHz
4. Metal (Fe)	Very good	30 Hz to 19 kHz

7.3.1   Miscellaneous points on tapes

1. The best choice for tape length is either 60 or 90 minutes since 30 minute tapes soon

create storage problems and are not so readily available. The 120 minute tapes do not deliver the same performance because of the necessary reduction in tape thickness. Thin tape can lead to snagging in machines (most manufacturers do not recommend them for this reason alone).

2. The ACR must be capable of operation with the type of tape chosen.

3. When not in the machine, tapes should be stored in their library case to protect them from dust, and stored in an area free from direct sunlight and dampness.

7.4 VIDEO CASSETTE RECORDERS

Video cassette recorders offer an attractive alternative to audio cassette recorders. The ability to record both Doppler signal and sonogram on one tape is a convenient feature since the signal and sonogram may be reviewed separately from the analyser. In addition, since patient identification and examination details can be superimposed on the image of the sonogram, the need for audio dubbing can be obviated.

Another way in which the VCR may be used is to record the Doppler signal and an associated ultrasonic real-time scan image.

Colour recording with a VCR is essential for Doppler flow mapping. Slow replay and frame pause facilities enable flow patterns which can be difficult to appreciate at the normal cardiac cycle rate to be identified.

Standard VCRs have a limited audio frequency range (typically 30 Hz to 12 kHz). If a better high-frequency response is required, so-called HiFi VCRs should be considered. Typical specifications for the audio channels of HiFi VCRs are shown in Table 7.3.

Table 7.3 Typical specification of the HiFi audio channels of a video cassette recorder

1. Frequency response	20 Hz to 20 kHz
2. Signal-to-noise ratio	80 dB
3. Harmonic distortion	0.1%
4. Input signal range	20 mV to 600 mV
5. Wow and flutter	Not quoted
6. Crosstalk	70 dB

The picture quality from a VCR, for all modes of operation (fast preview, slow replay, frame freeze, etc.), is normally reviewed in subjective terms and therefore the best method of selection from a picture quality view point is to utilize the wealth of reviews carried out by various video magazines and booklets.

The miscellaneous points related to the purchase and use of audio cassette recorders also apply to VCRs.

7.5 LOUDSPEAKERS

Loudspeakers in Doppler equipment often seem to have been selected so as to fit the box rather than to give faithful reproduction of the Doppler signals. No single speaker is capable of reproducing the entire audio spectrum. For low-frequency Doppler signals obtained from veins, for example, it would be worth considering a speaker suited for mid-to low-frequency use. Speakers in Doppler equipment are often small in size and only suited to reproducing high frequencies.

If it is decided to employ a quality speaker system, a HiFi power amplifier would normally also be required. The vast majority of amplifiers from well known manufacturers in the mid-price range would more than suffice. The same applies to loudspeakers.

7.6   HARD COPY

Although on-line analysis of Doppler signals can minimize the need for hard copy, such as film or photographic print, a strong argument can be put forward for its use in the recording of sonograms. Unmerited confidence is often placed on simple indices considered in isolation. An accompanying picture of the sonogram from which the indices were derived can convey a more realistic level of confidence.

The following factors should be borne in mind when a hard copy unit is being selected:

1. Initial cost
2. Running cost
3. Image quality
4. Ease of use
5. Additional equipment required (developers etc.)

The following options are available:

1. X-ray film
2. Photographic paper in multi-imager
3. Polaroid film
4. Thermal video printer
5. Fibre optic linescan recording (as used in echocardiography)

The first two offer excellent quality and reproduction but have a high initial or running cost. The third and fourth, while much less expensive to buy than (1) and (2), have slightly reduced image quality (few grey shades and poorer spatial resolution). Option 4 is likely to outsell the others in new installations because of its low running costs, relative cheapness and ease of use.

Although many Doppler instruments produce colour images, the production of colour hard copy in a reasonable time is restricted to the use of colour polaroid. Other colour reproduction techniques (colour printers, electrostatic plotters, etc.), are essentially computer peripherals and are unsuitable because of limited capabilities (limited colours, poor resolution, slow speed and high cost). Improved colour thermal video printers are now becoming available.

The fibre optic linescan recorder as widely encountered in ultrasonic M-scan recording has many attractions for Doppler recording. It is cheap to run, many cardiac cycles can be accommodated and other physiological signals can be simultaneously recorded. An associated ultrasonic real-time image may also be printed next to the Doppler record. When used with UV paper, the final record is usually unsatisfactory since the number of grey shades is limited. Higher-contrast photographic paper solves this problem at increased cost.

7.7 COMMON PROBLEMS WITH DOPPLER RECORDING SYSTEMS

Any system consisting of several interconnected units is liable to suffer from small problems which can prevent successful application. Table 7.4 is supplied to assist with the identification and resolution of such problems.

Table 7.4

Problem	Possible reason/remedy
1. Cannot record	'Standard' audio cables not making proper connections between instruments
2. Poor high frequency response	Recorder heads dirty Use readily available head cleaning kits
3. Mains hum superimposed on signal	Probably due to loops in the earth wire connections Try rearranging plugs or signal connections; where possible reduce system complexity to identify the cause
4. Left/right channels crossed over	Interchange left/right leads
5. Interference	Attempt to determine source, for example switching off any nearby equipment

7.8 SUMMARY

It is highly desirable to be able to record Doppler signals accurately. In this chapter the strengths and weaknesses of several recording techniques are reviewed. In particular, audio and video cassette recorders are examined. The options for hard copy recording of sonograms are briefly discussed. The successful recording of Doppler signals requires careful attention to each item in the composite system.

7.9 REFERENCE

Smallwood RH (1985) Recording Doppler blood flow signals on magnetic tape. Clin Phys Physiol Meas 6, 357–359.

Chapter 8

THE ORIGIN OF THE DOPPLER POWER SPECTRUM

8.1 INTRODUCTION

The simple Doppler equation for a single target passing through an infinitely wide ultrasound beam has already been presented in Chapters 1 and 3. The Doppler shift frequency f_d is given by:

$$f_d = f_t - f_r = (2f_t v \cos \theta)/c \qquad 8.1$$

where f_t and f_r are the transmitted and received ultrasound frequencies, v the velocity of the target, c the velocity of sound in tissue, and θ the angle between the ultrasound beam and the direction of motion of the target. When ultrasound is used to interrogate the flow within a blood vessel there are of course numerous targets in the ultrasound field with a range of velocities, and the Doppler shift signal therefore contains not just a single frequency, as shown by eqn 8.1, but rather a spectrum of frequencies which varies in shape as the velocity distribution within the vessel changes with time.

The Doppler shift frequency is proportional to velocity, and under ideal uniform sampling conditions the power in a particular frequency band of the Doppler spectrum is proportional to the number of erythrocytes moving with velocities that produce frequencies in that band, and therefore the Doppler power spectrum should have the same shape as a velocity distribution plot for the flow in the vessel. The velocity distribution corresponding to a variety of realistic velocity profiles (as derived for Fig. 2.8) are shown in Fig. 8.1. The spectra corresponding to flat velocity profiles have much of the power concentrated in a relatively small range of frequencies, whilst those corresponding to parabolic profiles are almost flat. In turbulent flow, the velocities of the targets in the ultrasound field fluctuate rapidly with time (Section 2.3.3) and this causes a broadening of the spectrum that would otherwise be obtained from flow with the same temporal average velocity profile. Gross haemodynamic disturbances such as large vortices may cause irregular spectra with large isolated forward and reverse components.

The variation in the shape of the Doppler power spectrum as a function of time is usually presented in the form of a sonogram (Fig. 8.2). In this type of display, time is plotted along the horizontal axis, frequency along the vertical axis, and the power at a particular frequency and time as the intensity of the corresponding pixel. Thus a single line on the sonogram corresponds to a single power spectrum, much in the same way as a line on an ultrasound B-scan corresponds to a single A-scan.

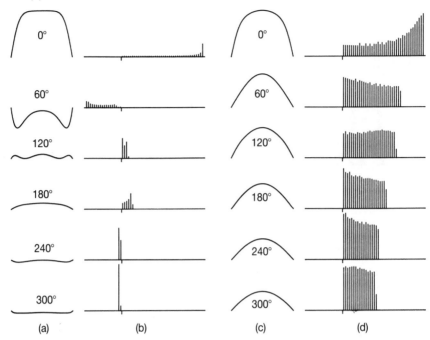

Figure 8.1 Series of velocity profiles for a common femoral artery (a) and common carotid artery (c), together with corresponding velocity distribution histograms. The peak of forward flow has been arbitrarily called 0° and the maximum velocities have been scaled to have the same amplitude

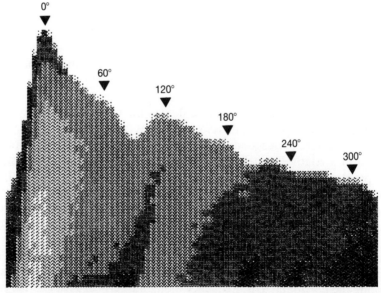

Figure 8.2 Sonogram of the Doppler signal from the common carotid artery whose velocity profiles are given in Fig. 8.1c. Time is represented along the horizontal axis, Doppler shift frequency along the vertical axis, and spectral amplitude by the blackness of the paper. The time slices corresponding to the histograms in Fig. 8.1d are marked with small arrowheads

There are a number of factors that distort the power spectra and which may limit the accuracy with which the velocity distribution in the vessel can be determined. In this chapter we consider the source of scattering of the ultrasound and the way in which the Doppler power spectrum is influenced by various physical and electronic mechanisms.

8.2 BLOOD AS A TARGET

Despite its appearances, blood is not a homogeneous liquid, but a suspension of cells and other particles in a clear straw-coloured fluid called plasma. Because of this microscopic structure ultrasound is scattered by blood, and Doppler shift measurements of blood velocity are possible.

The so-called formed elements of blood consist of the erythrocytes (red blood cells), the leukocytes (white blood cells), and platelets. The relative size and concentrations of these components are summarized in Table 8.1. It is generally believed that scattering of ultrasound by blood is almost entirely due to erythrocytes because they are much more numerous than the slightly larger leukocytes, and significantly larger than the platelets. Reid et al (1969) found that the scattering cross-section of platelets is approximately 10^{-3} times that of the erythrocytes and that at normal concentrations their contribution is undetectable.

Table 8.1 The sizes and concentrations of the major formed elements of blood

	Concentration (particles/mm^3)	Dimensions (μm)	% total of blood volume
Erythrocytes	5×10^6	7.2×2.2	45
Leukocytes	8×10^3	9–25	~0.8
Platelets	2.5×10^5	2–3	~0.2

The behaviour of erythrocytes as targets depends on their size, concentration and acoustic properties, and the acoustic properties of the embedding media, the plasma.

Erythrocytes are flexible biconcave discs with a diameter of 7.2 μm and a thickness of 2.2 μm, and their concentration is such that they normally occupy between about 36% and 54% of the total blood volume (the haematocrit is thus said to be betweeen 36% and 54%). Even at the highest ultrasound frequencies used for blood velocity measurements the diameter of the erythrocytes is much smaller than an ultrasound wavelength and therefore in low (unphysiological) concentrations they act as a random distribution of point targets.

In the high concentrations found in normal whole blood, however, they cannot be treated as a purely random distribution because their positions are no longer independent of each other; Shung et al (1976) have pointed out that when the haematocrit is 45% the average distance between two red cells is only about 10% of their diameter. This significantly complicates the behaviour of blood as a scattering target, but Shung and his colleagues were able to bring an analysis of scattering by blood to a successful conclusion by introducing the heuristic 'hole' first used by Beard et al (1967). Another approach that has been used with some success by Angelsen

(1980) is to treat the blood as an isotropic continuum, the source of scattering being fluctuations in the compressibility and mass density of the continuum.

The scattering properties of blood may be altered by the flow conditions. If it is allowed to stand, the erythrocytes may aggregate into multicellular clumps known as rouleaux; however, the bonding between such cells is so weak that the rouleaux disperse when exposed to the shear gradients found in normal arterial (but not necessarily venous) blood flow, and therefore are more troublesome in in-vitro experiments than in-vivo. Turbulence may also have a significant effect on the scattering of ultrasound by blood (Shung et al 1976) because local accelerations in the velocity field cause a separation between the plasma and erythrocytes due to their different mass densities, and a consequent increase in the fluctuations in the local cell concentrations (Angelson 1980). Caution must therefore be exercised when interpreting the results from the turbulent jets found distal to stenoses in both arteries and heart valves.

The acoustic properties of the blood constituents that influence the scattering of ultrasound are their densities and adiabatic compressibilities, and these are summarized in Table 8.2 (Urick 1947). The attenuation of ultrasound by blood is relatively low with a value of approximately $0.15 \, \mathrm{dB \, cm^{-1} \, MHz^{-1.2}}$ (Narayana et al, 1984), and the contribution of scattering towards this is negligible for frequencies of less than 15 MHz (Shung et al 1976).

Table 8.2 The density and compressibility of the major components of blood. Data based on Urick (1947)

	Density $(\mathrm{kg \, m^{-3}})\,(\rho)$	Adiabatic compressibility $(\mathrm{m^2 \, N^{-1}})\,(\beta)$
Erythrocytes	1.091×10^3	3.41×10^{-10}
Plasma	1.021×10^3	4.09×10^{-10}

8.2.1   Scattering of ultrasound from blood

The scattering of waves by particles that are small in comparison to the wavelength was first studied by Lord Rayleigh in 1871, and such scattering is therefore usually referred to as Rayleigh scattering. Two of its important characteristics are that the shape of the scatterers is unimportant, and that the scattered power is proportional to the fourth power of frequency. As mentioned in the last section, the behaviour of a concentrated ensemble of particles is rather different from that of individual scatterers, but Shung et al (1976) showed both theoretically and experimentally that the scattering of ultrasound by blood *is* proportional to the fourth power of frequency. This has the practical implication that the performance of a Doppler system falls off with frequency less rapidly than that of a pulse echo system, because the increase in scattered power with frequency partially offsets the increased attenuation of ultrasound by the intervening tissue.

In the same paper Shung et al also studied the effect of haematocrit on scattering and showed that the scattering coefficient is not proportional to haematocrit over a large range, as would be predicted by simple theory, but reaches a maximum for values of haematocrit of about 24–30%. The expression they derived for the scattering coefficient α_s is reproduced as eqn 8.2:

$$\alpha_s \simeq 0.21 \, k^4 a^3 W_0 (1 - W_0)(1 - 1.72 W_0)\left(\left|\frac{\beta_e - \beta_p}{(1 - W_0)\beta_p + W_0\beta_e}\right| + \left|\frac{\rho_e - \rho_p}{\rho_e}\right|\right)^2 \quad 8.2$$

where k is the wave number $(= 2\pi/\lambda)$, a the radius of the scatterers, W_0 the volume concentration of scatterers $(= \text{haematocrit}/100)$ and β_e, β_p, ρ_e and ρ_p are the adiabatic compressibility and density of the erythrocytes and plasma (see Table 8.2). Equation 8.2, together with the experimental results obtained by Shung et al at frequencies of 5.2 MHz, 8.5 MHz and 15.2 MHz, are illustrated in Fig. 8.3. It is seen that α_s does increase linearly with frequency up to a haematocrit value of about 10%, but from there on the interaction between the blood cells becomes important.

The same authors have also addressed the problem of the angular dependence of scattering of ultrasound from blood. They compared their experimental measurements with two theoretical descriptions of scattering, one given by Rschevkin (1963) and Morse and Ingard (1968), and another due to Ahuja (1972). The two theories lead to similar results, but that of Ahuja is more complete in that it recognizes that the embedding media is not frictionless and that therefore shear waves are generated by the oscillating particles. The experimental measurements of Shung et al (1977) agreed well with both theories, but particularly closely with that of Ahuja. Because the Rschevkin model is much simpler and performs reasonably well, its essential result is reproduced here, and the reader is referred to the paper of Shung et al (1977, eqn 6) for the more exact theoretical description. Rschevkin (1963) showed that the power received by a receiving transducer $(P(\phi))$ may be written:

$$P(\phi) = \frac{nVk^4 a^6 I_i A}{9s^2}\left(\frac{\beta_e - \beta_p}{\beta_p} + \frac{3\rho_e - 3\rho_p}{2\rho_e + \rho_p}\cos\phi\right)^2 \quad 8.3$$

where n is the concentration of particles, V the observation volume, I_i the incident intensity, A the receiver aperture, and s the distance between the receiver and the volume V. This function is plotted in the form of a polar diagram in Fig. 8.4. It may be seen that the scattered power is at a minimum when ϕ, the scattering angle, is $0°$, and a maximum when ϕ is $180°$.

Thus far we have considered only the time-averaged value of the scattered power, but it is well known that the power returning from blood fluctuates as a function of time. These fluctuations have been explored by Atkinson and Berry (1974) who showed that not only does the power vary with time but with the lateral displacement of the transducer. Figure 8.5 is redrawn from Atkinson and Berry's paper and shows typical pressure envelope signals as functions of time and displacement.

Atkinson and Berry argued that, if the blood cells are randomly distributed, the number contained in different small volumes of the same size V will not simply be given by nV, where n is the overall number density, but will fluctuate about that value. Furthermore, if the linear dimensions of V are approximately equal to the ultrasonic wavelength or the pulse length, or the transverse dimensions of the beam, these fluctuations will cause variations in scattering power throughout the specimen. They went on to develop a statistical diffraction theory to account for the observed fluctuations, and derived expressions for the mean rates of fluctuation due to these effects. The mathematics involved was fairly complex, but the results remarkably

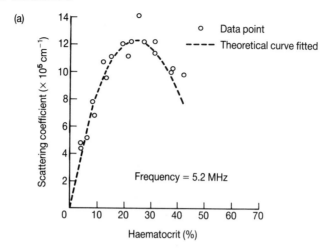

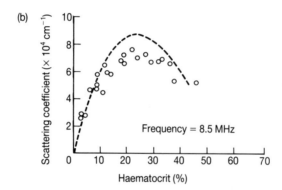

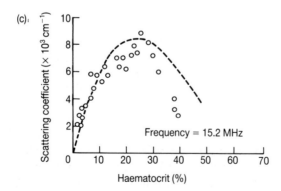

Figure 8.3   Scattering coefficient of erythrocytes suspended in plasma plotted against haematocrit at (a) 5.2 MHz, (b) 8.5 MHz, and (c) 15.2 MHz. Note that the vertical scale of each graph is different. The dots are the experimental points and the lines are plotted using eqn 8.2. (Reproduced with permission from Shung et al (1976), copyright © 1976, IEEE)

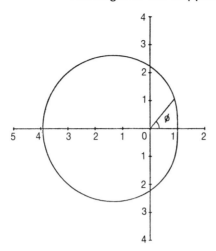

Figure 8.4 Normalized polar diagram of the scattered power from blood as a function of the scattering angle ϕ, calculated using eqn 8.3. Note that the backscattered power is about 6 dB greater than at forward angles

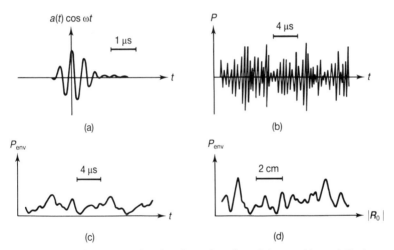

Figure 8.5 Fluctuations in the Doppler signal as a function of time and lateral displacement: (a) is the incident pressure wave, (b) the time dependence of the back-scattered echo, (c) the time dependence of the echo envelope, and (d) the dependence of the envelope on the lateral displacement of the source-receiver. (Reprinted with permission from Atkinson and Berry (1974), IOP Publishing)

simple, and agree very closely with experimental data. In particular they showed that the temporal fading rate N_t, defined as the average number of crossings of the pressure envelope p_{env} through its mean value $\langle p_{env} \rangle$ during unit time t with the source-receiver held fixed is given by:

$$N_t = \frac{e^{-\pi/4}}{T_L\sqrt{2}} = 0.322/T_L \qquad 8.4$$

where T_L is a measure of the pulse length. Similarly the lateral fading rate N_d, the mean number of crossings of p_{env} through the value $\langle p_{env} \rangle$ at a fixed time delay during a unit lateral displacement of the source-receiver is given by:

$$N_d = 0.322/R_B \qquad\qquad 8.5$$

where R_B is a measure of the ultrasound beam radius. The same theory failed to predict accurately the absolute scattered power, and this was thought to be a result of assumptions made to simplify the theory, and practical problems of preventing clotting and rouleaux formation during the experimental work.

8.3   CONTINUOUS WAVE DOPPLER SPECTRA

There are two distinct types of Doppler shift velocimeter (see Chapter 4), continuous wave (CW) and pulsed wave (PW). CW units, which are the subject of this section, both transmit and receive continuously; PW units transmit short bursts of ultrasound at regular intervals and switch to a receive mode during the inter-burst period. The Doppler spectra resulting from these two types of instrument may differ significantly and they are therefore treated separately.

8.3.1   Physical principles of CW Doppler

The simplest Doppler devices are those that employ CW operation (Fig. 8.6). The Doppler probe contains a pair of transducers, one of which is used to continuously transmit a monochromatic (single-frequency) ultrasound wave into the tissue, the other to continuously receive ultrasound waves as they are reflected or scattered from targets (both stationary and moving) within the tissue. Because transmission and reception are continuous there is no depth resolution except in the sense that signals originating from close to the transducer experience less attenuation than those from distant targets, and are therefore stronger.

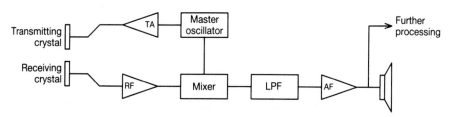

Figure 8.6   Block diagram of a simple CW Doppler system. TA, transmitting amplifier; RF radio frequency amplifier; LPF low pass filter; AF, audio frequency amplifier

The returning ultrasound signals are amplified and mixed with one or more reference signal from the master oscillator and the Doppler difference frequencies extracted. Further processing may then be used to display the signal in any one of a number of ways. Doppler transducers and the fields they produce are discussed in Chapter 5, and the various aspects of generating and demodulating ultrasound signals and handling the Doppler difference frequency are tackled in Chapters 6 and 9. As stated in the introduction to this chapter, the spectrum of Doppler shift frequencies will under 'ideal' circumstances have the same shape as the velocity distribution plot for the vessel of interest. In the sections that follow (8.3.2–8.3.7) a number of factors that distort the spectra from CW instruments are discussed.

8.3.2   Non-uniform insonation

Uniform insonation of blood vessels (particularly larger ones) is not easy to achieve. The spatial variation of sensitivity of a CW transducer depends on both the transmitting and receiving crystals and their relative positions and orientations. Evans and Parton (1981) and Douville et al (1983) have published studies of the directional characteristics of a number of twin-crystal CW transducers, and both groups noted a significant difference between probes of apparently identical manufacture. The near field, particularly close to the transducer, may be quite complex. In the far field there is usually, but not always, a single principal maximum at each range, and although side lobes may be present they are not of great significance. The width and rate of decay of the main lobe is highly dependent on transducer geometry.

If the sensitivity of the probe is not uniform across the diameter of a vessel from which measurements are to be made, some parts of the vessel will be preferentially sampled, and the frequencies corresponding to the velocities in the most sensitive part of the beam over-represented in the Doppler power spectrum.

Ultrasound transducers are usually manipulated to give the 'best' signal, and the axis of the ultrasound beam is likely to pass through, or close to, the centre of the vessel. If the vessel is small, the ultrasound beam may be sufficiently flat across the vessel for no undue distortion of the Doppler spectrum to occur, but if the vessel is large it is likely that the beam will effectively insonate only the middle portion of the vessel and the signal from the central streamlines will be over-represented. A number of authors have performed calculations on the effects of partial sampling of the blood vessel. Powalowski et al (1975) and Evans (1982a) considered models in which rectangular beams of ultrasound of varying size passed through the centre of vessels containing parabolic velocity profiles. Cobbold et al (1983) considered rectangular and Gaussian beams, both on-axis and off-axis, and also made some allowances for tissue attenuation. Evans (1982b) demonstrated the effect of using a narrow beam to insonate vessels containing complex velocity profiles, and later (Evans 1985) discussed the effect of non-uniform insonation on the measurement of mean velocity in vessels containing complex profiles.

In the simple case of a rectangular beam passing through the centre of a blood vessel it can be shown that the fraction, F, of the lamina at radius r which is intersected by an ultrasound beam of width w is given by (Evans 1982a):

$$F = (2/\pi)\sin^{-1}(w/2r) \qquad \text{for } w < 2r \qquad 8.6$$

If the radial distribution of velocities in the vessel is known, eqn 8.6 may be written in terms of velocity rather than radius. A particularly interesting case is that of a parabolic velocity profile which under conditions of uniform insonation yields a flat power spectrum (Section 2.3.1). Such a profile may be written:

$$v(r) = v_{max}(1 - r^2/R^2) \qquad 8.7$$

where v_{max} is the maximum velocity found at the centre of the vessel of radius R.

Substituting for r in eqn 8.6 leads to:

$$F = (2/\pi)\sin^{-1}[w/2R(1 - v/v_{max})^{1/2}]$$

$$\text{for } w < 2R(1 - v/v_{max})^{1/2} \qquad\qquad 8.8$$

A graph of this function (Fig. 8.7) shows the way in which the Doppler power spectrum from a parabolic velocity profile is distorted by incomplete vessel sampling.

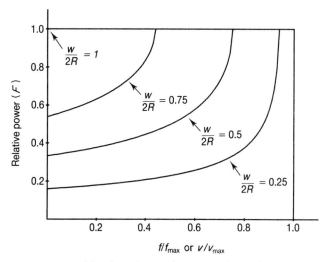

Figure 8.7   Power spectra resulting from insonating a vessel containing steady laminar flow with ultrasound beams of different relative widths

By applying eqn 8.8 to the two halves of the vessel separately it is also possible to deduce the effect of using an ultrasound beam which is displaced to one side. Alternatively the same results may be calculated numerically using a simple model such as that suggested by Cobbold et al (1983). Although such an approach is unnecessarily complicated for such simple situations it allows the investigation of the effects of more complex ultrasonic beam profiles, and the differential attenuation between blood and soft tissue. The results obtained by Cobbold et al for a parabolic velocity profile interrogated by a square ultrasound beam are shown in Fig. 8.8.

Velocity profiles found in arteries vary both from vessel to vessel and throughout the cardiac cycle, and therefore the effects of partial sampling may also vary widely. The results of incomplete sampling of some complex velocity profiles (those shown in Fig. 8.1) are illustrated in Fig. 8.9.

It is by now clear that non-uniform vessel sampling may severely distort the shape of the Doppler spectrum; this may significantly affect spectral broadening indices (Section 10.4.5), the output of Doppler signal processors (Chapter 9) and the measurement of volumetric flow (Chapter 11).

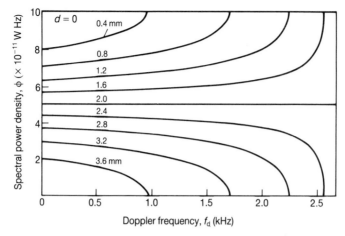

Figure 8.8 Spectral density graphs for a parabolic velocity profile interrogated by a square ultrasound beam with a width equal to the vessel diameter (4 mm). The results are plotted for successive 0.4 mm displacements of the beam central axis. (Reproduced with permission from Cobbold et al (1983), copyright © 1983, IEEE)

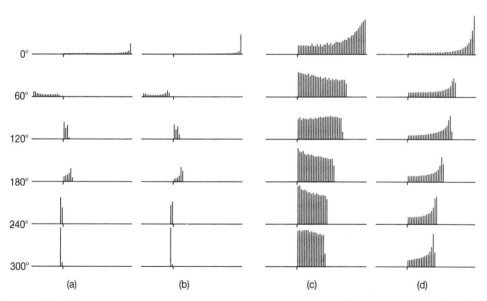

Figure 8.9 Effects of partial sampling of the complex velocity profiles shown in Fig. 8.1. The spectra shown in (a) and (c) are those resulting from uniform sampling and are identical in shape to the velocity histograms shown in Figs 8.1b and 8.1d. The spectra shown in (b) and (d) are those resulting from interrogating the vessel with a uniform beam which is only 25% of the width of the vessel. Each series of spectra has been normalized in terms of both maximum frequency and maximum power, but the magnitude relationships within each series have been maintained

8.3.3 Attenuation

Ultrasound waves are attenuated by a variety of mechanisms as they propagate through the body (Section 3.3.3). Attenuation rates in soft tissue (approximately 0.8 dB MHz^{-1} cm^{-1}) are much greater than those in blood (approximately 0.2 dB MHz^{-1} cm^{-1}) and therefore echoes returning from different parts of a blood vessel may experience different amounts of attenuation if they traverse different acoustic pathways. In particular, signals from the centre of the vessel will be stronger than those from its lateral edges where the ultrasound traverses more soft tissue and less blood, and this has much the same effect as reducing the effective width of the ultrasound beam, and exacerbates the effects of non-uniform insonation discussed in the last section. The effect becomes greater as the angle θ decreases and the path lengths in tissue and blood become more disparate, and is more pronounced at higher frequencies where the rates of attenuation are greater. Cobbold et al (1983) have documented this effect for a number of specific cases, and the results of one of their studies is reproduced in Fig. 8.10. This figure is directly comparable with Fig. 8.8 with the exception that attenuation effects have been considered in the calculations. The angle θ was taken to be $60°$, and the transmitted frequency 8 MHz.

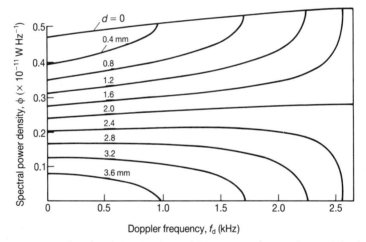

Figure 8.10 Spectral density graphs for a parabolic velocity profile interrogated by a square ultrasound beam with a width equal to the vessel diameter. These results differ from those shown in Fig. 8.8 because attenuation effects have been considered. (The assumed values are 6.4 dB cm^{-1} in tissue and 1.44 dB cm^{-1} in blood.) (Reproduced with permission from Cobbold et al (1983), copyright © 1983, IEEE)

8.3.4 Non-uniform target distribution

The scattering of sound from blood has been discussed in Section 8.2.1. In general ultrasound is equally scattered from the entire cross-section of a vessel, but it has been found that turbulence increases the amount of scattering that occurs (see Section 8.2). The amount by which the scattering is increased is as yet undetermined, but it is a common subjective observation that signals from turbulent jets are relatively strong. Velocity spectra recorded from highly turbulent flow must therefore be treated with caution.

8.3.5 Intrinsic spectral broadening

Equation 8.1 relates the Doppler shift frequency from a single target passing through an infinitely wide ultrasound beam to its velocity. In practice the ultrasound beam has only a finite width, and for reasons that will be explained subsequently even a single target produces a spectrum of Doppler shift frequencies rather than a single frequency. When there are many targets passing through a finite ultrasound beam each contributes a spectrum of Doppler shift frequencies to the overall spectrum, which will therefore be broader and more smeared than would otherwise have been the case. This type of spectral broadening, which is due to the properties of the measurement system rather than the nature of the system being measured is referred to as intrinsic spectral broadening (ISB).

ISB can be explained either in terms of the range of angles available to the incident and back-scattered radiation as the target traverses the ultrasound beam (so-called geometrical broadening) or in terms of the amplitude modulation caused by the finite transit time of the target through the ultrasound beam (so-called transit-time broadening). The equivalence of these two mechanisms was not appreciated in early publications and this has led to some confusion. Green (1964) identified what he considered to be three independent mechanisms of spectral spreading, unrelated to velocity gradients, in Doppler shift fluid flowmeters. The first, that due to Brownian motion of the individual scatterers, he showed to be negligible, but he found both geometrical and transit-time broadening to be significant. Griffith et al (1976) and Newhouse et al (1976) studied transit-time broadening and Newhouse et al (1977a) geometrical broadening, but the latter group of workers (Newhouse et al 1977b, 1980) were subsequently able to show the two effects to be equivalent. It is interesting that the same mistakes concerning the apparent sources of spectral broadening were made by a number of workers in the Doppler laser field until the equivalence of the two effects was asserted by Angus and his colleagues (1971). Both explanations for ISB are described here because although they are equivalent each may have computational advantages under certain circumstances.

8.3.5.1 Geometrical explanation

The Doppler shift frequency, f_d, has been shown to be proportional to the cosine of the angle θ between the ultrasound beam and the direction of flow. In practice, because of the finite size of the transducer, the velocity vector of each target in an ultrasound field subtends not a single angle, but a range of angles (Fig. 8.11) and therefore each target contributes a range of Doppler frequencies to the overall Doppler spectrum. Newhouse et al (1977a, 1980) have shown that the magnitude of the spectral broadening may under some circumstances be calculated simply by considering the 'extreme angle rays', i.e. the range of angles over which back-scattered ultrasound is received by the transducer.

8.3.5.2 Transit time explanation

Each individual target that traverses an ultrasound beam scatters ultrasound for a limited period of time, i.e. for as long as it takes to cross from one edge of the beam to

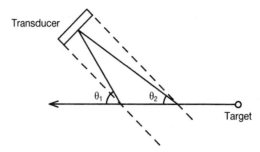

Figure 8.11   Each target passing through an ultrasound beam subtends a series of angles (θ_1 to θ_2) with the centre of the transducer face

the other. Even whilst the target is within the beam the intensity of ultrasound returning to the transducer will change as a result of the non-homogeneous ultrasound field. Although the spectrum of a continuous sine wave contains only a single frequency, the spectrum of a sinusoidal burst, or 'amplitude modulated' sine wave, contains a whole spectrum of frequencies, and therefore even a single target must result in a spectrum of frequencies being received by the transducer. The wider and more homogeneous the ultrasound beam, the longer the returning ultrasound pulse from each target, and the narrower the frequency spectrum. The frequency representation of a sinusoidal burst is a spectrum with a relative width ($\delta f/f$) given approximately by the reciprocal of the number of oscillations in the burst (the exact relationship depends on the shape of the modulating function and the definition of spectral width), i.e.

$$\delta f/f \simeq 1/\text{number of oscillations in received burst} \qquad 8.9$$

So far the explanation has involved only a single target. In practice there are many targets within the ultrasound beam at any given time, and as explained in Section 8.2.1 the scattered power from a random distribution of erythrocytes fluctuates quite dramatically. The rate of fluctuation is determined by the width of the ultrasound beam, and the consequent amplitude modulation of the signal produces spectral broadening similar to that described for individual targets.

Early calculations (e.g. Newhouse et al 1977a) suggested that in the near field (Fresnel zone) transit-time spectral broadening was much smaller in magnitude than geometrical spectral broadening, and this strengthened the view that the two effects were independent. In fact the broadening calculated by the transit-time approach was erroneously small because no account was taken of the strong modulation of the ultrasound signal in the extremely complicated near field.

8.3.5.3   Calculation of magnitude

It has already been mentioned that a number of authors (Green 1964, Griffith et al 1976, Newhouse et al 1976, 1977a, 1977b, 1980) have studied theoretical aspects of spectral broadening. In each case the assumed transducer geometries and beam shapes were idealized, and much simpler than those produced by practical CW Doppler units (Evans and Parton 1981, Douville et al 1983). The only way to calculate the effects of real CW Doppler transducers with their twin-crystal arrangement would be numerically; nevertheless the formulae that have been derived in analytical studies provide us with an insight into the magnitude of ISB effects, and are reproduced here for that reason. Recently Bascom et al (1986) have published some numerical calculations on the effect of ISB on the received Doppler spectra from flat and parabolic velocity profiles interrogated by square and circular transducers, but these too assumed the transmitting and receiving crystals to be coincident.

Newhouse et al (1980) analysed and experimentally verified the ISB due to constant laminar flow across the waist of a focused beam, and across the intermediate and far field of unfocused beams. For the case of the focused beam they showed that the Doppler fractional broadening $\delta f_d/f_d$ could be written:

$$\delta f_d/f_d \simeq \tan \theta \, \Delta \phi \qquad 8.10$$

where θ is the usual Doppler angle, and $\Delta \phi$ the angle subtended by the transducer face at the focal spot. For the case of an unfocused beam they showed that in the far field the fractional broadening was given by:

$$\delta f_d/f_d \simeq \kappa(\lambda/D)\tan \theta \qquad 8.11$$

where λ is the ultrasound wavelength, D the transducer diameter and κ a constant with a value of approximately 2 or 3.

There are two important facts to be gleaned about ISB from the preceding mathematics. The first is that for CW Doppler, ISB is proportional to the tangent of the angle between the ultrasound beam and the flow axis and therefore the smaller the angle the smaller is the effect of ISB. (This is intuitively obvious, since the smaller the angle the larger the sample length.) The second observation is that ISB can be significant in CW applications. For example for a 3 MHz transducer ($\lambda=0.5$ mm) of diameter 10 mm and a beam to vessel angle of 45°, the spectral broadening is approximately 10%. Equation 8.11 has been evaluated and plotted in Fig. 8.12 for a range of values of λ/D likely to be encountered in practice, and it may be seen that particularly at large angles $\delta f_d/f_d$ is quite appreciable.

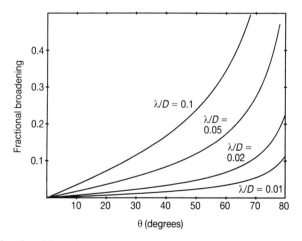

Figure 8.12 The fractional broadening in CW applications calculated from eqn 8.11. Each curve represents the effect for a given ratio between the ultrasound wavelength and the transducer diameter

The practical effect of spectral broadening on a complex Doppler spectrum is to blur its shape, and in particular to smooth out sharp changes. The effect of different degrees of ISB on a number of power spectra is illustrated in Fig. 8.13. The importance of ISB depends on the use to be made of the spectral information. Even large degrees of ISB will not influence the measurement of mean velocity, since the broadening is symmetric, but the maximum velocity becomes difficult to define precisely and may well be overestimated if too low a threshold value is chosen (see Fig. 9.4). Information derived from the spectral shape itself must clearly be treated with caution if there is a possibility of significant ISB.

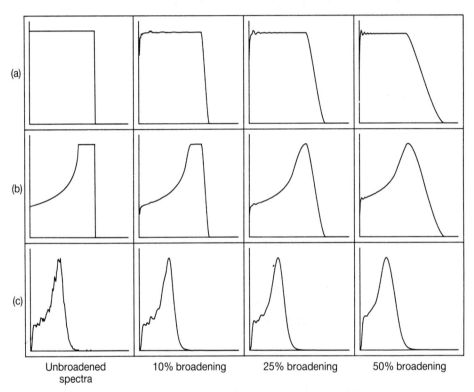

Figure 8.13   Effect of different degrees of spectral broadening on three different power spectra: (a) the spectrum resulting from uniform insonation of parabolic flow; (b) the spectrum from partially sampled uniform flow – cf. Fig. 8.7; (c) the averaged spectrum recorded from a common carotid artery at peak systole

8.3.6   Filtering

CW Doppler is incapable of precise range resolution and therefore, in addition to the Doppler shifted signal arising from blood flowing through the ultrasonic field, there is inevitably a component of the Doppler signal that arises from other moving structures such as the vessel wall. Fortunately most of these signals are of low frequency, but they may have amplitudes which are considerably greater than those from blood. It is therefore necessary to incorporate high pass filters into a Doppler unit, and in addition to removing the unwanted clutter signal they will remove the signal returning from blood which has a low velocity. This leads to a 'gap' in the power spectrum around the frequencies that represent zero flow, which may be particularly troublesome during diastole, and when recordings of very low flow velocities (such as those often observed from fetuses and low birthweight babies) are to be made. The cut off frequency of the high pass filters (often called wall thump filters) is usually user adjustable and they should always be set as low as is compatible with recording a 'noise-free' spectrum. Unfortunately it is not always possible to reduce the filter frequency adequately, either because the equipment manufacturer has not allowed sufficient adjustment, or because low-frequency interference signals are so strong that they prevent the use of a very low frequency cut off. The effect of filters on the performance of frequency followers is discussed in Chapter 9.

8.3.7 Spectral analysis limitations

In order to extract the power spectrum from a Doppler signal it must be transformed into the frequency domain (Fourier transformed). Such a procedure can only yield an estimate of the true underlying spectrum, and because of the nature of the Doppler signal individual estimates may appear very noisy. An example of this is provided by Fig. 8.14a, which is a single estimate of the power spectrum from a test rig containing steady flow.

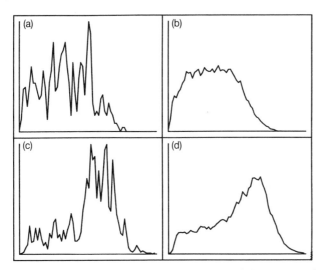

Figure 8.14   Individual and averaged spectral estimates: (a) individual spectral estimate for steady flow in a tube; (b) Bartlett estimate of power spectrum for the same flow using 64 individual estimates; (c, d) individual and averaged spectral estimates taken from the human carotid artery at peak systole

Oppenheim and Schafer (1975, pp 545–548) have shown that for a Gaussian random process (to which a Doppler signal may be approximated) the variance of the spectral estimate may be written:

$$\text{var}[P'(\omega)] = P(\omega)^2 \left[1 + \left(\frac{\sin \omega N}{N \sin \omega} \right)^2 \right] \qquad 8.12$$

where $P'(\omega)$ is the estimated power spectral density (PSD) as a function of ω the angular frequency $(=2\pi f)$; $P(\omega)$ is the true underlying PSD; and N the number of points in the Fourier transform. Even for large values of N the variance remains finite and tends to the square of the spectrum, and therefore the estimated PSD will always fluctuate wildly about the true spectrum value. For large N we may write:

$$\text{var}[P'(\omega)] \simeq P(\omega)^2 \qquad\qquad 8.13$$

It is possible to reduce the variance of the estimated PSD by averaging over a number of independent estimates, a method usually known as Bartlett's procedure (Oppenheim and Schafer 1975). In this case the variance may be written:

$$\text{var}[P'(\omega)] = \frac{1}{K} P(\omega)^2 \left[1 + \left(\frac{\sin \omega N'}{N' \sin \omega} \right)^2 \right] \qquad\qquad 8.14$$

where K is the number of estimates averaged and N' the number of points in each Fourier transform. For large values of N' this reduces to:

$$\text{var}[P'(\omega)] \simeq P(\omega)^2 / K \qquad\qquad 8.15$$

Thus, the spectral variance of the Bartlett estimate is inversely proportional to the number of raw estimates that are averaged together. Figure 8.14b is a Bartlett estimate of the PSD from the same rig as Fig. 8.14a with 64 successive individual estimates having been averaged, and shows a dramatic reduction in spectral variance.

The spectral resolution of the Fourier transform (i.e. the width of each spectral component) is determined solely by the length of the data sample used in the transform, T_a, and may be written:

$$\Delta f_a = 1/T_a \qquad\qquad 8.16$$

For a given length of data, T_x, it is possible to trade off spectral resolution against spectral variance. The entire sample may be used for a single estimate, in which case the resolution will be given by $1/T_x$ and the spectral variance by $P(\omega)^2$, or the sample may be split into K segments in which case the resolution will be given by K/T_x and the variance by $P(\omega)^2/K$.

It is an implicit assumption of the Fourier transform that the data are stationary, that is to say their statistical properties do not change over the sample period. This must be the case, otherwise the PSD will vary during the sample and so it will not be possible to estimate it with any accuracy. It is the stationarity condition that limits the length of data T_x that may be used to produce a spectral estimate of a Doppler signal (whether using a single transform of the entire segment or averaging several transforms). Arterial Doppler signals may not be considered stationary for periods of greater than about 10–20 ms, and sometimes less, and therefore it is impossible to obtain a frequency resolution of better than 50–100 Hz, even if the Doppler shift signal has a low frequency. The resolution will become even lower if Bartlett's procedure is used to reduce spectral variance, but it is possible to maintain the resolution and decrease the variance by averaging the transforms of the corresponding portions of signals from a number of heartbeats. This is possible because a longer data segment has been analysed without violating the stationarity condition. Figure 8.14c shows a single spectral estimate taken from a human carotid artery at peak systole, whilst Fig. 8.14d is an average of 64 estimates taken from 64 consecutive heartbeats.

In the preceding discussion it has been implicitly assumed that the data window is rectangular. In practice spectral analysers often multiply the data by a non-rectangular weighting function in order to reduce spectral leakage. The reason for this, and the details, are not important in the context of this discussion and may be found in any book on digital signal processing (e.g. Oppenheim and Schafer 1975, pp 239–250, Rabiner and Gold 1975, pp 88–105), although the use of such windows does slightly modify the theory given above. Windowing reduces the effective data length and therefore the spectral resolution without decreasing spectral variance. The reduction in the effective data length does, however, mean that it is possible to average spectral estimates from overlapping segments in order to reduce spectral variance, and this is a technique employed in some spectral analysers. This method was first described by Welch (1967), who showed that by using a window similar to a hanning window, a reduction in the spectral variance of 11/18 could be achieved using a 50% data overlap.

Many of the limitations discussed in this section could be overcome by using modern parametric spectral analysis techniques such as the autoregressive (AR) modelling method (Kay and Marple 1981) rather than the FFT, and some preliminary work on the application of this technique to Doppler signals has been reported by Kitney and Giddens (1986) and Kaluzynski (1987). A major drawback of the AR method has been that it could not be implemented on-line, but recently Schlindwein and Evans (1988) have described a real-time system for just this purpose.

8.4   PULSED WAVE DOPPLER SPECTRA

A drawback of CW Doppler is that the Doppler shift signal may originate from anywhere along the ultrasound beam. This limitation may be overcome using a pulsed Doppler technique, which allows the Doppler signal to be range gated.

PW Doppler is dealt with separately from CW Doppler in this chapter because the Doppler spectra resulting from the two types of instrument may differ considerably. PW units may be roughly classified into two (overlapping) groups; those that use long gate times sufficient to interrogate an entire vessel, and those (including multigate instruments) that use very short gates so that the flow in only a small part of the vessel is interrogated. Many of the characteristics of these two types of operation are similar, but they differ in other respects so that long gate PW operation may actually be more like CW operation than short-gate PW operation from the standpoint of the spectra they produce. Wherever possible it will be made clear if, and how, short- and long-gate operation differ. Details of the generation and demodulation of PW signals are given in Chapters 4, 5 and 6.

8.4.1   Physical principles of PW Doppler

Unlike CW Dopplers, PW Dopplers usually contain only a single crystal which serves as both a transmitter and a receiver. They emit short bursts of ultrasound several thousand times every second, usually at regular intervals. After each pulse has been transmitted there is a delay before one or more gates in the receiving circuit are opened for a short period of time to admit signals returning from a small volume of tissue. The delay between transmission and opening the gate may be altered by the operator to determine the depth from which the signals are gathered, whilst the time for which the gate is left open (and to some extent the receiver bandwidth), taken together with the length of the transmitted pulse, determines the length of the sample volume. Specifically, the distance from the transducer to the beginning of the range cell, Z_1, will be given by:

$$Z_1 = c(t_d - t_p)/2 \qquad\qquad 8.17$$

where c is the velocity of the ultrasound in tissue, t_p the pulse length, and t_d the time delay between the start of transmission and the moment at which the receiver gate opens. The distance from the transducer to the end of the range cell, Z_2, will be given by:

$$Z_2 = c(t_d + t_g)/2 \qquad\qquad 8.18$$

where t_g is the time for which the gate is open. The length of the range cell may therefore be written:

$$Z_r = Z_2 - Z_1 = c(t_g + t_p)/2 \qquad\qquad 8.19$$

8.4.1.1   Maximum velocity limit

With CW Doppler there is no practical limit on the maximum velocity that can be measured. This is not so with PW Doppler because of the finite sampling rates employed. In order to extract a Doppler shift from the ultrasound signal the velocimeter compares the phase relationships between a signal from a reference oscillator and each successive returning ultrasound pulse. The maximum phase change that can be observed between two pulses is limited to the range $-\pi$ to $+\pi$ radians (since angular measurements repeat themselves every 2π radians), and therefore if the target moves a distance of more than $\lambda/4$ between samples (equivalent to a round trip distance of $\lambda/2$ for the ultrasound pulse) its velocity may be interpreted incorrectly. This limitation is simply an expression of the sampling theorem (Shannon 1949) which states that it is necessary to sample a signal at at least twice the highest frequency present in the signal to avoid ambiguity. Mathematically this may be stated as:

$$f_d(\text{max}) = f_s/2 \qquad\qquad 8.20$$

where $f_d(\text{max})$ is the maximum Doppler shift that can be unambiguously detected, and f_s is the pulse repetition frequency or sampling frequency. The critical frequency $f_s/2$ is commonly known as the Nyquist frequency. The maximum velocity that may unambiguously detected, V_{max}, may be found by substituting eqn 8.1 into eqn 8.20, i.e.

$$V_{\text{max}} = f_s c/(4f_t \cos\theta) \qquad\qquad 8.21$$

The effect of aliasing (the incorrect interpretation of frequencies above the Nyquist limit) is illustrated in Fig. 8.15a. As soon as the Doppler shift frequency exceeds $f_s/2$ it is interpreted as being $-f_s/2$ and the top of the sonogram appears at the bottom of the reverse channel. Although the Doppler shift is normally interpreted as being between $-f_s/2$ and $+f_s/2$ it may be interpreted as any other range of frequencies provided the total range is only f_s, for example from $-f_s/4$ to $3f_s/4$ or from 0 to f_s. In the case of the sonogram shown in Fig. 8.15a a shift of the range to $-0.1f_s$ to $+0.9f_s$ results in the sonogram shown in Fig. 8.15b, which is in this case the correct interpretation of the signal. Some analysers have the ability to shift sonograms graphically in this way; they cannot change the permissible range of frequencies, only the way in which they map onto the frequency scale.

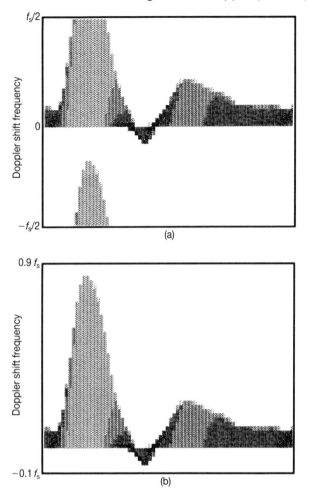

Figure 8.15   Effect of transmitted pulse repetition frequency aliasing on a sonogram. (b) Correction of aliasing by changing the interpretation of the information derived in (a)

8.4.1.2   Range ambiguity

In the previous section it was stated that the position of the range gate in a PW system is determined by the time delay between transmission and the commencement of signal acquisition. In fact there is a degree of range ambiguity since the signals arriving at the transducer at a given time may be echoes from the last transmitted pulse, the previous pulse, or even earlier pulses. Signals are therefore collected from ranges located around Z_n given by:

$$Z_n = (c/2)(t_d + nt_s) \qquad 8.22$$

where t_s is the time between subsequent pulse transmissions and n zero or any non-negative integer. In practice, because of attenuation, the signals returning from deeper tissue are weaker than signals from the most superficial tissues, and if f_s is low they may be negligible. If f_s is high, two or more significant gates may exist and it is for this reason that f_s cannot be increased at will to overcome the maximum velocity limit discussed in the last section. Another way of viewing this is that, as f_s is increased, the PW case tends to the CW case, where there is no maximum velocity limit but neither is there any depth resolution.

If a flowmeter is to use pulses that return during the same transmission cycle and thus avoid ambiguity, the maximum range z_{max} at which it can operate is given by:

$$z_{max} = c/2f_s \qquad\qquad 8.23$$

If, in addition, there is to be no velocity ambiguity, then eqn 8.21 must be satisfied, and therefore there is a maximum range–velocity limit given by:

$$z_{max} V_{max} = c^2/8f_t \cos \theta \qquad\qquad 8.24$$

In practice, some range ambiguity is not usually a serious problem, except possibly in the heart, because provided that the sites of the range gates are known care can be taken to ensure that only one of them encompasses a vessel. Some commercial scanners have high PRF modes which deliberately introduce range ambiguity so as to be able to measure high velocities without resorting to CW operation.

8.4.2 Non-uniform insonation

Whilst CW transducers often use separate elements for transmitting and receiving, it is usual in PW applications to use a single crystal or group of crystals, for both purposes. Hence the field shape produced by a PW unit differs from that from a CW unit both radially (due to crystal geometry) and longitudinally (due to finite pulse length).

As with CW Doppler, it is essential that the field from a long-gate Doppler unit is substantially uniform over the entire lumen of the vessel if the Doppler power spectrum is to be representative of the velocity distribution in the vessel, and so, in addition to the ultrasound beam being sufficiently wide to encompass the whole vessel, the gate must be open for a sufficiently long period of time. Short-gate Doppler units do not insonate the vessel uniformly and would ideally reflect the velocity distribution of the erythrocytes in the small sample volume they interrogate. Unfortunately spectral broadening mechanisms become important in such instances (see the next section) and considerable care must be taken with the interpretation of such spectra.

8.4.3 Intrinsic spectral broadening

In CW applications ISB is determined by the length of the bursts of ultrasound reflected by targets as they traverse the ultrasound beam (Section 8.3.5.2). In PW applications the length of each transmitted pulse may be so short that the target travels only a part of the way across the ultrasound beam during each transmission cycle. In this case the length of the reflected pulses will be the same as that of the transmitted pulses, and the ISB will arise solely from this source. If there is uniform flow within the sample volume, the Doppler spectrum will be a replica of the square of the spectrum of the transmitted ultrasound pulse, but scaled down in frequency by a factor of f_d/f_t (Newhouse et al 1976) (Fig. 8.16). The ISB for the short pulse case is therefore entirely determined by the bandwidth to

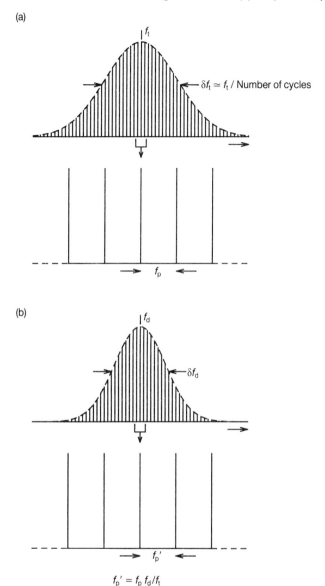

Figure 8.16 Spectrum of ultrasound power transmitted by a PW Doppler unit, and (b) the corresponding audio frequency replica derived by demodulation. In each case the lower part of the diagram represents an expanded version of the centre of the upper diagram

centre frequency of the transmitted pulse and is approximately equal to the reciprocal of the number of cycles in the transmitted pulse, i.e.

$$\delta f_d / f_d \simeq 1 / t_p f_t \qquad\qquad 8.25$$

where t_p is the length of the ultrasound pulse. Considerable spectral broadening therefore occurs where very short pulses are used, and it is difficult to distinguish between broadening caused by a true distribution of velocities in the sample volume and the intrinsic broadening. Hatle and Angelsen (1982, p 48) have counselled caution in the interpretation of reports concerning the ability to resolve spectral broadening due to turbulence using high spatial resolution measurements.

The use of pulses with large bandwidths creates additional problems in that, as both scattering and attenuation are frequency dependent, the shape of an ultrasound pulse and hence its Doppler replica can be considerably distorted is it travels through the body and is scattered by the blood cells. This is discussed further in the next two sections.

8.4.4   Frequency-dependent scattering

Scattering of ultrasound by blood is proportional to the fourth power of frequency (Shung et al 1976), and therefore the higher frequency components of a pulse of ultrasound are more strongly scattered than the lower frequencies. This results in a similar distortion in the demodulated audio frequency signal so that the higher Doppler shifts are accentuated and the mean frequency is displaced upwards. Calculations reported by Newhouse et al (1977c) suggest that this becomes important only for very large bandwidth pulses, and even a centre frequency-to-bandwidth ratio of three leads to an error of less than 5% in the modal Doppler frequency measured from uniform flow.

8.4.5   Attenuation

As in the case of CW Doppler (Section 8.3.3) attenuation may effectively modify the shape of a PW Doppler beam, but this will normally only be of importance with long-gate Doppler systems where the objective is to insonate the vessel uniformly.

For short-gate Dopplers it is the frequency-dependent nature of attenuation rather than attenuation itself that is important in distorting the Doppler power spectrum. Attenuation of ultrasound in tissue increases with increasing frequency, and therefore the higher frequency components of the transmitted ultrasound pulse are attenuated more rapidly than the lower frequencies. As with frequency-dependent scattering, this leads to a distortion of the demodulated audio frequency signal, but in this instance it is the lower frequencies that are accentuated and the mean frequency is displaced downwards. Once again Newhouse et al (1977c) have treated this distortion mathematically, and shown that it is only important for very short pulses which experience large degrees of attenuation. To some extent the increase in the mean Doppler shift frequency due to scattering is compensated for by the decrease due to frequency dependent attenuation, and Reid and Klepper (1984) have pointed out that the operating frequency of a Doppler unit could be chosen to minimize the net error over a finite range if the attenuation of the tissue were known, and that the same choice would maximize the average echo power.

8.4.6   Filtering

The effects of filtering on the Doppler spectrum have been discussed in Section 8.3.6 where it was explained that high pass filters are necessary to remove high-amplitude low-frequency signals arising from tissue movement and especially 'vessel wall thump'. Exactly the same considerations apply to long-gate PW Doppler, but for short-gate PW the sample may be placed entirely within the blood vessel of interest, and therefore many of the low-frequency signals will be rejected by the range gating. In these circumstances it is possible to reduce the filtering requirements.

8.4.7   Spectral analysis limitations

The interplay between spectral resolution and the variance of the spectral estimate has been explained in Section 8.3.7. It has also been shown that the spectral broadening for short-gate Doppler instruments can be very great (Section 8.4.3).

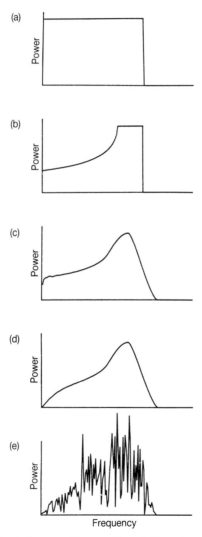

Figure 8.17   The progressive influence of a number of distorting mechanisms on the idealized Doppler power spectrum from steady laminar flow: (a) idealized spectrum; (b) the effect of partial sampling of the vessel – Section 8.3.2; (c) the effect of spectral broadening – Section 8.3.5; (d) the effect of high pass filtering – Section 8.3.6; (e) the effect of spectral estimate variance – Section 8.3.7

Therefore when such an instrument is used the spectral resolution may be limited not by the length of the data sample, T_a, but by the intrinsic spectral broadening. If the stationarity period of the signal is greater than the reciprocal of the intrinsic spectral width $(1/\delta f)$ then, since spectral resolution cannot be improved by using a larger data segment, it is better to reduce the spectral variance by using Bartlett's averaging procedure (Section 8.3.7).

8.5 SUMMARY

Under ideal uniform sampling conditions the Doppler power spectrum would have the same shape as a histogram of the velocity distribution of the erythrocytes within the Doppler sample volume. In practice this shape is modified by a number of mechanisms including non-uniform sampling, attenuation, intrinsic spectral broadening, filtering and the limitations of spectral estimation techniques. Figure 8.17 illustrates the way in which the ideal spectrum from steady laminar flow might be progressively influenced by these mechanisms.

There are significant differences between CW and PW Doppler, due both to the effects of sampling and to the wide signal bandwidths used by some of the latter type of device. In the case of short-gate PW Doppler these differences may profoundly alter the shape of the Doppler power spectrum.

8.6 NOTATION

a	Radius of scatterer
A	Aperture of receiver
c	Velocity of ultrasound in tissue
D	Diameter of a transducer
f	Centre frequency of an ultrasound burst
f_d	Doppler shift frequency
$f_d(\text{max})$	Maximum Doppler shift frequency that can be unambiguously detected
f_r	Received frequency
f_s	Pulse repetition frequency
f_t	Transmitted frequency
F	Weighting factor
I_i	Incident intensity
k	Wave number $(= 2\pi/\lambda)$
K	Number of spectral estimates averaged together
n	Concentration of particles
N	Number of points in a Fourier transform
N'	Number of points in a (different) Fourier transform
N_d	Lateral fading rate
N_t	Temporal fading rate
P_{env}	Pressure envelope
$P(\phi)$	Power as a function of scattering angle
$P(\omega)$	Power spectral density
$P'(\omega)$	Estimated power spectral density

r	Radial coordinate
R	Radius of a blood vessel
R_B	Measure of ultrasound beam radius
s	Distance from observation point to volume V
t_d	Delay between transmitting an ultrasound pulse and opening the receiver gate
t_g	Time for which a gate is open
t_p	Length of an ultrasonic pulse
t_s	Period between pulse transmissions ($=1/f_s$)
T_a	Length of a Fourier transformed data sample
T_L	Measure of pulse length
T_x	Total length of a data sample
v	Velocity of a target
v_{max}	Maximum velocity in a blood vessel
V	Sample volume
V_{max}	Maximum velocity shift that can be unambiguously detected
w	Width of an ultrasound beam
W_0	Volume concentration of scatterers
z_{max}	Maximum unambiguous range of a PW Doppler
Z_1	Beginning of a Doppler range cell
Z_2	End of a Doppler range cell
Z_n	Nominal range of multiple range gates
Z_r	Length of Doppler range cell
α_s	Scattering coefficient
β_e	Adiabatic compressibility of erythrocyte
β_p	Adiabatic compressibility of plasma
δf	Spectral width of an ultrasound burst
δf_a	Spectral resolution of a Fourier transform
$\Delta\phi$	Angle subtended by a transducer at its focal spot
θ	Angle between ultrasound beam and blood flow
κ	A constant
λ	Wavelength of ultrasound
ρ_e	Density of erythrocytes
ρ_p	Density of plasma
ϕ	Scattering angle
ω	Angular frequency

8.7 REFERENCES

Ahuja AS (1972) Effect of particle viscosity on propagation of sound in suspensions and emulsions. J Acoust Soc Am 51, 182–191.

Angelsen BAJ (1980) A theoretical study of the scattering of ultrasound from blood. IEEE Trans Biomed Eng BME-27, 61–67.

Angus JC, Edwards RV, Dunning JW (1971) Signal broadening in the laser Doppler velocimeter. AIChE J 17, 1509–1510.

Atkinson P, Berry MV (1974) Random noise in ultrasonic echoes diffracted by blood. J Phys A: Math Nuc Gen 7, 1293–1302.

Bascom PAJ, Cobbold RSC, Roelofs BHM (1986) Influence of spectral broadening on continuous wave Doppler ultrasound spectra: a geometric approach. Ultrasound Med Biol 12, 387–395.

Beard CI, Kays TH, Twersky V (1967) Scattering by random distributions of spheres versus concentration. IEEE Trans Antennas Propag AP-15, 99–118.

Cobbold RSC, Veltink PH, Johnston KW (1983) Influence of beam profile and degree of insonation on the CW Doppler ultrasound spectrum and mean velocity. IEEE Trans Sonics Ultrasonics SU-30, 364–370.

Douville Y, Arenson JW, Johnston KW, Cobbold RSC, Kassam M (1983) Critical evaluation of continuous-wave Doppler probes for carotid studies. J Clin Ultrasound 11, 83–90.

Evans DH (1982a) Some aspects of the relationship between instantaneous volumetric blood flow and continuous wave Doppler ultrasound recordings. I. Ultrasound Med Biol 8, 605–609.

Evans DH (1982b) Some aspects of the relationship between instantaneous volumetric blood flow and continuous wave Doppler ultrasound recordings. III. Ultrasound Med Biol 8, 617–623.

Evans DH (1985) On the measurement of the mean velocity of blood flow over the cardiac cycle using Doppler ultrasound. Ultrasound Med Biol 11, 735–741.

Evans DH, Parton L (1981) The directional characteristics of some ultrasonic Doppler blood-flow probes. Ultrasound Med Biol 7, 51–62.

Green PS (1964) Spectral broadening of acoustic reverberation in Doppler-shift fluid flowmeters. J Acoust Soc Am 36, 1383–1390.

Griffith JM, Brody WR, Goodman L (1976) Resolution performance of Doppler ultrasound flowmeters. J Acoust Soc Am 60, 607–610.

Hatle L, Angelson B (1982) Doppler ultrasound in cardiology. Physical principles and clinical applications, 2nd edn, Lea and Febiger, Philadelphia.

Kaluzynski K (1987) Analysis of application possibilities of autoregressive modelling to Doppler blood flow signal spectral analysis. Med Biol Eng Comp 25, 373–376.

Kay SM, Marple SL (1981) Spectrum analysis – a modern perspective. Proc IEEE 69, 1380–1419.

Kitney RI, Giddens DP (1986) Linear estimation of blood flow waveforms measured by Doppler ultrasound. In: Medinfo 86 (Eds R Salamon, B Blum, M Jørgensen), Elsevier, Amsterdam.

Morse PM, Ingard KU (1968) Theoretical acoustics, pp 427, McGraw-Hill, New York.

Narayana PA, Ophir J, Maklad NF (1984) The attenuation of ultrasound in biological fluids. J Acoust Soc Am 76, 1–4.

Newhouse VL, Bendick PJ, Varner LW (1976) Analysis of transit time effects of Doppler flow measurement. IEEE Trans Biomed Eng BME-23, 381–387.

Newhouse VL, Varner, LW, Bendick PJ (1977a) Geometrical spectrum broadening in ultrasonic Doppler systems. IEEE Trans Biomed Eng BME-24, 478–480.

Newhouse, VL, Johnson FG, Furgason ES (1977b) Transit time and geometrical broadening in Doppler ultrasound spectra. Proc 30th ACEMB, p 353.

Newhouse VL, Ehrenwald AR, Johnson GF (1977c) The effect of Rayleigh scattering and frequency dependent absorption on the output spectrum of Doppler blood flowmeters. In: Ultrasound in medicine 3B (Eds D White, RE Brown), pp 1181–1191, Plenum Press, New York.

Newhouse VL, Furgason ES, Johnson GF, Wolf DA (1980) The dependence of ultrasound Doppler bandwidth on beam geometry. IEEE Trans Sonics Ultrasonics SU-27, 50–59.

Oppenheim AV, Schafer PW (1975) Digital signal processing, Prentice-Hall, Englewood Cliffs, New Jersey.

Powalowski T, Borodzinski K, Nowicki A (1975) Effect of ultrasonic beam width on blood flow estimation by means of CW Doppler flowmeter. Scripta Medica Univ Brno 48, 97–103.

Rabiner LR, Gold B (1975) Theory and application of digital signal processing. Prentice-Hall, Englewood Cliffs, New Jersey.

Reid JM, Klepper J (1984) Frequency errors in pulse Doppler systems. Ultrasonic Imaging 6, 212.

Reid JM, Sigelmann RA, Nasser MG, Baker DW (1969) The scattering of ultrasound by human blood. Abst 8th Int Congr Med Biol Eng, pp 10–7.

Rschevkin SN (1963) A course of lectures on the theory of sound, p 374, Pergamon Press, Oxford.

Schlindwein FS, Evans DH (1988) A real time autoregressive spectrum analyzer for Doppler ultrasound signals. Ultrasound Med Biol (in press).

Shannon CE (1949) Communications in the presence of noise. Proc IRE 37, 10–21.

Shung KK, Sigelmann RA, Reid JM (1976) Scattering of ultrasound by blood. IEEE Trans Biomed Eng BME-23, 460–467.

Shung KK, Sigelman RA, Reid JM (1977) Angular dependence of scattering of ultrasound from blood. IEEE Trans Biomed Eng BME-24, 325–331.

Urick RJ (1947) A sound velocity method for determining the compressibility of finely divided substances. J Appl Phys 18, 983–987.

Welch PD (1967) The use of fast Fourier transform for the estimation of power spectra: a method based on time averaging over short, modified periodograms. IEEE Trans Audio Electroacoust AU-15, 70–73.

Chapter 9

DOPPLER SIGNAL PROCESSORS: THEORETICAL CONSIDERATIONS

9.1 INTRODUCTION

The Doppler shift signal that results from the demodulation process contains a wealth of information about blood flow occurring within the sample volume of the Doppler velocimeter. The most complete way to display this information is to perform a full spectral analysis and present the results in the form of a sonogram (see for example Fig. 8.2). Whilst such a display may be of great value for assessing the general quality of the signal and for making qualitative statements about disease state, it contains so much information that some feature reduction must take place before quantitative statements may be made. It is therefore usual to extract some kind of envelope signal from the full Doppler shift spectrum, either using analogue processing methods on the raw Doppler signal or using digital methods to process the Fourier transformed Doppler data. The most useful envelope detectors are probably the mean frequency and maximum frequency processors, but a number of others have been used and may even have some advantages.

It has already been shown in Chapter 8 that the Doppler power spectrum is influenced by a number of factors that are unrelated to the velocity distribution in the target blood vessel, and the same must necessarily be true of envelope signals derived from the Doppler spectrum. In this chapter we compare the theoretical merits of a number of envelope processors and consider the effects that various distorting factors may have on their output. We start, however, with a brief discussion of spectral analysis.

9.2 SPECTRAL ANALYSIS TECHNIQUES

Spectral analysis of the Doppler signal has in the past been performed using a variety of equipment including parallel filter analysers and swept filter analysers of both the analogue and digital time-compression variety (Atkinson and Woodcock, 1982), but virtually all new instruments now use the fast Fourier transform (FFT) method. The time interval histogram (Daigle and Baker 1977) which has been used as a method for estimating the spectral content of Doppler signals has been shown by Angelsen (1980) and Burckhardt (1981) to give only an estimate of the average frequency and the width of the spectrum, with no detailed information about the shape of the spectrum.

9.2.1 The Fourier transform analyser

The fast Fourier transform is simply a high-speed algorithm for computing the Fourier transform (spectral content) of a discrete (sampled) signal. The method was first described by Cooley and Tukey (1965) and because it is so rapid it has revolutionized spectral analysis using digital methods. The details of the implementation of the FFT are beyond the scope of this book, and have been discussed in detail elsewhere (Cochran et al 1967, Rabiner and Gold 1975 pp 356–437, Brigham 1974). Recently Schlindwein et al (1988) have published details of an FFT analyser for Doppler signals which uses one of the new digital signal processing chips.

The important properties of the Fourier transform analyser have already been discussed in Section 8.3.7. These may be summarized as follows:

1. The best spectral resolution obtainable from a transform is given by the reciprocal of the data segment length used. If, for example, the transform is carried out on a 5 ms data segment, the resolution can be no better than 200 Hz. In practice the data is usually 'windowed' to prevent spectral leakage (see Section 8.3.7) and this reduces the effective data length and hence spectral resolution further. The maximum data segment length that may be legitimately used is determined by the stationary period of the signal (the period for which none of its statistics changes significantly).
2. The maximum frequency component that can be detected by analysers operating on a real signal is half the sampling frequency of the analyser. Frequencies above this value must be removed at the input by an analogue low pass filter, otherwise they will be 'aliased' down into the working frequency range (i.e. they are misinterpreted as falling in the analysis range). In the case of analysers operating on complex signals (i.e. using two quadrature components) the frequency of neither signal must exceed the sampling rate.
3. The Fourier transform of a random signal is merely an estimate of the true underlying spectrum and in particular has a large variance. This variance may be reduced by averaging a number of independent estimates using Bartlett's procedure (see Section 8.3.7).

Most of the limitations discussed above are common to all the analysers mentioned at the beginning of this section; analogue systems, however, do not suffer from aliasing problems.

Figures 8.14a and 8.14c are individual spectral estimates of the Doppler signal recorded from steady flow in a tube, and from a human carotid artery at peak systole, calculated using a Fourier transform technique. Sonograms (Section 8.1) produced using FFT analysers are to be found throughout this book (see for example Fig. 10.2).

9.3 MEAN FREQUENCY PROCESSORS

Perhaps the most obvious envelope signal to extract from a Doppler audio signal is the intensity weighted mean frequency, $\bar{f}(t)$, defined by:

$$\bar{f}(t) = \frac{\int_f P(f)f\,df}{\int_f P(f)\,df} \qquad 9.1$$

Here $P(f)$ is the Doppler power spectrum. Under ideal conditions this derived frequency is proportional to the mean velocity of flow in the target blood vessel. If the angle between the ultrasound beam and the blood vessel is known, the output from a mean frequency follower may be calibrated directly in terms of flow velocity; if in addition the cross-sectional area of the vessel is known, the output may be calibrated in terms of volumetric flow.

Several methods of deriving the intensity-weighted mean frequency using analogue techniques have been described, of which the best known are those of Arts and Roevros (Arts and Roevros 1972, Roevros 1974) and DeJong (DeJong et al 1975, Gerzberg and Meindl 1977, 1980a, 1980b). An alternative approach is to evaluate eqn 9.1 numerically using a real-time spectrum analyser and a microcomputer (see for example Kontis 1987, Schlindwein et al 1988). Details of the Arts and Roevros circuit can be found in Section 6.5.1.

9.3.1   Effect of non-uniform insonation

The effect of non-uniform insonation on the Doppler power spectrum was discussed in Section 8.3.2, and illustrated for a number of specific situations in Figs 8.7–8.10. Clearly such dramatic changes will have a considerable effect on the output of mean frequency followers, and most importantly on the estimation of mean velocity. The magnitude of these errors has been investigated for only a few simple idealized geometries (Evans 1982, Cobbold et al 1983, Evans 1985a), but the results show that use of an ultrasound beam which is much narrower than a blood vessel containing a parabolic velocity profile may cause an overestimation of mean velocity of up to 33%. Interestingly the error in the time-averaged mean velocity over the cardiac cycle is dependent only upon the shape of the *mean* component of the velocity profile and the way in which this is sampled by the ultrasound beam (Evans 1985a). The shape of the mean component depends on a number of factors (see Evans 1985a and Chapter 2) but will be parabolic at sites that are sufficiently far from curves, branches, bifurcations and other sources of haemodynamic disturbance. Figure 9.1 illustrates the error in the measurement of mean velocity in a vessel containing parabolic flow using a rectangular ultrasound beam of various sizes with various degrees of offset from the vessel centre.

9.3.2   Effect of attenuation

The differential rates of attenuation in soft tissue and blood have the effect of slightly reducing the effective width of an ultrasound beam (Section 8.3.3) and therefore of accentuating the higher velocities at the centre of the vessel. Cobbold et al (1983) have published diagrams illustrating the effect of attenuation on the error in mean velocity measurements when square and Gaussian beams interact with a parabolic velocity profile (Fig. 9.2).

9.3.3   Effect of intrinsic spectral broadening

Intrinsic spectral broadening (Section 8.3.5), per se, does not affect the mean frequency of a spectrum because the broadening is symmetrical. However, if very short pulses with large bandwidths are used, then frequency dependent scattering will

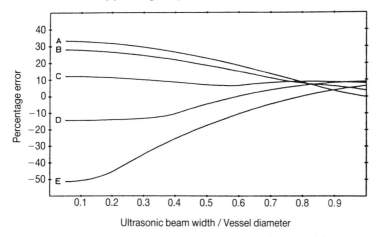

Figure 9.1   The error that results if a mean frequency processor is used with an ultrasound beam that is narrower than the blood vessel. Curve A has been calculated for a centralized beam, curves B–E each represent the behaviour of offset beams. The values of the offsets (in fractions of the vessel radius) are: B, $R/5$; C, $2R/5$; D, $3R/5$; E, $4R/5$. (Reprinted with permission from Evans (1985a), © Pergamon Journals Ltd)

tend to increase the mean frequency of the received spectrum (Section 8.4.4), whilst frequency-dependent attenuation will tend to decrease the mean frequency (Section 8.4.5). The effects of these frequency-dependent mechanisms on the measured Doppler frequency have been explored theoretically by Newhouse et al (1977), and experimentally by Holland et al (1984). Unfortunately the experimental and theoretical results are not completely compatible and it seems that further studies of this complex phenomenon are needed. It seems likely, however, that for CW or long-gate PW Doppler measurements of blood flow, frequency-dependent scattering and attenuation cause only minor inaccuracies.

9.3.4   Effect of filters

As has been explained in Section 8.3.6, Doppler devices contain high pass filters designed to reject high-amplitude low-frequency signals arising from stationary and near stationary tissue. In addition to rejecting unwanted noise they also reject low frequency signals, and this has the effect of biasing the mean frequency estimate in an upwards direction. Gill (1979) has studied the error that this introduces and shown that it is influenced both by the cutoff frequency of the filter, f_{HP}, and the velocity profile.

For velocity profiles of the form:

$$v = v_{max}[1 - (r/R)^\beta]$$
9.2

Gill shows the error to be given by:

$$f_{error} = 2f_{HP}/(2 + \beta)$$
9.3

For a parabolic profile, β has a value of 2; for blunt profiles a value of greater than 2; and for plug flow β tends to infinity.

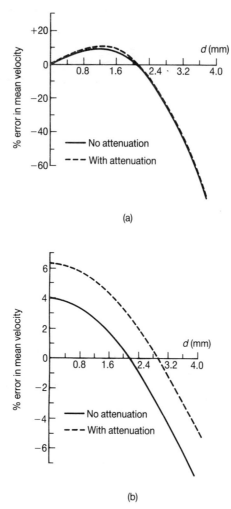

Figure 9.2   The effect of attenuation on the error in mean velocity calculated from the power density spectrum, for a parabolic velocity profile, for various beam displacements from the centre: (a) represents the results from a square beam of 4 mm width insonating a vessel of the same diameter, (b) represents the results from a Gaussian beam with a standard deviation of 2 mm insonating the same vessel as in (a). (Reprinted with permission from Cobbold et al (1983), copyright © 1983, IEEE)

Equation 9.3 must be applied to the instantaneous velocity profiles, and not the time-averaged velocity profile, and therefore it is not easy to correct for this error when pulsatile flow is present.

9.3.5   Effect of signal-to-noise ratio

The signal-to-noise ratio of the Doppler signal is often not as great as might be hoped. The mean frequency processor has no way of distinguishing between signal and noise within the acceptable band for Doppler signals, and therefore noise will bias the output.

Gerzberg and Meindl (1977) have shown that in the presence of band-limited stationary noise, the output of a mean frequency follower is given by:

$$\bar{f}' = \bar{f}_d/(1+1/S) + \bar{f}_n/(1+S) \qquad 9.4$$

where $\bar{f}_d$ is the mean Doppler shift frequency, $\bar{f}_n$ the centroid of the noise spectrum, and S the signal-to-noise power ratio. The effects of various levels of additive white noise on the output of a mean follower (calculated from eqn 9.4) is shown in Fig. 9.3; these results have been experimentally verified by Gerzberg and Meindl.

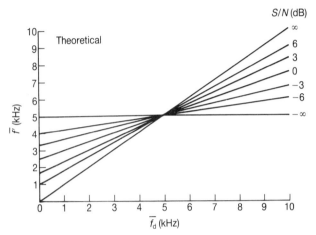

Figure 9.3   Predicted centroid $\bar{f}'$ of a signal $\bar{f}_d$ accompanied by band-limited stationary noise whose centroid is located at 5 kHz. (Reprinted with permission from Gerzberg and Meindl (1977), Plenum Publishing Corp)

If the characteristics of the noise are known, it is theoretically possible to eliminate the effects of this error (Gerzberg and Meindl 1977, Gill 1979, Gerzberg and Meindl 1980b).

9.4   MAXIMUM FREQUENCY PROCESSORS

A second popular envelope detector is the maximum frequency processor. This method was first used to derive a single-valued waveform from sonograms produced off-line with Kay spectrographs, simply by drawing around the darkened area of the sonogram (Gosling et al 1969). When the signal is unidirectional, the definition of maximum is self-evident; however, if there is bidirectional flow and particularly simultaneous forward and reverse flow, then a 'composite maximum frequency' must be defined. The most usual method of producing a composite maximum frequency envelope is to follow both the absolute maximum frequency and the absolute minimum frequency and to sum these two envelopes (absolute maximum is not allowed to fall below zero, nor absolute minimum to exceed zero). Thus if only forward flow is present the composite maximum corresponds to the highest forward velocity, and when both are present to the velocity differences between maximum forward and maximum reverse.

Several methods of extracting the maximum frequency envelope automatically have now been described; some use analogue processing methods on the Doppler audio signal (Sainz et al 1976, Skidmore and Follet 1978) whilst others use digital methods to process the output of a spectrum analyser (Gibbons et al 1981, Prytherch and Evans 1985, D'Alessio 1985). The digital techniques are particularly valuable for reasons that will be explained later (Section 9.4.6).

There are also a group of 'maximum' frequency followers which actually follow the 7/8th or 15/16th power or total amplitude levels for each spectrum, and which perform in a similar way to maximum frequency followers. These followers are less immune to the distorting factors discussed in Sections 9.4.2–9.4.6 than are the maximum frequency followers that depend on threshold levels.

9.4.1   Relationship to mean frequency

The relationship between the instantaneous mean and maximum velocities in an artery depends entirely on the shape of the velocity profile. For plug flow the mean and maximum velocities are equal, whilst for parabolic flow the maximum velocity is exactly twice the mean. In pulsatile flow the velocity profiles may change quite dramatically during the cardiac cycle and thus the ratio of maximum to mean frequency in the Doppler signal will also vary widely. Despite this the shapes of maximum and mean frequency envelopes derived from the same signals are often superficially similar (Evans and Macpherson 1982).

Under certain flow conditions (such as those found in vessels supplying low-impedance vascular beds) the maximum velocity in the vessel is in its centre throughout the cardiac cycle. If this is the case, the ratio of the time-averaged maximum Doppler frequency to the time-averaged mean Doppler frequency is the same as the ratio of the maximum velocity to the mean velocity in the time-averaged mean component of the velocity profile (Evans 1985a). This means that if the shape of the mean velocity profile is known then the time averaged mean velocity can be calculated from the time averaged maximum velocity. In the case where the mean velocity profile is parabolic, as it will be at sufficient distances from geometry changes, then the time averaged maximum velocity will be twice the time averaged mean velocity (Evans 1985a).

9.4.2   Effect of non-uniform insonation

One of the great advantages of the maximum frequency processor is that its output is not significantly influenced by ultrasonic beam shape or the ratio of the beam width to vessel diameter, provided some part of the ultrasound beam passes through the part of the vessel containing the maximum velocity.

9.4.3   Effect of attenuation

Since ultrasonic beam shape is not critical for the correct functioning of the maximum-frequency follower, differential attenuation between blood and tissue is of little consequence.

9.4.4 Effect of intrinsic spectral broadening

The effect of ISB is to blur the Doppler power spectrum so that, rather than falling off rapidly at the frequency corresponding to maximum flow, there is a gradual fall-off which depends on the amount of ISB present. This effect is illustrated in Fig. 9.4. If severe degrees of ISB are present, such as might be obtained with a very short gate Doppler unit, the derived maximum frequency may be critically dependent on the threshold setting.

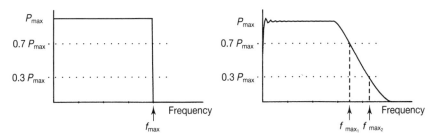

Figure 9.4   Effect of intrinsic spectral broadening on the output of a maximum frequency follower: (a) the Doppler spectrum from parabolic flow in the absence of ISB: changes in threshold have no effect on the derived maximum frequency; (b) the spectrum from parabolic flow with ISB. The derived maximum frequency is influenced by the threshold level

9.4.5 Effect of filters

Provided the maximum frequency in the Doppler signal does not fall below the cutoff frequency of the high pass (wall thump) filter, the maximum frequency follower is totally unaffected by its action. If the maximum frequency does fall below this value, the follower output will drop to zero. This problem can only be alleviated by using the lowest practical filter cutoff frequency for each application.

9.4.6 Effect of signal-to-noise ratio

A major advantage of maximum frequency followers is their noise immunity, but full advantage of this virtue can only be realized if the maximum frequency envelope is superimposed on a real-time sonogram (Fig. 9.5). Using this method it is possible for an operator to monitor the performance of the follower, and if necessary change its threshold level to ensure that it follows the required signal. A practical method of implementing this has been described by Prytherch and Evans (1985). Experience has shown that the threshold level becomes critical only when there is a very poor signal-to-noise ratio, and that even then it is usually possible by careful adjustment to 'pull out' the required signal. Figure 9.5 also illustrates that the system can separate even the signals from two vessels that have been simultaneously insonated. The threshold has been set to extract the higher frequency signals but, by suitable manipulation of the analyser gain and threshold level, the maximum frequency of the second vessel could be followed. This could not be achieved using any of the other followers described in this chapter. Other sources of noise such as venous flow and vessel wall thump are easily recognized on the sonogram and are usually easy to eliminate.

152

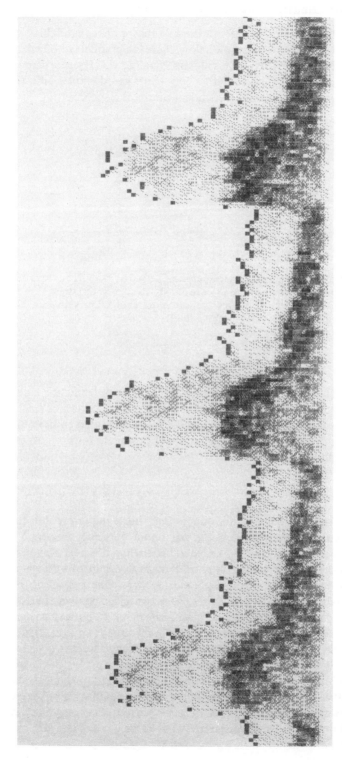

Figure 9.5 Sonogram with superimposed maximum frequency envelope. This particular display resulted from accidental simultaneous insonation of two vessels in the neck of a normal subject. The threshold of the microcomputer has been set to pick out the vessel giving the larger Doppler shifts, but could equally well have been set to follow the maximum frequency outline of the other vessel. The maximum frequency follower is the only processor that will allow such separation, and it is only possible with a system that superimposes the maximum frequency envelope directly on the sonographic display. (Reproduced from Evans (1985b), Courtesy of Elsevier Science Publishing Co)

Without simultaneous sonographic display the maximum follower is better at coping with noise than other followers; used in conjunction with a spectral display it is very much better.

9.5   ZERO-CROSSING PROCESSORS (RMS FOLLOWERS)

Zero-crossing processors were the first type of envelope follower to be successfully employed with Doppler ultrasound velocimeters (Franklin et al 1961) and are still in widespread use today in simple commercial devices. For the case of band-limited noise (such as the Doppler audio signal) the output of such a device is proportional to the root mean square (RMS) frequency of the input signal (Rice 1944). These devices have retained their position because of their simplicity, but are subject to many errors (Flax et al 1970, 1973, Lunt 1975, Johnston et al 1977).

9.5.1   Statistical noise and output filters

Flax et al (1970) have studied the statistical noise generated in the zero-crossing detector both theoretically and experimentally and have highlighted the conflicting conditions necessary for both a low noise level and an adequate frequency response at the output. They calculated the relative standard deviation at the output of a low pass filter receiving three types of pulse trains, and showed that for band-limited white noise the ratio of the standard deviation to the mean of the signal at the output is given by:

$$\frac{\sigma}{\mu} = \left(\frac{1 - 2\alpha\lambda}{2\lambda\beta}\right)^{1/2} \qquad\qquad 9.5$$

where λ is the mean zero-crossing density, β the filter time constant and α the time interval after the occurrence of a pulse, during which the probability of the occurrence of a second pulse is considered to be zero. They also showed that α could be approximated by $0.48/\lambda$ and eqn 9.5 may therefore be rewritten:

$$\frac{\sigma}{\mu} = \frac{0.2}{(2\lambda\beta)^{1/2}} \qquad\qquad 9.6$$

For low values of λ, that is for portions of the cardiac cycle where there is a low flow velocity, the signal-to-noise ratio at the output may be very poor unless the filter time constant is long. Unfortunately a long time constant will smooth the output waveform and prevent the accurate observation of rapid flow changes, such as those that occur at the beginning of systole: therefore some compromise between required frequency response and tolerable signal-to-noise ratio must be reached.

9.5.2   Relationship to mean frequency

The relationship between the mean frequency and RMS frequency of the Doppler power spectrum depends on the shape of the velocity profile. If the profile is flat, the two frequencies will be equal; if the profile is parabolic (and uniformly insonated), the RMS frequency will be 15% higher than the mean frequency (Woodcock et al 1972, Evans 1982). In pulsatile flow the velocity profile changes during the cardiac cycle and therefore so does the relationship between the output of a zero-crossing detector and the mean frequency. As Johnston et al (1977) have pointed out, a number of authors have tested the linearity of the zero-crossing detector using steady flow regimens, and have erroneously concluded that its output accurately follows changes in mean velocity.

9.5.3 Effect of non-uniform insonation

The Doppler spectrum may be dramatically distorted by non-uniform insonation of the blood vessel (Section 8.3.2), and this affects the behaviour of the RMS follower. Evans (1982, 1985a) has discussed the changes in the output of such devices resulting from partial insonation of vessels containing parabolic velocity profiles.

9.5.4 Effect of attenuation, spectral broadening and filters

The effects of attenuation, spectral broadening and filters on the output of RMS followers are qualitatively similar to their effects on the output of mean followers. These have not been investigated in detail, presumably because of other more significant limitations of RMS followers.

9.5.5 Effect of signal-to-noise ratio

A major problem with zero-crossing detectors is that they require a high signal-to-noise ratio (S/N) to operate properly. The ideal zero-crossing detector would function by generating an output pulse each time the signal crossed its own mean level. Such a detector is however impracticable because even in the absence of signal the electrical noise from the system would generate a large output signal. This problem is overcome in practice by using a SET–RESET system (Lunt 1975), which produces a pulse when two trigger levels placed either side of the zero line are passed in succession (Fig. 9.6). Provided the peak-to-peak amplitude of the noise is less than the difference between the two trigger levels, it cannot trigger the detector. Note, however, that in the example shown an 'ideal' zero-crossing detector would produce more pulses from the noise alone than from the noise plus signal.

For the simple signal shown in Fig. 9.6 the zero-crossing detector is insensitive to amplitude (provided it exceeds the difference between the SET and RESET levels), but for the complex noisy signals encountered from blood flow this is not so and the number of output pulses depends on the signal amplitude and threshold levels (Fig. 9.7). If the S/N at the input of the detector is high, the detector output will be independent of signal amplitude over a wide range, and will be proportional to the RMS frequency of the input; if the S/N is low, there may be no plateau region where the output is independent of input signal amplitude, and the output will be unreliable.

9.6 OTHER SIGNAL LOCATION ESTIMATORS

Mean, maximum and RMS frequency processors are at present much more common than any other type of signal location estimator, but a number of others have been suggested from time to time. Two frequency followers which may have some potential are the first moment follower and the mode frequency follower.

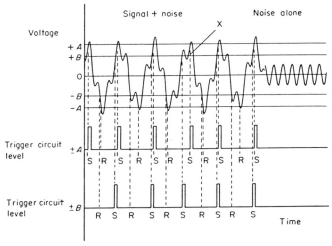

Figure 9.6 The SET–RESET system with noise. The output is correct with trigger levels $\pm A$ and $\pm B$. Note that the noise alone has more 'true' zero-crossings than the signal + noise. (Reprinted with permission from Lunt (1975), © Pergamon Journals Ltd)

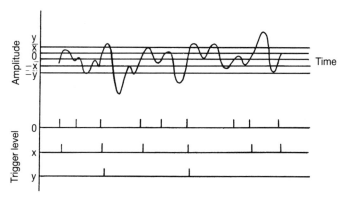

Figure 9.7 The effect of signal amplitude on the output of a zero-crossing detector. With a complex signal the number of pulses depends on the trigger level, or, if the signal level is kept constant, on the mean amplitude of the signal. (Reprinted with permission from Lunt (1975), © Pergamon Journals Ltd)

9.6.1 First moment followers

The first moment of the Doppler power spectrum is a particularly interesting quantity because it is theoretically more closely related to instantaneous volumetric flow than is the intensity-weighted mean frequency (Saini et al 1983).

The intensity-weighted mean frequency of the Doppler power spectrum $\bar{f}(t)$ was defined in eqn 9.1 as its first moment, M_1, divided by its zeroth moment M_0 where:

$$M_1 = \int_f P(f) f \, df \qquad 9.7$$

and

$$M_0 = \int_f P(f)\,\mathrm{d}f \qquad\qquad 9.8$$

Note that M_0 represents the total power in the Doppler spectrum and that for conditions of uniform insonation this is proportional to the cross-sectional area of the target blood vessel.

Under ideal conditions the instantaneous mean velocity of flow is proportional to $\bar{f}(t)$ and may be calculated using the standard Doppler equation (eqn 1.1). The instantaneous volumetric flow may then be calculated by multiplying the instantaneous mean velocity by the cross-sectional area of the target blood vessel. If the vessel cross-section is relatively constant throughout the cardiac cycle, a single area measurement will suffice; if, however, the vessel is very pulsatile, then ideally this should be measured continuously. Since M_0, the zeroth moment of the power spectrum, is proportional to vessel cross-sectional area, multiplying $\bar{f}(t)$ by M_0 (or in practice refraining from dividing M_1 by M_0) results in a quantity proportional to instantaneous flow.

The value of the first moment follower as a means of monitoring mean flow has been investigated by Saini et al (1983), and it has also been used in waveform analysis applications by Maulik et al (1982) and Thompson et al (1985). It is slightly easier to implement a first moment follower than a mean frequency follower, whether using analogue or digital techniques, because less processing is required.

It is, as yet, too early to assess the likely place of the first moment follower. One cause for concern, however, must be that the total power in the Doppler spectrum is affected by factors other than the vessel cross-section, and in particular may be reduced very significantly during some parts of the cardiac cycle by the high pass filters designed to remove the high-amplitude low-frequency power from static and nearly static structures.

9.6.2   Mode frequency followers

An off-line mode frequency follower has been described by Greene et al (1982). The algorithm first finds the region of the spectrum containing the signal by locating the median frequency which exceeds a fixed threshold, and then searches its immediate vicinity for the signal mode. This estimator ignores the high-amplitude narrow-bandwidth noise around the zero flow region, and is claimed to be insensitive to most cases of background noise and quadrature channel crosstalk. Such a follower may be of use because of its noise-resistant properties.

9.6.3   Median frequency followers

A median frequency follower which operates off-line on the output of a spectrum analyser has been described by Thomson (1980). During a first pass an arithmetic logical unit functions as an accumulator and computes the total amplitude of the spectrum. This amplitude is divided by two, and then during a second pass the amplitude of each component of the spectrum is subtracted from this value until it reaches zero. The bin in which this occurs is stored as the median. Strictly speaking it would seem more correct to operate on the intensity of each bin rather than the amplitude.

9.7 SIGNAL AVERAGING TECHNIQUES

The possibility of reducing spectral variance by averaging has been discussed in Section 8.3.7. The signal-to-noise ratio (S/N) of the Doppler envelope waveform may similarly be improved by using a coherent averaging technique. The principle of such a technique is that if the waveforms are properly aligned (by time locking them to a suitable feature) and summed, the signal amplitude will increase more rapidly than noise, which has a random time relationship to the signal. If there is perfect alignment and the noise is of a random Gaussian nature, the signal amplitude will rise in proportion to the number of waveforms summed, N, whilst the noise will increase only as the square root of N, leading to an improvement in the S/N of root N (English and Woollons 1982). Thus, for example, if 64 Doppler waveforms are coherently averaged there will be an improvement in the S/N of 800%. An example of the improvement achieved by summing only 17 waveforms is shown in Fig. 9.8.

The feature used to time-lock the averager can be part of the signal itself, for example the foot of the systolic up-slope (Evans et al 1985), or an associated external event such as the R-wave of the ECG. In practice physiological waveforms are never exactly repetitive, and

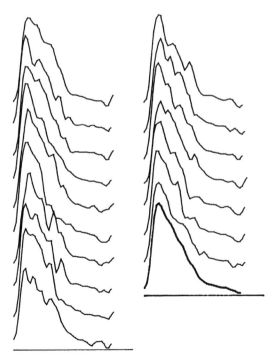

Figure 9.8 Maximum frequency envelope derived from a consecutive series of Doppler recordings from the anterior cerebral artery of a neonate. The final waveform is an ensemble average and shows a considerable improvement in S/N over the individual components

the interval between the R-wave and the flow pulse is slightly variable; nevertheless considerable improvement in S/N is possible.

> Kitney and Giddens (1983) have reported the use of phase shift averaging with Doppler velocity waveforms. This elegant method helps to reduce phase variation due to physiological variability and trigger errors. An initial ensemble average is calculated by using the R-wave as the fiducial point. Each individual waveform is then cross-correlated with this average, and a new ensemble average calculated by aligning the waveforms on the correlation maxima. The process may then be repeated until there is no further change in the ensemble average or until some other convergence criterion is satisfied.

9.8   CHOICE OF PROCESSING TECHNIQUE

Each of the processing techniques discussed has a number of advantages and disadvantages, and their utility must to some extent depend on their intended use.

For volumetric blood flow measurements, the mean frequency processor has traditionally been the method of choice, but both the first moment processor and the maximum processor may offer advantages. Whilst the output of the mean processor is proportional to instantaneous mean velocity, the output of the first moment processor is (ideally) proportional to instantaneous volumetric flow, and thus may be more accurate when there are significant changes in the vessel cross-section over the cardiac cycle (although absolute calibration is not easy). The maximum frequency processor has the advantage that its output is essentially independent of the shape and size of the ultrasound beam and may be of value for flow measurement when the velocity profile is flat, or where the flow is fully established and relatively non-pulsatile.

The relative merits of different processing techniques when used prior to waveform analysis have not been studied in any detail even though the use of different methods may considerably modify the waveform shape (Evans 1985b), and there is little evidence that any particular processing technique improves the sensitivity of any of the waveform analysis techniques discussed in the next chapter. It is, however, very important that each laboratory establishes normal and abnormal ranges for each technique they intend to use as these may be significantly affected by the choice of ultrasound transducer and envelope extraction technique. Since there is little evidence to suggest the use of any particular processor, it seems expedient to use a method that is relatively insensitive to noise and other distorting influences. Both the maximum follower and the mode follower are valuable in this respect, and the former is particularly useful because it is easy to monitor its performance by superimposing its output on a sonogram.

9.9   SUMMARY

There are several methods available for extracting an envelope signal from the full Doppler signal. Methods in widespread use include those that extract the instantaneous mean, maximum and RMS frequencies from the power spectrum. More recently, first moment and mode frequency followers have been suggested as being valuable. Each of the followers has a number of advantages and disadvantages and none is ideal for all purposes.

Under ideal conditions the output of the mean frequency follower is proportional to instantaneous mean velocity, and the output of the first moment follower to instantaneous volumetric flow.

The maximum frequency follower is probably the method of choice in waveform shape applications and under some circumstances may be used for flow measurement. Its major advantages are that it is resistant to noise and to measures designed to reduce noise such as high pass filters, and that its output is fairly independent of the way in which the ultrasound beam samples the blood vessel. For the best results, its output should be superimposed on a real-time sonogram display, so that an operator can ensure that it is following the desired part of the spectrum.

The zero-crossing detector (or RMS follower) is cheap and simple to construct but has little else to recommend its use.

9.10   REFERENCES

Angelsen BAJ (1980) Spectral estimation of a narrow-band Gaussian process from the distribution of the distance between adjacent zeros. IEEE Trans Biomed Eng BME-27, 108–110.

Arts MGJ, Roevros JMJG (1972) On the instantaneous measurement of blood flow by ultrasonic means. Med Biol Eng 10, 23–34.

Atkinson P, Woodcock JP (1982) Doppler ultrasound and its use in clinical measurement, Academic Press, London.

Brigham EO (1974) The fast Fourier transform. Prentice-Hall, Englewood Cliffs, New Jersey.

Burckhardt CB (1981) Comparison between spectrum and time interval histogram of ultrasound Doppler signals. Ultrasound Med Biol 7, 79–82.

Cobbold RSC, Veltink PH, Johnston KW (1983) Influence of beam profile and degree of insonation on the CW Doppler ultrasound spectrum and mean velocity. IEEE Trans Sonics Ultrasonics SU-30, 364–370.

Cochran WT, Cooley JW, Favin DL et al (1967) What is the fast Fourier transform? IEEE Trans Audio Electroacoust AU-15, 45–55.

Cooley JW, Tukey JW (1965) An algorithm for the machine calculation of complex Fourier series. Math Comp 19, 297–301.

Daigle RE, Baker DW (1977) A readout for pulsed Doppler velocity meters. ISA Trans 16, 41–44.

D'Alessio T (1985) 'Objective' algorithm for maximum frequency estimation in Doppler spectral analysers. Med Biol Eng Comp 23, 63–68.

DeJong DA, Megens PHA, DeVlieger M, Thon H, Holland WPJ (1975) A directional quantifying Doppler system for measurement of transport velocity of blood. Ultrasonics 13, 138–141.

English MJ, Woollons DJ (1982) Basic methods – preprocessing and signal averaging. In: Digital signal processing (Ed. NB Jones), pp 76–89, Peter Peregrinus.

Evans DH (1982) Some aspects of the relationship between instantaneous volumetric blood flow and continuous wave Doppler ultrasound recordings. I. Ultrasound Med Biol 8, 605–609.

Evans DH (1985a) On the measurement of the mean velocity of blood flow over the cardiac cycle using Doppler ultrasound. Ultrasound Med Biol 11, 735–741.

Evans DH (1985b) Doppler signal processing. In: Cardiovascular ultrasonic flowmetry (Eds SA Altobelli, WF Voyles, ER Greene), pp 239–261, Elsevier, New York.

Evans DH, Macpherson DS (1982) Some aspects of the relationship between instantaneous volumetric blood flow and continuous wave Doppler ultrasound recordings. II. Ultrasound Med Biol 8, 611–615.

Evans, DH, Archer LNJ, Levene MI (1985) The detection of abnormal neonatal cerebral haemodynamics using principal component analysis of the Doppler ultrasound waveform. Ultrasound Med Biol 11, 441–449.

Flax SW, Webster JG, Updike SJ (1970) Statistical evaluation of the Doppler ultrasonic blood flowmeter. Biomed Sci Instrum 7, 201–222.

Flax SW, Webster JG, Updike SJ (1973) Pitfalls using Doppler ultrasound to transduce blood flow. IEEE Trans Biomed Eng BME-20, 306–309.

Franklin DL, Schlegel W, Rushmer RF (1961) Blood flow measured by Doppler frequency shift of back-scattered ultrasound. Science 134, 564–565.

Gerzberg L, Meindl JD (1977) Mean frequency estimator with applications in ultrasonic Doppler flowmeters. In: Ultrasound in medicine 3B (Eds D White, RE Brown), pp 1173–1180, Plenum Press, New York.

Gerzberg L, Meindl JD (1980a) Power-spectrum centroid detection for Doppler systems applications. Ultrasonic Imaging 2, 232–261.

Gerzberg L, Meindl JD (1980b) The root f power-spectrum centroid detector: system considerations, implementation and performance. Ultrasonic Imaging 2, 262–289.

Gibbons DT, Evans DH, Barrie WW, Cosgriff PS (1981) Real-time calculation of pulsatility index. Med Biol Eng Comp 19, 28–34.

Gill RW (1979) Performance of the mean frequency Doppler modulator. Ultrasound Med Biol 5, 237–247.

Gosling RG, King DH, Newman DL, Woodcock JP (1969) Transcutaneous measurement of arterial blood velocity by ultrasound. In: Ultrasonics for industry 1969, conference papers, pp 16–23.

Greene FM, Beach K, Strandness DE, Fell G, Phillips DJ (1982) Computer based pattern recognition of carotid arterial disease using pulsed Doppler ultrasound. Ultrasound Med Biol 8, 161–176.

Holland SK, Orphanoudakis SC, Jaffe CC (1984) Frequency-dependent attenuation effects in pulsed Doppler ultrasound: experimental results. IEEE Trans Biomed Eng BME-31, 626–631.

Johnston KW, Maruzzo BC, Cobbold RSC (1977) Errors and artifacts of Doppler flowmeters and their solutions. Arch Surg 112, 1335–1341.

Kitney RI, Giddens DP (1983) Analysis of blood velocity waveforms by phase shift averaging and autoregressive spectral estimation. J Biomech Eng (Trans ASME) 105, 398–401.

Kontis S (1987) Algorithms for fast computation of the intensity weighted mean Doppler frequency. Med Biol Eng Comp 25, 75–76.

Lunt MJ (1975) Accuracy and limitations of the ultrasonic Doppler blood velocimeter and zero crossing detector. Ultrasound Med Biol 2, 1–10.

Maulik D, Saini VD, Nanda NC, Rosenzweig MS (1982) Doppler evaluation of fetal haemodynamics. Ultrasound Med Biol 8, 705–710.

Newhouse VL, Ehrenwald AR, Johnson GF (1977) The effect of Rayleigh scattering and frequency dependent absorption on the output spectrum of Doppler blood flowmeters. In: Ultrasound in medicine 3B (Eds D White, RE Brown), pp 1181–1191, Plenum Press, New York.

Prytherch DR, Evans DH (1985) Versatile microcomputer-based system for the capture, storage and processing of spectrum-analyzed Doppler ultrasound blood flow signals. Med Biol Eng Comp 23, 445–452.

Rabiner LR, Gold B (1975) Theory and application of digital signal processing. Prentice-Hall, Englewood Cliffs, New Jersey.

Rice RO (1944) Mathematical analysis of random noise. Bell Syst Tech J 23, 282–332.

Roevros JMJG (1974) Analogue processing of CW Doppler flowmeter signals to determine average frequency shift momentaneously without the use of a wave analyzer. in: Cardiovascular applications of ultrasound (Ed. RS Reneman), pp 43–54, North-Holland, Amsterdam.

Saini VD, Maulik D, Nanda NC, Rosenzweig MS (1983) Computerized evaluation of blood flow measurement indices using Doppler ultrasound. Ultrasound Med Biol 9, 657–660.

Sainz A, Roberts VC, Pinardi G (1976) Phase-locked loop techniques applied to ultrasonic Doppler signal processing. Ultrasonics 14, 128–132.

Schlindwein FS, Smith MJ, Evans DH (1988) Spectral analysis of Doppler signals and computation of the normalized first moment in real-time using a digital signal processor. Med Biol Eng Comp 26, 228–232.

Skidmore R, Follett DH (1978) Maximum frequency follower for the processing of ultrasonic Doppler shift signals. Ultrasound Med Biol 4, 145–147.

Thompson RS, Trudinger BJ, Cook CM (1985) Doppler ultrasound waveforms in the fetal umbilical artery: quantitative analysis techniques. Ultrasound Med Biol 11, 707–718.

Thomson FJ (1980) Refreshed display of ultrasonic spectragrams and measurement of haemodynamic parameters. Med Biol Eng Comp 18, 33–38.

Woodcock J, Gosling R, King D, Newman D (1972) Physical aspects of blood velocity measurement by Doppler-shifted ultrasound. In: Blood flow measurement (Ed. VC Roberts), pp 19–23, Sector, London.

Chapter 10

WAVEFORM ANALYSIS AND PATTERN RECOGNITION

10.1 INTRODUCTION

In the previous chapter we considered various ways in which the Doppler shift signal may be processed to achieve either a flow velocity waveform or a Doppler power spectrum. In this chapter we will explore methods of extracting clinically useful information from these two types of output.

It may seem strange that the effort that has been expended over the last decade or more on attempts to quantify the shape of velocity waveforms has exceeded that on developing methods of measuring volumetric flow, but there are two reasons why this has been so. Firstly, volumetric flow is remarkably difficult to measure with any degree of accuracy (see Chapter 11) and requires a knowledge not only of the velocity waveform (derived under certain fairly stringent conditions) but also of the vessel dimensions and orientation and thus necessitates much more sophisticated hardware. Secondly, the shape of the waveform may provide information that a simple measurement of mean flow cannot. One example of this is in the assessment of arterial disease where it is well documented (Lee et al 1978, Farrar et al 1979) that the pulsatile components of a flow waveform are affected by lesser degrees of proximal stenosis than is mean flow. Even if mean flow is reduced in such circumstances it is impossible to tell whether this is a result of proximal or distal disease, but the shape of the waveform may give a clue as to where the problem lies.

10.2 PATTERN RECOGNITION PRINCIPLES

It is convenient to think of the interpretation of the shape of Doppler waveforms as a process of pattern recognition. The object of waveform analysis is to recognize those waveforms that are abnormal, even if the details of why a particular physiological or pathological change gives rise to a particular change in waveform shape is not fully understood.

The process of pattern recognition may be split into three stages: transduction, feature extraction and classification (see Fig. 10.1). Essentially transduction consists of deriving some type of pattern vector (for example the velocity waveform or the Doppler power spectrum) from the blood flow in the artery; feature extraction consists of extracting and combining salient features of the pattern vector into a feature vector (for example pulsatility index or an index of spectral broadening); and classification consists of deciding whether such a vector was obtained from a normal or abnormal artery.

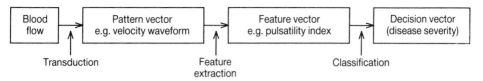

Figure 10.1 Schematic representation of the complete pattern recognition process

The transduction process has largely been dealt with in previous chapters, and is further considered in the next section. The rest of this chapter is devoted to feature extraction and classification. It is important to note that each stage in the process is dependent on the previous stages, and that therefore different methods of interrogating the blood flow, different signal processing techniques and different types of envelope detection will all influence the details, and perhaps the performance of the rest of the recognition sequence.

10.3 PRE-PROCESSING

Feature extraction may be carried out on raw waveforms or spectra, but various pre-processing strategies may be used to compress the data or improve its signal-to-noise ratio before this stage is commenced. Two sophisticated examples of pre-processing can be found in the studies of Greene et al (1982) and Thompson et al (1985) who analysed the Doppler signals from carotid arteries and fetal umbilical arteries respectively. Greene and his colleagues (Greene et al 1982, Knox et al 1982a) collected the spectra from 20 heartbeats conforming to criteria intended to eliminate the effects of cardiac arrhythmias, and formed an ensemble average (Section 9.7), using the ECG R-wave as the fiducial point, over a period extending from -100 ms to $+600$ ms with respect to the R-wave. They then found the mode frequency for each 2.5 ms time slice using a signal location estimator, and the 3 dB and 9 dB down points both above and below the mode. All further processing was carried out using the mode and four contour frequencies. For features based on the time relationships they derived a frequency–time waveform from an average of these five frequencies weighted according to their relative amplitudes, whilst for features based on frequency relationships they averaged each contour over a ± 12.5 ms window centred about a specified point in time relative to peak systole. The peak of systole itself was found using the median smoothed first derivative of the mode waveform.

Thompson et al (1985) performed their ensemble averaging on the maximum velocity, first moment, amplitude sum, and mean velocity waveforms by aligning each individual waveform with respect to a point corresponding to the beginning of each maximum velocity waveform as identified using an autocorrelation method. They then fitted an analytical function to each ensemble average waveform using a least squares method, and carried out their further analysis on this function.

In the vast majority of feature extraction methods described in this chapter little or no pre-processing has been attempted, but clearly it could be of considerable benefit.

10.4 FEATURE EXTRACTION

It is useful to treat the sequence of N numbers that comprise the velocity waveform or power spectrum as the components of a vector in N-dimensional 'pattern space', and the problem of separating normal and abnormal waveforms or spectra as one of separating the tips of these vectors in pattern space. In general, N is too large to tackle this problem directly and therefore it is necessary to extract a limited number of significant features from the pattern vectors and thus produce a 'feature vector' of considerably lower dimensionality. Some feature extractors achieve this by combining individual features of the waveform (for example, the maximum amplitude, minimum amplitude, mean amplitude, maximum acceleration) into one or more indices, whilst others make use of either a model or a transform to describe the overall wave shape.

Over the years many methods of feature extraction have been tried and recommended in many applications, but only a very limited number are in widespread use. The glossary of methods that follows is not exhaustive, but includes the most commonly used methods and many that are seldom used (at least at present). These are included to illustrate the variety of methods of tackling the problem.

10.4.1   Subjective interpretation

The human brain is remarkably adept at pattern recognition and an experienced observer can tell a great deal about blood flow simply by listening to a Doppler shift signal, or by looking at the sonogram of the signal. It may not be possible for such an observer to explain exactly what characteristics influence judgement, but there are many, some obvious and some more subtle, which may be subconsciously taken into account. Some of the more obvious features are listed in Table 10.1.

Table 10.1   Some major factors that may influence subjective interpretation of the sonogram

1. Was the signal more difficult to obtain than is usual for a patient of a particular build?
2. Was there any flow during diastole? If so, how much?
3. Was there any reverse flow?
4. Is the height of the sonogram roughly as expected for the combination of transmitted frequency, site and angle?
5. Is there a window under the sonogram? If there is spectral broadening, when does it occur and for how long?
6. Are there any vortical spikes on the sonogram?
7. Are there any shoulders on the sonogram?

The normal sonogram does of course vary from site to site and not all the features listed in the table are of value at every site; furthermore, a feature considered pathological at one site (for example reverse flow in the cerebral arteries) may be a sign of normality at another site (for example in the femoral artery). Figure 10.2 shows sonograms which were taken from the common femoral arteries of a normal volunteer and three patients with peripheral vascular disease, and which exhibit a variety of abnormal features.

Subjective interpretation may be very powerful in experienced hands (Walton et al 1984), but objective methods do not rely on the expertise of the user, are not subject to observer bias, allow methods to be transported between centres, and may be able to distinguish much more subtle changes in the waveforms. At present, however, most objective methods concentrate on one particular aspect of the sonogram (for example the

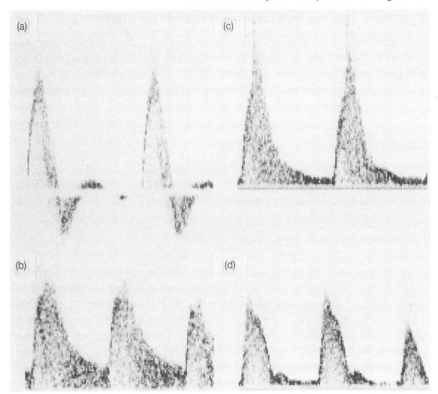

Figure 10.2   Sonograms recorded from the common femoral arteries of a normal subject and three patients with peripheral vascular disease. (a) Normal triphasic sonogram. (b) Damped sonogram resulting from an upstream stenosis. (c) Sonogram with a vortical spike originating from an upstream flow disturbance. (d) Sonogram with a shoulder on the downslope of the systolic peak associated with an occluded superficial femoral artery

outline shape) and may thus sometimes ignore features that may be obvious to the human observer.

10.4.2   Maximum frequency

When blood flows through a stenosis its velocity must increase to maintain the same flow rate as in the pre- and post-stenotic region of the vessel. The detection of the high velocities within a stenosis is the basis of a test described by Spencer and Reid (1979) and subsequently widely adopted for the detection of internal carotid artery stenosis (Zwiebel et al 1982, Johnston et al 1982a, Manga et al 1986). The Doppler shift is of course proportional not only to the velocity, but also to the frequency of the transmitted ultrasound and the cosine of the angle between the blood vessel and the ultrasound probe (eqn 1.1), and these variables must be allowed for. Furthermore, a very tight stenosis will reduce the volumetric flow through the artery, and the increase in peak velocity is therefore less than would be predicted on purely geometrical grounds. A more sophisticated test based on the comparison of the velocities within and beyond the stenosis is described in Section 10.4.10.3.

10.4.3   Simple single-site normalized frequency indices

Several of the most popular feature extraction techniques are based on finding the ratio of the height of one feature of a waveform to that of another. The advantage of taking such a ratio is that both the numerator and the denominator include the cosine of the angle between the Doppler probe and the blood vessel. The cosine term therefore cancels, and the index is independent of angle.

10.4.3.1   Pulsatility index

Perhaps the most widely used index of all is pulsatility index (*PI*). This index was originally defined (Gosling et al 1971) as the total oscillatory energy in the flow-velocity waveform divided by the energy in the mean flow-velocity over the cardiac cycle, i.e.

$$PI_{\text{orig}} = \sum_{n=1}^{\infty} a_n^2 / M^2 \qquad\qquad 10.1$$

where a_n is the amplitude of the nth harmonic, and M the mean value of amplitude over the cardiac cycle. The purpose of this index was to summarize the degree of pulse-wave damping at different arterial sites; the smaller the *PI* the greater the degree of damping.

This index requires the calculation of the Fourier transform of the velocity waveform, which at that time presented some computational difficulties and it was soon superseded (Gosling and King 1974) by a similar but simpler *PI* calculated by dividing the maximum vertical excursion of the waveform by its mean height (Fig. 10.3), i.e.

$$PI = (S - D)/M \qquad\qquad 10.2$$

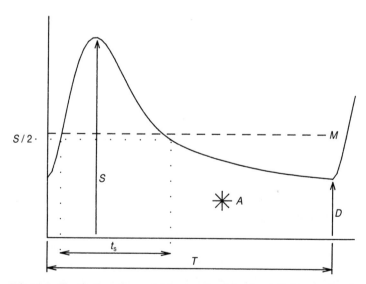

Figure 10.3   Diagram illustrating the variables involved in the definitions of pulsatility index, resistance index, S/D ratio, constant flow ratio and height width index. S is the maximum height of the waveform, D the minimum height, M the mean height over the cardiac cycle, T the length of the cardiac cycle, A the area under the curve, and t_s the duration of the systolic peak (measured between half amplitude points)

where S and D are the maximum and minimum values of amplitude during the cardiac cycle. (For pulsatile waveforms such as that shown in Fig. 10.2a, D may be negative and may not occur at end diastole.)

The two PI values are correlated (Gosling and King 1974, Johnston et al 1978) but are not numerically equal and therefore cannot be used interchangeably. The simplicity of PI as defined in eqn 10.2 allows it to be calculated on-line using a simple microprocessor-based system (Gibbons et al 1981, Johnston et al 1982b), and it has been widely accepted as one of the standard ways of quantifying the 'pulsatility' of a Doppler waveform. It could be argued that of the two definitions the earlier has more physical meaning, and that, since on-line Fourier analysis is a relatively simple task with modern microcomputer-based systems, this method should be revived in the future.

PI is a useful objective description of the pulsatility of a waveform, but great care must be taken with its interpretation in the clinical setting as it is influenced by many factors including proximal stenosis (Johnston and Tarashuk 1976, Evans et al 1980, Johnston et al 1983), distal stenosis (Thiele et al 1983, Junger et al 1984, Macpherson et al 1984), and peripheral resistance (Evans et al 1980).

10.4.3.2   *Pourcelot's resistance index*

A second widely used index of pulsatility is Pourcelot's resistance index (RI) which is defined (Fig. 10.3) as:

$$RI = (S - D)/S \qquad\qquad 10.3$$

This index was first used on waveforms from the common carotid artery (Planiol and Pourcelot 1973, Pourcelot 1976) as an indicator of the circulatory resistance beyond the measurement point, and has subsequently been widely used for the study of neonatal cerebral haemodynamics (Perlman 1985).

Unfortunately considerable confusion has arisen between PI as defined in eqn 10.2 and RI as defined in eqn 10.3, and the latter is now generally termed pulsatility index in the North American neonatal literature. The term resistance index is preferred here to avoid ambiguity, but even this term is not ideal since the value of RI may be influenced by many factors of which distal resistance is only one. RI is often considered to vary between 0 and 1, and all waveforms with reverse flow are reported as having an RI of unity. There is, however, no reason why negative values of D cannot be inserted in eqn 10.3 to produce values of RI of greater than unity.

RI is extremely simple to calculate and has been used to detect waveform changes in a wide variety of pathological conditions including internal carotid stenosis (Pourcelot 1976), birth asphyxia (Bada et al 1979), intraventricular haemorrhage (Bada et al 1979), patent ductus arteriosus (Perlman et al 1981, Lipman et al 1982), pneumothorax (Hill et al 1982) and hydrocephalus (Hill and Volpe 1982).

10.4.3.3   S/D *ratio*

A slight variation of the RI has been used by a number of workers in the obstetric field to describe changes that occur in the shape of the velocity waveform in the umbilical artery with gestational age (Stuart et al 1980, Trudinger et al 1985a) and in some high-risk pregnancies (Trudinger et al 1985a). This index, which is often called the AB or A/B ratio, is simply calculated by dividing the maximum systolic height (S) by the end diastolic height (D) (Fig. 10.3). This ratio is called the S/D ratio here to distinguish it from the A/B ratio which has been used for the evaluation of common carotid waveforms and is described in Section 10.4.3.5.

S/D is in fact a simple transformation of RI and may be rewritten:

$$S/D = 1/(1 - RI) \qquad 10.4$$

One drawback of the S/D ratio is that, as D tends to zero, S/D tends to infinity, and there is a discontinuity at $D=0$ with S/D equal to infinity at $D=0^{+}$ and to minus infinity at $D=0^{-}$.

One further variation of the RI, the D/S ratio, has been used by Trudinger and his colleagues (1985b) to describe the velocity waveform in the uteroplacental circulation. This quantity is simply related to RI by eqn 10.5:

$$D/S = 1 - RI \qquad 10.5$$

This index avoids the discontinuity at $D=0$.

10.4.3.4   The constant flow ratio

Thompson et al (1985) have suggested the 'constant flow ratio' (CFR) as a method of describing fetal umbilical artery waveforms. They define this quantity (Fig. 10.3) as:

$$CFR = DT/A \qquad 10.6$$

where D is the minimum height of the waveform, T the length of the waveform and A the area under the curve. The ratio A/T is the mean height of the waveform over the cardiac cycle (M), and therefore eqn 10.6 may be written:

$$CFR = D/M \qquad 10.7$$

This index is related to PI and RI by eqn 10.8:

$$CFR = PI/RI - PI \qquad 10.8$$

10.4.3.5   The A/B ratio

The A/B ratio was introduced as a means of characterizing the shape of the signals from the common carotid and supraorbital arteries (Gosling 1976). The waveforms from both these sites exhibit two peaks (A and B) during systole (Fig. 10.4), and the ratios of their amplitudes change with both age and disease. Tests for internal carotid disease using a combination of the A/B ratios in the common carotid and supraorbital arteries (Baskett et al 1977) and in the common carotid and supratrochlear arteries (Prichard et al 1979) have been described.

10.4.4   Simple single-site indices that include time

Changes in the circulation affect not only the height of the recognizable features of a waveform, but also the time relationship between those features. The upslope of the waveform is particularly susceptible to changes in the cardiac impulse and the circulation

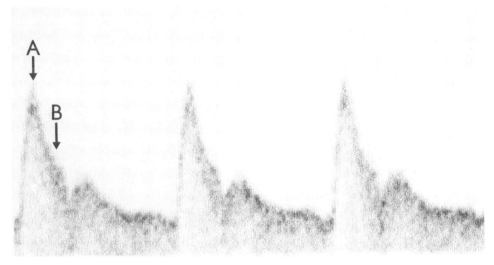

Figure 10.4   Sonogram recorded from the common carotid artery of a normal subject showing the two systolic peaks, A and B

proximal to the site of measurement, whilst the decay of the velocity after peak systole is more influenced by the distal circulation. The time relationships within the waveform are very easy to measure, and have been used both alone and in combination with the heights of discernible features to characterize the shape of the waveform.

10.4.4.1   Systolic decay time index

Few indices have been reported that use only the time relationships within a single cardiac cycle, but the systolic decay time index (*SDTI*) reported by Thompson et al (1985) is one such. It describes the ratio of the normalized rise and decay slopes close to peak systole and is defined (Fig. 10.5) as:

$$SDTI = (t_r/t_1)/(t_d/t_2) \qquad\qquad 10.9$$

where t_1 is the time from the beginning of systole to peak systole, t_2 $(=T-t_1)$ the time from peak systole to the end of diastole, t_r the rise time from $0.75S$ to S, and t_d the decay time from S to $0.75S$. Thompson et al show it to be of some use in detecting abnormal fetal umbilical artery waveforms.

10.4.4.2   Height width index

The height width index (*HWI*) suggested by Johnston and his colleagues (Johnston et al 1984) combines information about both the pulsatility of, and the time relationships within, the arterial waveform. Waveforms recorded distal to arterial stenoses have relatively small pulsatile components and relatively wide systolic peaks, and both of these serve to decrease the value of *HWI* which is defined (Fig. 10.3) as:

$$HWI = [(S-D)/M]\,(T/t_s) \qquad\qquad 10.10$$

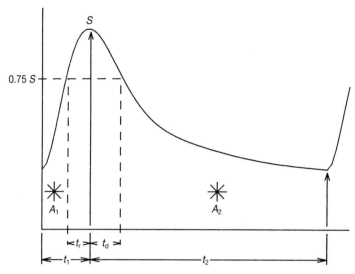

Figure 10.5   Diagram illustrating the variables involved in the definition of systolic decay time index and relative flow rate index. t_1 is the time from the beginning of systole to peak systole, t_2 the time from peak systole to the end of diastole, t_r the rise time from $0.75S$ to S, t_d the decay time from S to $0.75S$, A_1 the area under the curve before peak systole and A_2 the area under the curve after peak systole

where t_s is the duration of the systolic peak measured between the half amplitude points. The expression in the square bracket is equal to pulsatility index and so eqn 10.10 may be rewritten:

$$HWI = PI(T/t_s) \qquad\qquad 10.11$$

10.4.4.3   Path length index

A further index of pulsatility introduced by Johnston et al (1984) is the path length index (*PLI*). This exploits the decrease in the total path length traced out over the cardiac cycle as a waveform becomes more damped. It is normalized to remove angle dependence and heart rate, and is defined (Fig. 10.6) as:

$$PLI = \sum_{i=0}^{N-1} [(f_{i+1}-f_i)^2/M^2 + (t_{i+1}-t_i)^2/T^2]^{1/2} \qquad\qquad 10.12$$

where f_i is the *i*th value of the waveform amplitude and t_i the *i*th value of time.

10.4.4.4   Relative flow rate index

The relative flow rate index (*RFRI*) describes the ratio of the average flow rate before the systolic peak to the average flow rate during the rest of the cardiac cycle, and was introduced by Thompson et al (1985) for use with fetal umbilical artery waveforms. It is defined (Fig. 10.5) as:

$$RFRI = (A_1/A_2)/(t_1/t_2) = M_1/M_2 \qquad\qquad 10.13$$

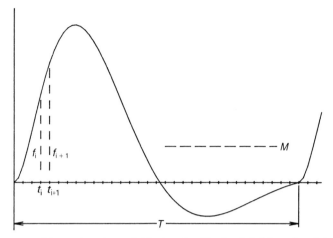

Figure 10.6 Diagram illustrating the variables involved in the definition of path length index. M is the mean height of the waveform, T the cycle length, f_i the ith value of the waveform amplitude, and t_i the ith value of time

where A_1 and A_s are the areas under the curve before and after peak systole, t_1 is the time from the beginning of systole to peak systole, t_2 the time from peak systole to end diastole, M_1 the mean height of the curve before the systolic peak, and M_2 the mean height of the curve during the rest of the cardiac cycle.

10.4.5 Spectral broadening indices

Each of the indices mentioned so far has been concerned with the shape of the velocity waveform (whether it be maximum, mean or rms). These indices ignore the information contained in the power spectrum concerning the distribution of velocities within the ultrasound beam, which may sometimes be of diagnostic importance. In particular it has been found that the shape of the spectrum measured from the internal carotid artery at or around peak systole is influenced by quite moderate degrees of proximal disease (Barnes et al 1976, Blackshear et al 1979, Reneman and Spencer 1979). Sonograms recorded from normal vessels exhibit a clear window under the systolic peak, resulting from a relatively flat velocity profile at this point in the cardiac cycle (Fig. 10.7a), whilst those from diseased vessels show a degree of spectral broadening (Fig. 10.7b) which is related to the degree of proximal stenosis, and is believed to be the result of disturbed non-axial flow.

One of the earliest attempts to quantify spectral broadening was that of Bodily et al (1980) who defined fractional broadening to be s_f/f_{mean}, where f_{mean} is the mean frequency at peak systole and s_f the standard deviation of frequency at the same time. The mean frequency was estimated to be half way between the minimum and maximum frequency, and the standard deviation assumed to be proportional to the difference between the two frequencies. This led to the empirical relationship for a spectral broadening index (SBI):

$$\text{Fractional broadening } (SBI(1)) = k \frac{f_{max} - f_{min}}{f_{max} + f_{min}} \qquad 10.14$$

where f_{max} and f_{min} are the maximum and minimum frequency at peak systole, and k is an experimentally derived constant found to be equal to 0.47.

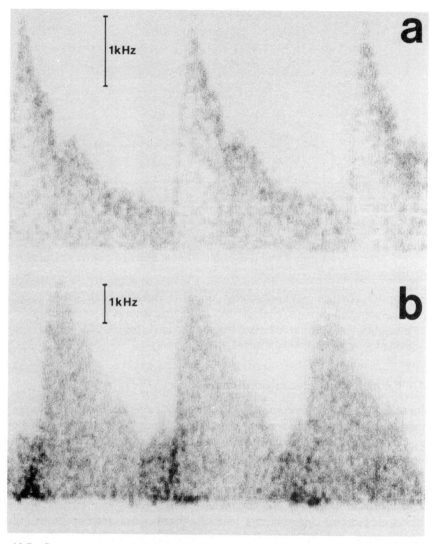

Figure 10.7 Sonograms recorded from the internal carotid arteries of (a) a normal subject exhibiting a clear window under the systolic peak, and (b) a patient with disease at the origin of the internal carotid artery showing spectral broadening. Note that the frequency scale of (b) is twice that of (a).

Since that time many indices have been tried both *in vitro* and *in vivo* with varying degrees of success, and a selection of these are enumerated in Table 10.2. A problem with many of these indices is the rather loose way in which the minimum, mean and maximum frequencies have been defined, which makes it difficult to repeat the original studies. There is also a tremendous variation in the time period over which the data have been collected, the way in which data from different time slices have been combined, the amount of averaging that has been used, and the size of the sample volume. This last factor is particularly important as it may change the Doppler spectrum completely (Evans 1982, Knox et al 1982b, Van Merode et al 1983).

Table 10.2 Selection of spectral broadening indices—see Section 10.7 for notation

Reference	Definition	Comments
Bodily et al 1980	$SBI(1) = 0.47(f_{max} - f_{min})/(f_{max} + f_{min})$	PW
Johnston et al 1981	$SBI(2) = (f_{max} - f_{min})/f_{max}$	CW; same information content as $SBI(1)$
Woodcock et al 1982	$SBI(3) = (f_{max} - f_{min})/f_{mean}$	PW
Brown et al 1982	$SBI(4) = (f_{max} - f_{mean})/f_{max}$	CW: f_{mean} calculated using analogue frequency follower
Woodcock et al 1983	$SBI(5) = (f_{max} - f_{min})/f_{median}$	PW
Rittgers et al 1983	$SBI(6) = f_{min}/f_{max}$	CW; same information content as $SBI(1)$; calculated from 12 dB down points from t_0 to $t_{0+100 \text{ ms}}$
Sheldon et al 1983	$SBI(7) = f_{max}/f_{mean}$	PW; same information content as $SBI(4)$; averaged over 16–32 cycles
Kalman et al 1985	$SBI(8) = s_f/f_{mean}$	CW; four lines of spectral data from around peak systole averaged
Kalman et al 1985	$SBI(9) = m3_f/(s_f)^3$	As for $SBI(8)$
Kalman et al 1985	$SBI(10) = (m4_f/(s_f)^4) - 3$	As for $SBI(8)$

10.4.6 Stepwise feature selection algorithms

The feature extraction methods discussed so far have all been based to some degree on experience or intuition, have resulted in a one-dimensional feature vector, and have concentrated on one particular attribute of the Doppler signal (either the shape of the velocity waveform or the degree of spectral broadening). Another approach is to evaluate a large number of features which may assist with the classification process, and to select a set of these on the basis of their ability to discriminate between disease states. This approach has been adopted by Rutherford et al (1977) who selected five out of nine features from the common carotid artery waveform, and Greene and his colleagues (Greene et al 1982, Knox et al 1982a, Langlois et al 1984) who selected two or three out of 94 features for each of three binary decisions concerning disease severity. The advantage of such a strategy is that it is able to highlight features or combinations of features which are unexpectedly good discriminators. An example of this is 'feature I.3' of Greene et al (1982) which was a measure of the relative increase in spectral width between peak systole and peak systole + 100 ms, and proved to be very important for separating normal and diseased carotid arteries. Feature selection cannot be entirely divorced from classification with these methods since the selection of features is based upon their ability, to discriminate between disease classes. The classifier ultimately chosen may however be independent of the selection algorithm.

The first step in feature selection is to choose a number of candidate features that may potentially assist in the classification process. These may be simple values such as the height of the waveform at different times during the cardiac cycle, derived features such as pulsatility index or fractional broadening, or any combination of these.

Simple analyses tend to concentrate on features known or thought to be most closely correlated with disease, but more complicated analyses may evaluate several dozen features. It is then necessary to eliminate a proportion of the features, either because they are poor discriminators on their own or because they share the same discriminating information with other features and are thus redundant. The analysis proceeds in a step-wise manner; the feature that provides the greatest univariate discrimination is first selected and then paired with all the remaining features to ascertain the combination that produces the greatest discrimination. The procedure is then repeated to find the best triplet and so forth until the addition of further features no longer improves the discrimination or a predetermined number of features has been selected.

Although the method described produces an optimal set of discriminating features it may not necessarily be the best combination, and various strategies that combine 'forward' selection (adding one feature at a time) and 'backward' selection (dropping one feature at a time) can be used to improve the final selection (Klecka 1980).

10.4.7 Laplace transform analysis

A method of feature extraction based on a description of the shape of the entire blood velocity waveform in terms of a Laplace transform was introduced by Skidmore and Woodcock (1978) and explored in depth in a series of papers (Skidmore and Woodcock 1980a,b, Skidmore et al 1980). The envelope signal is first transformed into the frequency domain, and then fitted to the third-order Laplace equation:

$$H(S) = 1/(S^2 + 2\delta\omega_0 S + \omega_0^2)(S + \gamma) \qquad 10.15$$

where $S = j\omega$. Equation 10.15 can be represented in graphical form by plotting its poles on an Argand diagram. The three poles are found by equating the denominator to zero and are given by:

$$S_{1,2} = -\delta\omega_0 \pm j\omega_0(1 - \delta^2)^{1/2} \qquad 10.16$$

and

$$S_3 = -\gamma \qquad 10.17$$

The S-plane representation of eqn 10.15 is shown in Fig. 10.8. The particular virtue of the Laplace transform model is that in health, at least, each of the terms ω_0, γ and δ appears to be related to a different aspect of the circulation, specifically arterial stiffness, distal impedance and proximal lumen size. This is clearly advantageous as it allows the effects of distal and proximal disease on the Doppler waveform to be deconvolved. Several studies have evaluated δ as a measure of proximal stenosis (Baird et al 1980, Johnston et al 1984, Macpherson et al 1984, Junger et al 1984).

Although Skidmore and his colleagues calculated the coefficients ω_0, γ and δ by fitting the Fourier transform of the waveform to a third-order Laplace equation, an equally valid technique is to fit the waveform to the inverse Laplace transform of eqn 10.15, i.e.

$$f(t) = \frac{\gamma(\alpha^2 + \beta^2)}{(\gamma - \alpha)^2 + \beta^2}\left[e^{-\gamma t} + e^{-\alpha t}\left(\frac{\gamma - \alpha}{\beta}\sin\beta t - \cos\beta t\right)\right] \qquad 10.18$$

where $\alpha = \omega_0\delta$ and $\beta^2 = \omega_0^2 - \alpha^2$. This was the method used by Johnston et al (1984).

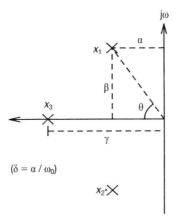

Figure 10.8   *S*-plane representation of eqn 10.15 showing the position of two complex and one real pole

10.4.8   Principal component analysis

Another technique used to describe the shape of the entire Doppler waveform is that of principal component analysis. This type of analysis has been applied to signals from the common carotid arteries (Martin et al 1980), the common femoral arteries (Macpherson et al 1984), and the anterior cerebral arteries of newborn infants (Evans et al 1985).

Principal component analysis is analogous to Fourier analysis in that the waveform is described in terms of the coefficients of a predetermined orthogonal set of waveforms, but rather than using sines and cosines the orthogonal set is chosen so that it describes the waveforms from the study population most efficiently, with the smallest number of terms. Because the transform is so efficient only two or three terms are required to reconstruct the original waveform to a high degree of accuracy, and therefore the waveform can be represented by two or three coefficients and plotted in two- or three-dimensional feature space (Fig. 10.9).

The methodology consists of two distinct stages, that of defining the principal components (PCs) of the study population, and that of calculating the coefficients of the PCs for each test waveform. The required PCs can be shown to be the eigenvectors of the covariance matrix of the population, which may be estimated from a sufficiently large and representative sample of waveforms. The first step is to calculate the sample mean record (*SMR*) which is simply the ensemble average of the sample waveforms, i.e.:

$$SMR_i = \sum_{n=1}^{N} f_{in}/N \qquad\qquad 10.19$$

where SMR_i is the *i*th element of the *SMR*, f_{in} is the *i*th element of the *n*th waveform, and *N* is the total number of waveforms in the sample. Each element of the covariance matrix can then be written:

$$C_{ij} = \sum_{n=1}^{N} (f_{in} - SMR_i)(f_{jn} - SMR_j)/(N-1) \qquad\qquad 10.20$$

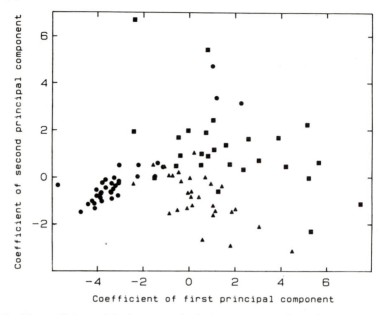

Figure 10.9   The coefficients of the first two principal components of waveforms recorded from the femoral arteries of patients with various disease patterns. The three groups were: severe aorto-iliac disease (●), probably normal aorto-iliac segment, but blocked femoral artery (▲), and probably normal aorto-iliac segment and patent superficial femoral artery (■). (Reprinted with permission from Evans (1984) J Biomed Eng 6, 275, © Butterworth & Co (Publishers) Ltd)

The eigenvalues and the corresponding eigenvectors of the covariance matrix can be found using a standard computer package. The first principal component is the eigenvector corresponding to the largest eigenvalue and so on, and it can be shown that, if only K PCs out of a possible P are used, the efficiency of the transform is given by:

$$E = \sum_{i=1}^{K} \lambda_i \bigg/ \sum_{i=1}^{P} \lambda_i \qquad \qquad 10.21$$

where λ_i is the ith eigenvalue.

Once the PCs have been defined, the calculation of their coefficients for each test waveform is straightforward. If b_k is the coefficient of the kth PC then:

$$b_k = \sum_{i=1}^{I} (f_i - SMR_i)(r_{ki}) \qquad \qquad 10.22$$

where f_i is the ith element of the test waveform and r_{ki} the ith element of the kth PC. The initial calculation of the PCs requires a fair amount of computing power, but the second stage is simple and can be carried out on-line on a small microcomputer system (Prytherch et al 1982).

Although principal component analysis has been applied mainly to the waveform shape it is possible to describe the entire sonogram in terms of a principal component series, and this approach has been explored by Martin and his colleagues (Martin et al 1981, Sherriff et al 1982).

10.4.9 Reactive hyperaemia tests

Doppler ultrasound has been used to quantify the hyperaemia that occurs after an artery has been artificially occluded for a short period of time. Both Fronek et al (1973) and Ward and Martin (1980) used this technique to detect aorto-iliac disease. They established baseline velocity readings, inflated a blood pressure cuff on the upper thigh to a suprasystolic pressure for a period of some minutes, and then recorded changes in the Doppler shift frequency following deflation. Fronek et al found a smaller velocity increase and a longer $T_{\frac{1}{2}}$ (the time in seconds for the velocity to return half-way to the control baseline) in patients with disease; Ward and Martin did not report on the percentage velocity increase, but found a similar augmentation in $T_{\frac{1}{2}}$.

10.4.10 Two site measurements

Most feature extraction techniques rely on the analysis of the Doppler signal recorded from a single point in the circulation; there are, however, a number of methods that involve the comparison of signals from two or more points. Several of these have been evaluated by Humphries et al (1980) and Baker et al (1986).

10.4.10.1 Damping factor and transit time

The first reported attempts to quantify disease severity by comparing Doppler signals from two sites were those of Woodcock and his colleagues (Woodcock 1970, Fitzgerald et al 1971, Woodcock et al 1972). They examined the effects of disease in the femoral arteries on both the damping of the Doppler waveform and its velocity as it propagated from the common femoral artery at the groin to the popliteal artery behind the knee. For this purpose they introduced two new quantities, the damping factor (DF) and the transit time (TT) (Fig. 10.10), which they defined as:

$$DF = PI_{\text{fem}}/PI_{\text{pop}} \qquad\qquad 10.23$$

and

$$TT = t_{\text{pop}} - t_{\text{fem}} \qquad\qquad 10.24$$

PI_{fem} and PI_{pop} are the pulsatility indices in the common femoral and popliteal arteries, and t_{fem} and t_{pop} the times of arrival of the 'feet' of the Doppler waveforms at these two sites. Increasing disease severity was shown to result in an increase in both TT and DF.

Normalized versions of TT and DF have since been described (Gosling 1976, Gosling and King 1978) both to eliminate site dependence and to allow for variations in mean blood pressure, which can significantly influence the pulse wave velocity.

10.4.10.2 Transit time ratio

Another index based on the transit time of the arterial pulse, the transit time ratio (TTR) was suggested by Craxford and Chamberlain (1977). This index was defined (Fig. 10.10) as:

$$TTR = T_a/T_g \qquad\qquad 10.25$$

where T_a is the time from the ECG R-wave to the arrival of the flow waveform at the

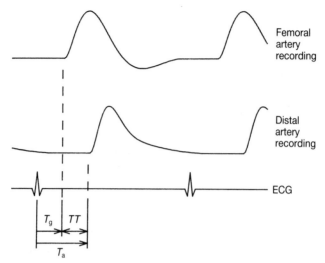

Figure 10.10   Diagram illustrating the variables involved in the definitions of transit time, damping factor and transit time ratio. T_g is the time lapse between the ECG R-wave and the arrival of the pulse at the groin, t_a the time lapse from the R-wave to the arrival of the pulse at the ankle, TT the time lapse between the arrival of the pulse at the two arterial sites

posterior tibial artery, and T_g the time from the R-wave to the arrival of flow at the common femoral artery. In both cases the reference point on the flow waveform was taken as the mid-way point on the upslope of the systolic pulse. It was suggested that this index was capable of separating normal subjects and patients with proximal, distal and mixed arterial disease.

10.4.10.3   Frequency ratio

A further method of using information gathered from two different sites is to compare the maximum Doppler shift frequencies recorded at different locations either in an unbranched segment of artery (Spencer and Reid 1979, Manga et al 1986), or in two separate arteries (Blackshear et al 1980). If the artery is unbranched, the flow at every point must be equal and therefore any decrease in the cross-sectional area of the vessel will cause an increase in velocity, and it should be possible to measure the degree of stenosis by comparing the highest Doppler shift frequency within the stenosis with that found in a normal segment of vessel. This method is better than simply measuring the maximum frequency (Section 10.4.2) because it is not upset if the stenosis is severe enough to reduce the flow through the artery. Spencer and Reid (1979) showed that the ratio f_2/f_1 (where f_1 is the Doppler shift frequency recorded from the stenosis and f_2 the downstream frequency) was more closely related to the change in radiological diameter of the vessel than the diameter squared, but this was thought to be a result of asymmetrical plaque development, and the intrinsic limitations of arteriography.

10.5   CLASSIFICATION

The final stage of the pattern recognition process is to assign the feature vectors obtained during feature extraction to one of a number of classes. There may be only two classes, usually normal or abnormal, or several consisting of different degrees or distributions of disease. The complexity of classification depends on the dimensionality of the feature vectors. If they are one-dimensional, as for pulsatility index and all the other indices discussed in Sections 10.4.3–10.4.5, it is only necessary to define one or more thresholds which will be used to decide the class. If the feature vectors are two or more dimensional, as for example those resulting from step-wise selection algorithms (Section 10.4.6) or principal component analysis (Section 10.4.8) then both the form and the position of the separating surfaces must be decided.

10.5.1   Sensitivity, specificity and ROC curve analysis

The simplest classification problem is that of separating one-dimensional feature vectors into two groups. In this situation the only choice that needs to be made is where to locate the decision threshold. If there is no overlap between the magnitudes of the vectors obtained from patients belonging to the two classes, the threshold can simply be chosen to separate the classes completely. In general, however, the results from the two classes do overlap and so depending on where the threshold is placed some signals from normal subjects will be adjudged abnormal and/or some signals from abnormals will be adjudged normal. The best choice of threshold will then depend on a number of factors including the consequences of making both types of false classification (false positive and false negative) and the prevalence of disease in the target population. Various aspects of this problem have been considered in depth by Metz (1978) and O'Donnell et al (1980).

There are two important measures of the performance of a diagnostic test; sensitivity (or true positive fraction) and specificity (or true negative fraction) which are defined as:

$$\text{Sensitivity } (TPF) = \frac{\text{Number of true positive decisions}}{\text{Number of actually positive cases}} \qquad 10.26$$

and

$$\text{Specificity } (TNF) = \frac{\text{Number of true negative decisions}}{\text{Number of actually negative cases}} \qquad 10.27$$

The two are not independent since they are both affected by the position of the decision threshold; and as the threshold is moved to increase sensitivity, so specificity falls. Two closely related quantities often quoted in the literature are positive and negative predictive values. The former is the probability that a positive result in a test indicates a genuine positive result, the latter that a negative result indicates a genuine negative result. These may be calculated as follows:

$$\text{Positive predictive value} = \frac{\text{Number of true positive decisions}}{\text{Total number of positive decisions}} \qquad 10.28$$

and

$$\text{Negative predictive value} = \frac{\text{Number of true negative decisions}}{\text{Total number of negative decisions}} \qquad 10.29$$

One further performance indicator sometimes used is accuracy which is defined as:

$$\text{Accuracy} = \frac{\text{Number of correct decisions}}{\text{Total number of cases}} \qquad 10.30$$

and is related to sensitivity and specificity by the expression:

$$\text{Accuracy} = \text{Sensitivity} \times \text{Fraction of the study population}$$
$$\text{that is actually positive}$$
$$+ \text{Specificity} \times \text{Fraction of the study population}$$
$$\text{that is actually negative} \qquad 10.31$$

Since the accuracy is determined by the prevalence of disease in the study population it is actually a very poor measure of performance for a diagnostic test.

The best method of assessing the value of a test and defining an appropriate decision threshold is to plot a receiver operating characteristic (ROC) curve for the test. Such a curve is derived by varying the decision threshold in small steps and determining the *TPF* and *TNF* for each new threshold value. In this way curves of the type shown in Fig. 10.11 are built up. Conventionally the false positive fraction (*FPF*) or (1 − specificity) is plotted along the abscissa, and *TPF* or sensitivity plotted along the ordinate. A good test (curve (a)) is one for which *TPF* rises rapidly and *FPF* hardly increases at all until *TPF* is high; a poor test (curve (c)) is one for which *TPF* and *FPF* increase at similar rates. Plotting ROC curves for two or more tests in this way enables their relative diagnostic values to be determined. It should, however, be noted that the locus of the ROC curve for a particular test is also affected by the 'gold standard' against which it is compared (O'Donnell et al 1980). If the gold standard is made more strict (in the sense that only severely diseased arteries are placed in the abnormal group) then, for a given threshold, the apparent sensitivity of the test will increase and the apparent specificity will fall.

The 'best' decision threshold for any test depends on the shape of the ROC curve, the prevalence of disease in the population to be studied, and the costs (both financially and in terms of medical consequences) of missing disease when it is present (false negatives), and of diagnosing disease when it is absent (false positives). If the disease is rare in the study

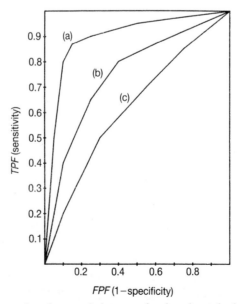

Figure 10.11   Receiver operating characteristic curves for three hypothetical tests of varying utility

population (as for example in screening situations), the threshold should be strict, i.e. placed near the lower left portion of the ROC curve, otherwise almost all positive decisions will be false positives. If the disease is common, the threshold can be relaxed. If the cost of a false positive is much higher than that of a false negative (as for example when the treatment for a disease is potentially harmful to healthy patients, and of limited benefit to diseased patients) then once again the threshold should be relatively strict, whereas if the cost of a false negative is much greater than that of a false positive the threshold should be set towards the upper right portion of the ROC curve. Methods of calculating the optimal threshold are discussed in Metz (1978).

10.5.2   Classification in two or more dimensions

When the feature vectors are of two or more dimensions, the first step in the classification procedure is to find a suitable shape and orientation for the separating surfaces. It may then, depending on the procedure adopted, be possible to vary a threshold to produce ROC curves and pick an optimum operating point as for the one-dimensional problem discussed in the last section. There are many approaches to the N-dimensional classification problem (Andrews 1972, Tou and Gonzalez 1974), but there have as yet been few reports of their use with Doppler signals. Nicolaides et al (1976), Rutherford et al (1977) and Keagy et al (1982) have all used discriminant analysis applied to simple features extracted from the Doppler waveform. Greene et al (1982) compared the performance of five training algorithms for classifying the results of their step-wise feature-selection algorithms (Section 10.4.6), and Evans and Caprihan (1985) compared four techniques of classifying the results of principal component analysis (Section 10.4.8). In a further paper, Evans et al (1985) applied two classification techniques to principal component analysis data collected from neonatal cerebral arteries, and compared the results using ROC curve analysis. A detailed discussion of N-dimensional classification techniques is beyond the scope of this chapter, but a brief outline of three commonly used methods follows. Each is an example of 'supervised' learning, that is to say the classification algorithms are trained on a set of data each of which is of known class. Before applying any of these techniques the features should be transformed to have a population mean of zero and a variance of unity so as to remove their dimensions (Sebestyen 1962).

10.5.2.1   Nearest neighbour algorithms

The nearest neighbour (NN) method of classification requires no assumptions to be made about the statistical distribution of the data and is simple to implement. A training set of data, each of known class, is stored and serves as a reference against which new data are compared. The training set may consist of one or many vectors from each class, and the classification algorithm may assign the new vector to the class of its NN or the majority of its q NNs (where q is an integer greater than two, and usually odd for the two class problem). In the simplest case where there are only two classes, each represented by a single prototype, the NN rule leads to a separating function which is a linear hyperplane. Multiple prototypes lead to piecewise-linear separating functions. The NN technique is particularly valuable where the members of each class form distinct clusters, i.e. the intra-class distances are small compared with the inter-class distances. Pattern classification by distance function is dealt with in detail by Tou and Gonzalez (1974, Chapter 3), and Evans and Caprihan (1985) have illustrated the NN method with two sets of Doppler data.

10.5.2.2   Multivariate discriminant analysis

Discriminant analysis is an example of a statistical classification. Each class is assumed to be drawn from a population with a multivariate normal distribution, and the covariance matrices of all classes are assumed to be identical, or nearly so. The classification functions that result under these circumstances are simple linear combinations of the discriminating variables. Details of discriminant analysis and its implementation can be found in Thorndike (1978) and Klecka (1980), and the method has been used to classify Doppler data by Rutherford et al (1977) and Keagy et al (1982). Statistical techniques are potentially much more powerful than distribution-free methods, such as the nearest neighbour method, because they allow probabilities to be assigned to each classification, and the decision threshold to be varied and optimized for disease prevalence and the consequences of both correct and incorrect classifications (Section 10.5.1).

10.5.2.3   The Bayes method

A second method which uses the statistical properties of the data is the Bayes method. In this method the relative probabilities that a given feature vector belongs to each possible class are calculated from the probability density function and *a priori* probability of each class. The advantage of this method over that discussed in the previous section is that the probability density function may be estimated using either parametric or non-parametric methods, and that if a multivariate normal density function is used (as is most usual) there is no requirement that the covariance matrices of each class are identical or even similar. Because of this the Bayes method seems more appropriate for separating the feature vectors derived from Doppler signals as it is unlikely that the normal and abnormal classes will have similar multivariate normal distributions in N-dimensional feature space. Details of the Bayes method can be found in Tou and Gonzalez (1974, Chapter 4) and it has been used to classify Doppler data by Evans and Caprihan (1985) and Evans et al (1985).

10.6   SUMMARY

Waveform analysis is a powerful diagnostic tool which is complementary to, rather than a substitute for, volumetric blood flow measurement. Formally the pattern recognition process involved in waveform analysis may be split into transduction, feature extraction and selection, and classification. A variety of approaches have been tried for each of these stages, and the best combination is influenced by the recording site and the objectives of the analysis. The widespread introduction of microcomputers should allow the introduction of more sophisticated methods of feature selection and classification, and the provision of on-line diagnostic information.

10.7 NOTATION

a_n	Amplitude of the nth Fourier harmonic
A	Area under the curve, or height of the first peak of twin peaked waveforms
A_1	Area under Doppler curve up to the systolic peak
A_2	Area under Doppler curve from peak systole onwards
B	Height of the second peak of twin peaked waveforms
CFR	Constant flow ratio
C_{ij}	One element of a covariance matrix
D	Minimum height of a Doppler waveform
DF	Damping factor
E	Efficiency of a transform
f_i	The ith value of the waveform amplitude
f_{in}	The ith value of the nth waveform
f_{max}	Maximum frequency at peak systole
f_{mean}	Mean frequency at peak systole
f_{median}	Median frequency at peak systole
f_{min}	Minimum frequency at peak systole
f_1	Doppler frequency within a stenosis
f_2	Doppler frequency beyond a stenosis
FPF	False positive fraction
HWI	Height width index
$H(S)$	Laplace transform
j	Square root of -1
$m3_f$	Third moment of frequency about the mean
$m4_f$	Fourth moment of frequency about the mean
M	Mean value of the amplitude of a waveform over the cardiac cycle
M_1	Mean height of Doppler curve up to systolic peak
M_2	Mean height of Doppler curve from peak systole onwards
PI	Pulsatility index
PI_{fem}	Pulsatility index in common femoral artery
PI_{orig}	Original Fourier pulsatility index
PI_{pop}	Pulsatility index in popliteal artery
PLI	Path length index
r_{ki}	The ith element of the kth principal component
$RFRI$	Relative flow rate index
RI	Pourcelot's resistance index
s_f	Standard deviation of frequency
S	Maximum height of a Doppler waveform (or $j\omega$)
SBI	Spectral broadening index (numbered 1–10)
$SDTI$	Systolic decay time index
SMR	Sample mean record
t_0	Time of peak systole
t_{0+n}	n ms after peak systole
t_1	Time from beginning of systole to peak systole
t_2	Time from peak systole to the end of diastole
t_i	The ith value of time

t_d	Decay time from S to $0.75S$
t_{fem}	Time of arrival of waveform foot at common femoral site
t_{pop}	Time of arrival of waveform foot at popliteal site
t_r	Rise time from $0.75S$ to S
t_s	Duration of systolic peak (measured between the half amplitude points)
T	Total length of a waveform
TNF	True negative fraction
TPF	True positive fraction
TT	Transit time
TTR	Transit time ratio
δ	Coefficient derived from Laplace transform analysis related to proximal arterial narrowing
γ	Coefficient derived from Laplace transform analysis related to vasoconstriction/vasodilation
λ_i	The ith eigenvalue of a covariance matrix
ω_0	Coefficient derived from Laplace transform analysis related to arterial elastic modulus

10.8   REFERENCES

Andrews HC (1972) Introduction to mathematical techniques in pattern recognition. Wiley-Interscience, New York.

Bada HS, Hajjar W, Chua C, Sumner DS (1979) Non-invasive diagnosis of neonatal asphyxia and intraventricular haemorrhage by Doppler ultrasound. J Pediatr 95, 775–779.

Baird RN, Bird DR, Clifford PC, Lusby RJ, Skidmore R, Woodcock JP (1980) Upstream stenosis: its diagnosis by Doppler signals from the femoral artery. Arch Surg 115, 1316–1322.

Baker AR, Prytherch DR, Evans DH, Bell PRF (1986) Doppler ultrasound assessment of the femoro-popliteal segment: comparison of different methods using ROC curve analysis. Ultrasound Med Biol 12, 473–482.

Barnes RW, Bone GE, Reinertson J, Slaymaker EE, Hokanson DE, Strandness DE (1976) Non-invasive ultrasonic carotid angiography: Prospective validation by contrast arteriography. Surgery 80, 328–335.

Baskett JJ, Beasley MG, Murphy GJ, Hyams DE, Gosling RG (1977) Screening for carotid junction disease by spectral analysis of Doppler signals. Cardiovasc Res 11, 147–155.

Blackshear WM, Phillips DJ, Thiele BL, Hirsch JH, Chikos PM, Marinelli MR, Ward KJ, Strandness DE (1979) Detection of carotid occlusive disease by ultrasonic imaging and pulsed Doppler spectrum analysis. Surgery 86, 698–706.

Blackshear WM, Phillips DJ, Chikos PM, Harley JD, Thiele BL, Strandness DE (1980) Carotid artery velocity patterns in normal and stenotic vessels. Stroke 11, 67–71.

Bodily KC, Zierler RE, Marinelli MR, Thiele BL, Greene FM, Strandness DE (1980) Flow disturbances following carotid endarterectomy. Surg Gynaecol Obstet 151, 77–80.

Brown PM, Johnston KW, Kassam M, Cobbold RSC (1982) A critical study of ultrasound Doppler spectral analysis for detecting carotid disease. Ultrasound Med Biol 8, 515–523.

Craxford AD, Chamberlain J (1977) Pulse waveform transit ratios in the assessment of peripheral vascular disease. Br J Surg 64, 449–452.

Evans DH (1982) Some aspects of the relationship between instantaneous volumetric blood flow and continuous wave Doppler ultrasound. III. Ultrasound Med Biol 8, 617–623.

Evans DH, Caprihan A (1985) The application of classification techniques to biomedical data, with particular reference to ultrasonic Doppler blood velocity waveforms. IEEE Trans Biomed Eng BME-32, 301–311.

Evans DH, Barrie WW, Asher MJ, Bentley S, Bell PRF (1980) The relationship between ultrasonic pulsatility index and proximal arterial stenosis in a canine model. Circ Res 46, 470–475.

Evans DH, Archer LNJ, Levene MI (1985) The detection of abnormal neonatal cerebral haemodynamics using principal component analysis of the Doppler ultrasound waveform. Ultrasound Med Biol 11, 441–449.

Farrar DJ, Green HD, Peterson DW (1979) Noninvasively and invasively measured pulsatile haemodynamics with graded arterial stenosis. Cardiovasc Res 13, 45–47.

Fitzgerald DE, Gosling RG, Woodcock JP (1971) Grading dynamic capability of arterial collateral circulation. Lancet i, 66–67.

Fronek A, Johansen KH, Dilley RB, Bernstein EF (1973) Noninvasive physiologic tests in the diagnosis and characterization of peripheral arterial occlusive disease. Am J Surg 126, 205–214.

Gibbons DT, Evans DH, Barrie WW, Cosgriff PS (1981) Real-time calculation of ultrasonic pulsatility index. Med Biol Eng Comp 19, 28–34.

Gosling RG (1976) Extraction of physiological information from spectrum-analyzed Doppler-shifted continuous-wave ultrasound signals obtained non-invasively from the arterial system. In: IEE medical electronics monographs 18–22 (Eds DW Hill, BW Watson), Chapter 4, pp 73–125, Peter Peregrinus, Stevenage, Hertfordshire.

Gosling RG, King DH (1974) Continuous wave ultrasound as an alternative and complement to X-rays in vascular examination. In: Cardiovascular applications of ultrasound (Ed. RS Reneman), Chapter 22, pp 266–282, North-Holland, Amsterdam.

Gosling RG, King DH (1978) Processing arterial Doppler signals for clinical data. In: Handbook of clinical ultrasound (Eds M de Vlieger et al), John Wiley, London.

Gosling RG, Dunbar G, King DH, Newman DL, Side CD, Woodcock JP, Fitzgerald DE, Keates JS, McMillan D (1971) The quantitative analysis of occlusive peripheral arterial disease by a non-intrusive technique. Angiology 22, 52–55.

Greene FM, Beach K, Strandness DE, Fell G, Phillips DJ (1982) Computer based pattern recognition of carotid arterial disease using pulsed Doppler ultrasound. Ultrasound Med Biol 8, 161–176.

Hill A, Volpe JJ (1982) Decrease in pulsatile flow in the anterior cerebral arteries in infantile hydrocephalus. Pediatrics 69, 4–7.

Hill A, Perlman JM, Volpe JJ (1982) Relationship of pneumothorax to occurrence of intraventricular hemorrhage in the premature newborn. Pediatrics 69, 144–149.

Humphries KN, Hames TK, Smith SWJ, Cannon VA, Chant ADB (1980) Quantitative assessment of the common femoral to popliteal arterial segment using continuous wave Doppler ultrasound. Ultrasound Med Biol 6, 99–105.

Johnston KW, Taraschuk I (1976) Validation of the role of pulsatility index in quantitation of the severity of peripheral arterial occlusive disease. Am J Surg 131, 295–297.

Johnston KW, Maruzzo BC, Cobbold RSC (1978) Doppler methods for quantitative measurement and localization of peripheral arterial occlusive disease by analysis of the blood velocity waveform. Ultrasound Med Biol 4, 209–223.

Johnston KW, deMorais D, Kassam M, Brown PM (1981) Cerebrovascular assessment using a Doppler carotid scanner and real-time frequency analysis. J Clin Ultrasound 9, 443–449.

Johnston KW, Brown PM, Kassam M (1982a) Problems of carotid Doppler scanning which can be overcome by using frequency analysis. Stroke 13, 660–666.

Johnston KW, Kassam M, Cobbold RSC (1982b) On-line identifying and quantifying Doppler ultrasound waveforms. Med Biol Eng Comp 20, 336–342.

Johnston KW, Kassam M, Cobbold RSC (1983) Relationship between Doppler pulsatility index and direct femoral pressure measurements in the diagnosis of aortoiliac occlusive disease. Ultrasound Med Biol 9, 271–281.

Johnston KW, Kassam M, Koers J, Cobbold RSC, MacHattie D (1984) Comparative study of four methods for quantifying Doppler ultrasound waveforms from the femoral artery. Ultrasound Med Biol 10, 1–12.

Junger M, Chapman BLW, Underwood CJ, Charlesworth D (1984) A comparison between two types of waveform analysis in patients with multisegmental arterial disease. Br J Surg 71, 345–348.

Kalman PG, Johnston KW, Zuech P, Kassam M, Poots K (1985) In vitro comparison of alternative methods for quantifying the severity of Doppler spectral broadening for the diagnosis of carotid arterial occlusive disease. Ultrasound Med Biol 11, 435–440.

Keagy BA, Pharr WF, Thomas D, Bowes DE (1982) A quantitative method for the evaluation of spectral analysis patterns in carotid artery stenosis. Ultrasound Med Biol 8, 625–630.

Klecka WR (1980) Discriminant analysis. Sage university paper series on quantitative applications in the social sciences. Sage Publications, Beverly Hills and London.

Knox RA, Greene FM, Beach K, Phillips DJ, Chikos PM, Strandness DE (1982a) Computer based classification of carotid arterial disease: a prospective assessment. Stroke 13, 589–594.

Knox RA, Phillips DJ, Breslau PJ, Lawrence R, Primozich J, Strandness DE (1982b) Empirical findings relating sample volume size to diagnostic accuracy in pulsed Doppler cerebrovascular studies. J Clin Ultrasound 10, 227–232.

Langlois YE, Greene FM, Roederer GO, Jager KA, Phillips DJ, Beach KW, Strandness DE (1984) Computer based pattern recognition of carotid artery Doppler signals for disease classification: prospective validation. Ultrasound Med Biol 10, 581–595.

Lee BY, Assadi C, Madden JL, Kavner D, Trainor FS, McCann WJ (1978) Hemodynamics of arterial stenosis. World J Surg 2, 621–629.

Lipman B, Serwer GA, Brazy JE (1982) Abnormal cerebral hemodynamics in preterm infants with patent ductus arteriosus. Pediatrics 69, 778–781.

Macpherson DS, Evans DH, Bell PRF (1984) Common femoral artery Doppler waveforms: a comparison of three methods of objective analysis with direct pressure measurements. Br J Surg 71, 46–49.

Manga P, Dhurandhar RW, Stockard B (1986) Doppler frequency ratio and peak frequency in the assessment of carotid artery disease: a comparative study with angiography. Ultrasound Med Biol 12, 573–576.

Martin TRP, Barber DC, Sherriff SB, Prichard DR (1980) Objective feature extraction applied to the diagnosis of carotid artery disease using a Doppler ultrasound technique. Clin Phys Physiol Meas 1, 71–81.

Martin TRP, Sherriff SB, Barber DC, Lakeman JM (1981) Analysis of the total Doppler signal obtained from the common carotid artery. Ultrasonics 2, 269–276.

Metz CE (1978) Basic principles of ROC analysis. Sem Nuc Med 8, 283–298.

Nicolaides AN, Gordon-Smith IC, Dayandas J, Eastcott HHG (1976) The value of Doppler blood velocity tracings in the detection of aortoiliac disease in patients with intermittent claudication. Surgery 80, 774–778.

O'Donnell TF, Pauker SG, Callow AD, Kelly JJ, McBride KJ, Korwin S (1980) The relative value of carotid noninvasive testing as determined by receiver operator characteristic curves. Surgery 87, 9–17.

Perlman JM (1985) Neonatal cerebral blood flow velocity measurement. Clin Perinatol 12, 179–193.

Perlman JM, Hill A, Volpe JJ (1981) The effect of patent ductus arteriosus on flow velocity in the anterior cerebral arteries: ductal steal in the premature newborn infant. J Pediatr 99, 767–771.

Planiol T, Pourcelot L (1973) Doppler effect study of the carotid circulation. In: Ultrasonics in medicine (Eds M de Vlieger, DN White, VR McCready), pp 104–111, Elsevier, New York.

Pourcelot L (1976) Diagnostic ultrasound for cerebral vascular diseases. In: Present and future of diagnostic ultrasound (Eds I Donald, S Levi), pp 141–147, Kooyker, Rotterdam.

Prichard DR, Martin TRP, Sherriff SB (1979) Assessment of directional Doppler ultrasound techniques in the diagnosis of carotid artery diseases. J Neurol Neurosurg Psych 42, 563–568.

Prytherch DR, Evans DH, Smith MJ, Macpherson DS (1982) On-line classification of arterial stenosis severity using principal component analysis applied to Doppler ultrasound signals. Clin Phys Physiol Meas 3, 191–200.

Reneman RS, Spencer MP (1979) Local Doppler audio spectra in normal and stenosed carotid arteries in man. Ultrasound Med Biol 5, 1–11.

Rittgers SE, Thornhill BM, Barnes RW (1983) Quantitative analysis of carotid artery Doppler spectral waveforms: diagnostic value of parameters. Ultrasound Med Biol 9, 255–264.

Rutherford RB, Hiatt WR, Kreutzer EW (1977) The use of velocity waveform analysis in the diagnosis of carotid artery occlusive disease. Surgery 82, 695–702.

Sebestyen G (1962) Decision making processes in pattern recognition, Macmillan, New York.

Sheldon CD, Murie JA, Quin RO (1983) Ultrasonic Doppler spectral broadening in the diagnosis of internal carotid artery disease. Ultrasound Med Biol 9, 575–580.

Sherriff SB, Barber DC, Martin TRP, Lakeman JM (1982) Use of principal component factor analysis in the detection of carotid artery disease from Doppler ultrasound. Med Biol Eng Comp 20, 351–356.

Skidmore R, Woodcock JP (1978) Physiological significance of arterial models derived using transcutaneous ultrasonic flowmeters. J Physiol 277, 29–30P.

Skidmore R, Woodcock JP (1980a) Physiological interpretation of Doppler-shift waveforms. I. Theoretical considerations. Ultrasound Med Biol 6, 7–10.

Skidmore R, Woodcock JP (1980b) Physiological interpretation of Doppler-shift waveforms. II. Validation of the Laplace transform method for characterisation of the common femoral blood-velocity/time waveform. Ultrasound Med Biol 6, 219–225.

Skidmore, R, Woodcock JP, Wells PNT, Bird D, Baird RN (1980) Physiological interpretation of Doppler shifted waveforms. III. Clinical results. Ultrasound Med Biol 6, 227–231.

Spencer MP, Reid JM (1979) Quantitation of carotid stenosis with continuous wave Doppler ultrasound. Stroke 10, 326–330.

Stuart B, Drumm J, Fitzgerald DE, Duignan NM (1980) Fetal blood velocity waveforms in normal pregnancy. Br J Obstet Gynaecol 87, 780–785.

Thiele BL, Bandyk DF, Zierler RE, Strandness DE (1983) A systematic approach to the assessment of aortoiliac disease. Arch Surg 118, 477–481.

Thompson RS, Trudinger BJ, Cook CM (1985) Doppler ultrasound waveforms in the fetal umbilical artery: quantitative analysis technique. Ultrasound Med Biol 11, 707–718.

Thorndike RM (1978) Correlation procedures for research. Chapter 8 – Discriminant Analysis, Gardner, New York.

Tou JT, Gonzalez RC (1974) Pattern recognition principles, Addison-Wesley, Reading, Mass.

Trudinger BJ, Giles WB, Cook CM, Bombardieri J, Collins L (1985a) Fetal umbilical artery flow velocity waveforms and placental resistance: clinical significance. Br J Obstet Gynaecol 92, 23–30.

Trudinger BJ, Giles WB, Cook CM (1985b) Uteroplacental blood flow velocity–time waveforms in normal and complicated pregnancy. Br J Obstet Gynaecol 92, 39–45.

Van Merode T, Hick P, Hoeks APG, Reneman RS (1983) Limitations of Doppler spectral broadening in the early detection of carotid artery disease due to the size of the sample volume. Ultrasound Med Biol 9, 581–586.

Walton L, Martin TRP, Collins M (1984) Prospective assessment of the aorto-iliac segment by visual interpretation of frequency analysed Doppler waveforms – a comparison with arteriography. Ultrasound Med Biol 10, 27–32.

Ward AS, Martin TP (1980) Some aspects of ultrasound in the diagnosis and assessment of aortoiliac disease. Am J Surg 140, 260–265.

Woodcock JP (1970) The transcutaneous ultrasonic flowmeter and the significance of changes in the velocity–time waveform in occlusive arterial disease of the leg. PhD Thesis, University of London.

Woodcock JP, Gosling RG, Fitzgerald DE (1972) A new non-invasive technique for assessment of superficial femoral artery obstruction. Br J Surg 59, 226–231.

Woodcock JP, Shedden J, Skidmore R, Machleder H, Evans J, Wells PNT (1982) Doppler spectral broadening and anomalous vessel wall movement in the study of atherosclerosis of the carotid arteries. Ultrasound Med Biol 8, Suppl 1, 211.

Woodcock JP, Shedden J, Aldoori M, Skidmore R, Burns P, Evans J (1983) Doppler spectral broadening and anomalous vessel wall movement in the study of atherosclerosis of the carotid arteries. In: Ultrasound '82 (Eds RA Lerski, P Morley), pp 235–237, Pergamon Press, Oxford.

Zwiebel WJ, Zagzebski JA, Crummy AB, Hirscher M (1982) Correlation of peak Doppler frequency with lumen narrowing in carotid stenosis. Stroke 13, 386–391.

Chapter 11

VOLUMETRIC BLOOD FLOW MEASUREMENTS

11.1   INTRODUCTION

One of the most exciting applications of Doppler ultrasound is the measurement of volumetric blood flow, and there are now available a wide variety of commercial machines equipped to make such determinations. At present the majority of such machines are of the 'duplex' variety, i.e. they combine a CW or single-gate PW Doppler unit and a real-time ultrasonic B-scanner, but some multigate systems are also available. In addition a number of alternative techniques have been suggested or used to determine flow or one or more of its component parts.

Perhaps the greatest advantage of Doppler flow measurements is that they can be made without in the least interfering with flow. They can be made on conscious patients totally non-invasively and may therefore be repeated at will so as to monitor the progress of a disease process or the effect of a therapy.

In this chapter we first examine the most frequently used methods of flow measurement and some of the errors these may involve, and then describe some of the alternative approaches to the flow measurement problem.

11.2   FLOW MEASUREMENT WITH DUPLEX SCANNERS

Most ultrasonic duplex scanners now have facilities for calculating and displaying volumetric blood flow. The pulse echo system is used to image the blood vessel of interest, and allows the operator accurately to place the Doppler sample volume to totally encompass the vessel whilst avoiding signals from nearby structures, to measure the angle between the ultrasound beam and the axis of the blood vessel (θ), and to measure the diameter of the blood vessel (Fig. 11.1). The Doppler system is used to estimate the mean velocity of flow in the direction of the ultrasound beam (using a uniform insonation technique) which may then, with a knowledge of θ, be converted to the mean velocity parallel to the vessel axis. This in turn is multiplied by the vessel cross-section area to yield mean flow. The method which has been explored by many workers (e.g. Gill 1979, 1982, Avasthi et al 1984, Eik-Nes et al 1984, Gill et al 1984, Griffin et al 1985, Qamar et al 1985, Evans 1986, Lewis et al 1986) has considerable potential but it is important that users appreciate the large errors that may occur under unfavourable circumstances.

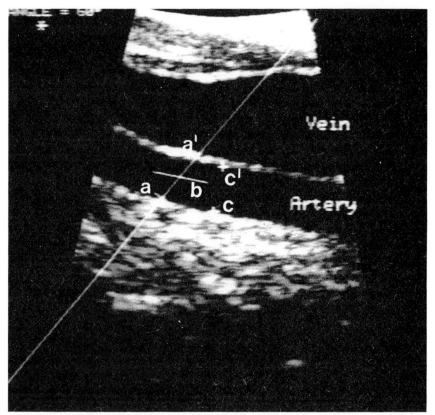

Figure 11.1 Duplex image of an artery and vein showing the placement of the sample gate (aa') to exclude the signal from a vein, the cursor used to measure the angle between the vessel axis and the ultrasound beam (b), and the cursors used to measure the vessel diameter (cc')

11.2.1 Theoretical aspects

The time-averaged volumetric flow through a vessel, $\bar{Q}$, is given by the time-averaged product of the cross-sectional area of the vessel $A(t)$ and the spatial mean velocity of flow within the vessel $\bar{v}(t)$, and may be written:

$$\bar{Q} = 1/T \int_{t=0}^{T} A(t)\,\bar{v}(t)\,dt \qquad\qquad 11.1$$

If the vessel is uniformly insonated, the mean velocity may be calculated from the mean Doppler shift frequency using the standard Doppler equation, i.e.:

$$\bar{v}(t) = \bar{f}_d(t)\,c/2f_t\,\cos\theta \qquad\qquad 11.2$$

where $\bar{f}_d(t)$ is the instantaneous mean Doppler shift, c the velocity of ultrasound in blood, f_t the transmitted zero-crossing frequency, and θ the angle between the ultrasound beam and the blood vessel axis. Substituting eqn 11.2 into eqn 11.1 leads to:

$$\bar{Q} = (c/2f_t\,\cos\theta) \int_{t=0}^{T} (A(t)\,\bar{f}_d(t)/T)\,dt \qquad\qquad 11.3$$

Ordinary commercial duplex machines have no method of following changes in cross-sectional area over the cardiac cycle and therefore it is implicitly assumed that such changes are insignificant or at least cancel out, and the equation that is actually evaluated is a simplified form of eqn 11.3, namely:

$$\bar{Q} = (cA/2f_t \cos \theta) \int_{t=0}^{T} (\bar{f}_d(t)/T) \, dt \qquad \qquad 11.4$$

where A is the 'effective vessel area'. The zero-crossing frequency of the transmitted signal is easily measured and the velocity of sound in blood is more or less constant, but each of the other terms in eqn 11.4 ($\bar{f}_d$, A and θ) must be measured.

11.2.2   Measurement of mean Doppler shift

The determination of mean Doppler shift frequency, and hence mean blood velocity, has already been dealt with at some length in Chapter 9. There it was shown that non-uniform insonation of the blood vessel (Section 9.3.1), differential attenuation between soft tissue and blood (Section 9.3.2), intrinsic spectral broadening (Section 9.3.3), high pass filters designed to reject high-amplitude low-frequency Doppler shifts (Section 9.3.4), and a poor signal-to-noise ratio (Section 9.3.5) could all affect the measured mean frequency.

The most important of these errors is probably that due to non-uniform vessel insonation (Evans 1985), and caution must be exercised when interpreting the mean velocity derived by duplex scanners if the ultrasound beam is significantly narrower than the vessel diameter. Since on average the velocity at the centre of a vessel is higher than at the periphery, a narrow ultrasound beam passing through the centre of the vessel leads to an overestimation of the mean velocity. The magnitude of this error has been considered in Section 9.3.1 and is summarized for a simple beam shape (rectangular) and velocity profile (parabolic) in Fig. 11.2.

11.2.3   Measurement of vessel area

Most duplex scanners offer a choice of two methods for determining vessel cross-section. The simpler and most usual method is to make a diameter measurement of the vessel using the same image used to place the Doppler sample volume (see Fig. 11.1), and then to calculate the area by assuming the vessel to be circular in cross-section. Alternatively the scan plane of the imaging device may be rotated around the vertical axis to produce a cross-sectional view of the vessel, whose area can then be measured directly. In either case the operation of most scanners encourages the user to make only a single measurement of a quantity which is known to vary over the cardiac cycle, and this can cause quite significant errors in addition to those intrinsic to the measurement of a single unchanging diameter.

Short of evaluating eqn 11.3 directly by constantly monitoring both vessel diameter and mean Doppler shift frequency (Hartley et al 1978, Furuhata et al 1978, Uematsu 1981), the error due to changing vessel size cannot be completely eliminated, but it may be reduced by measuring the vessel diameter a number of times and using the mean or median value in the flow calculations, and all users should be encouraged to do this if a vessel is at all pulsatile. In an investigation of the blood flow in the descending aortas of

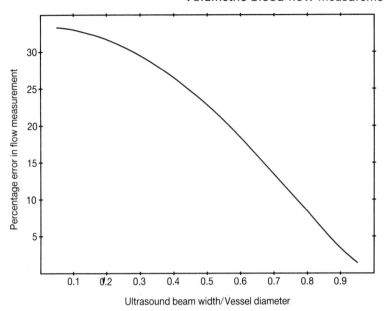

Figure 11.2   Percentage error in flow measurement due to partial insonation, by a centrally placed rectangular beam of ultrasound, of a blood vessel containing flow whose time-averaged velocity profile is parabolic. (Reprinted with permission from Evans (1986). © The Institute of Physical Sciences in Medicine)

fetal lambs, Struyk et al (1985) found that errors of between +9% and −19% were possible if the aortic diameter happened to be measured at its maximum or minimum value, but that the use of a time-averaged diameter led to a systematic error of −5%.

Other important sources of error in determining the vessel area include incorrect assumptions about the vessel shape (i.e. the cross-section may not be circular, possibly due to disease), the limited axial resolution of the pulse echo system, and incorrect caliper velocity settings. Axial resolution is important because echoes from the inner and outer surfaces of the blood vessel (and possibly from different layers within the wall) tend to merge together and therefore only the first echo from each wall is reliable. Because of this, workers in the field of fetal blood flow often measure from the outer aspect of the proximal wall to the inner aspect of the distal wall (Eik-Nes et al 1982, Teague et al 1985). This produces an overestimate of the vessel diameter, but this known systematic error is less serious than a similarly sized unknown random error. It is doubtful if a similar approach would be valid in adult arteries, the walls of which may have been affected by pathological changes.

Even if the axial resolution is very good, the best achievable accuracy will be of the order of a wavelength of the imaging ultrasound which may be quite significant when compared with the diameter of small vessels. Figure 11.3 shows the percentage error in measured flow plotted against vessel diameter for various errors in diameter measurement. Given that the wavelengths of ultrasound with frequencies of 3, 5 and 10 MHz are 0.5, 0.3 and 0.15 mm respectively, it can be seen that potential errors in measurements on vessels of less than 4 mm diameter may be very large indeed.

Caliper setting errors may arise because the velocity of ultrasound in blood

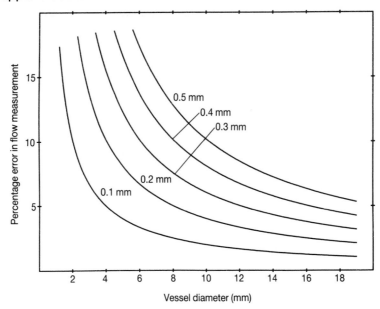

Figure 11.3   Percentage error in flow measurement due to the uncertainty in the diameter measurement (± 0.1 mm to ± 0.5 mm) plotted as a function of diameter. (Reprinted with permission from Evans (1986). © The Institute of Physical Sciences in Medicine)

(~ 1580 m s^{-1}) is significantly greater than that in soft tissue (~ 1540 m s^{-1}), and if this is overlooked the flow may be underestimated by about 5%.

11.2.4   Measurement of angle

The angle between the ultrasound beam and the blood vessel axis is usually found by rotating a dedicated cursor on the B-scan image to align it with the axis of the vessel (Fig. 11.1). Since it is the cosine of θ that determines the component of velocity measured by the Doppler probe, the accuracy of this measurement becomes much more critical at angles approaching 90°, where the cosine function varies very rapidly. The percentage error in the flow measurement due to a given error in the measurement of the angle θ is plotted as a function of the true value of θ in Fig. 11.4. With care, on a straight vessel, it is usually possible to measure θ to $\pm 2°$, and thus provided θ itself is kept below 60°, and preferably below 45°, errors from this source should not be excessive.

It has been assumed in the derivation of Fig. 11.4 that the plane of scan is coincident with the axis of the vessel; deviations from this will cause the flow to be underestimated, and therefore efforts should be made to ensure that a reasonable length of the vessel is in the plane of the scan wherever possible. Fortunately, for angles of less than 15° the error due to misalignment of the scan plane and vessel axis is less than 3%, but at 20°, 25° and 30° the errors are 6%, 9% and 13% respectively.

11.2.5   Practical limitations

There are numerous sources of error in volumetric flow measurements made using duplex scanners, but many of them can be practically eliminated by careful attention to

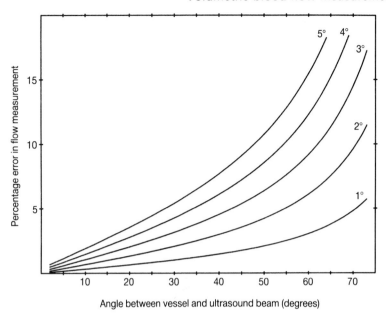

Figure 11.4   Percentage error in flow measurement due to the uncertainty in the measurement of the angle ($\pm 1°$ to $\pm 5°$) between the ultrasound beam and the axis of the blood vessel, plotted as a function of angle. (Reprinted with permission from Evans (1986). © The Institute of Physical Sciences in Medicine)

technique. The method is at its best when applied to medium-sized vessels of between about 4 mm and 8 mm diameter. For smaller vessels it is very difficult to make accurate determinations of the vessel size, whereas for larger vessels it becomes progressively more difficult to ensure uniform insonation.

11.3   FLOW MEASUREMENT WITH MULTIGATE SYSTEMS

A second type of flow measuring system that is available is the multigate flowmeter which, rather than relying on uniform insonation to sample all parts of the vessel equally, actually measures the shape of the instantaneous velocity profiles using a series of small sample volumes. In order to do this efficiently it is important that the ultrasound beam is considerably narrower than the vessel and that the sample length is such that several may be placed across the vessel lumen. These constraints limit the use of the method to fairly large superficial vessels.

The multigate technique may be applied using a single gate which is stepped across the vessel over a number of cardiac cycles (Peronneau et al 1972, Histand et al 1973) but is much better carried out with a true multigate system which may either consist of a large number of finite gates operating in parallel (Baker 1970, Keller et al 1976), or a single serial digital signal processor capable of behaving as a multigate system (Brandestini 1978, Hoeks et al 1981).

As with Duplex scanning, an independent measure of the angle θ between the ultrasound beam and the direction of blood flow is required, but, since multigate systems are often not linked to B-scan devices, some solution other than an imaging one may have to be found to this particular problem.

11.3.1   Theoretical aspects

If the cross-section of a blood vessel is split into a large number of small elements, ΔA_i, the instantaneous total flow through the vessel, $Q(t)$, is given by the sum of the flows through each of these elements, i.e.:

$$Q(t) = \sum_i \Delta A_i\, v_i(t) \qquad\qquad 11.5$$

where $v_i(t)$ is the velocity of the blood travelling through ΔA_i. If the velocity, $v_n(t)$, is uniform within a semi-annulus (see Fig. 11.5) then eqn 11.5 may be rewritten:

$$Q(t) = \pi \sum_n r_n\, \Delta r_n\, v_n(t) \qquad\qquad 11.6$$

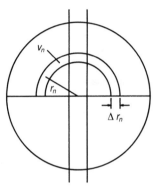

Figure 11.5   Schematic diagram of a narrow ultrasound beam passing through the centre of a blood vessel containing flow that is uniform within any small semi-annulus

where Δr_n is the thickness of the semi-annulus at distance r_n from the vessel centre. $v_n(t)$ may be found from the instantaneous mean Doppler shift frequency f_{dn} measured from the appropriate range gate of a multigate system, i.e.:

$$v_n(t) = f_{dn}(t)\, c/2f_t \cos\theta \qquad\qquad 11.7$$

Substituting eqn 11.7 into eqn 11.6, and remembering that in practice the Doppler gates are uniformly spaced and that the suffix n may therefore be dropped from Δr_n, we obtain:

$$Q(t) = (c\pi\, \Delta r/2f_t \cos\theta) \sum_n r_n f_{dn}(t) \qquad\qquad 11.8$$

Finally the mean flow may be determined by integrating eqn 11.8 over the cardiac cycle, i.e.:

$$\bar{Q} = (c\pi\, \Delta r/2f_t \cos\theta\, T) \int_{t=0}^{T} \sum_n r_n f_{dn}(t)\, dt \qquad\qquad 11.9$$

The velocity of ultrasound, c, is a known constant, and the transmitted zero-crossing frequency, f_t, for a given system is constant, and therefore in order to determine volumetric flow only θ, Δr, r_n and $f_{dn}(t)$ need be found.

11.3.2 Measurement of angle

Multigate systems do not usually incorporate a B-scan imager, but the angle θ may still be found in any of a number of ways (see Section 11.6.2). The MAVIS system (GEC Medical Ltd) uses a pair of cross-sectional Doppler images from either side of the flow measurement point to establish the axis of the vessel (Section 11.6.2.1), whilst the EBF system (Novamed) uses a second transducer to find the direction perpendicular to the flow axis both by measuring the size of the echoes from the vessel walls and by minimizing the Doppler shift frequency (Section 11.6.2.2).

Similar considerations concerning inaccuracies in flow measurement due to inaccuracies in angle measurement apply to multigate systems as to duplex systems (Section 11.2.4) and Fig. 11.4 is therefore appropriate for assessing such errors (but see the next section).

11.3.3 Measurement of Δr

The separation of the semi-annuli may be calculated in one of two ways, both of which are subject to errors. The first relies on the fact that the separation of the gates, s, for a particular system is known, and that the angle θ must be measured to calculate the blood flow velocity parallel to the axis of the vessel; thus Δr may be simply calculated by dividing s by sin θ (Fig. 11.6). The major problem with this method is that it is critically dependent on the accurate measurement of θ, particularly when θ is small. Fortunately this error is to some extent compensated for by the error in measuring velocity because, if θ is underestimated, the velocity is underestimated but the vessel diameter overestimated.

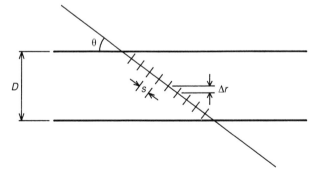

Figure 11.6 Schematic diagram of a multigate Doppler beam intersecting a blood vessel of diameter D at an angle θ

The second method is to measure the diameter of the vessel directly by means of an A-mode ultrasound transducer held perpendicular to the vessel. In this case Δr may be calculated by dividing the diameter of the vessel, D, by the number of Doppler gates that detect a blood flow signal. There are two sources of inaccuracy in this determination, those related to measuring the diameter accurately and the difficulty of defining precisely the edge of the vessel lumen.

11.3.4   Measurement of radius

The measurement of r_n is very closely related to the measurement of Δr, since r_n is an integral multiple of Δr. Therefore the same considerations apply to the accuracy of measuring r_n as to Δr, and since the two quantities are multiplied together the error in flow that results from an error in Δr is double in percentage terms. This is exactly analogous to the Duplex case where an error in measuring the vessel diameter leads to twice the percentage error in the flow measurement.

11.3.5   Measurement of velocity

In theory each sample volume of a multigate system should be so small that only a very narrow band of Doppler shift frequencies is present. This being the case, f_{dn} could be determined by any of a number of methods, but the ones normally chosen are to use a mean frequency follower or zero-crossing detector. In this particular application uniform insonation is not an issue, and so the major sources of error in the measurement of f_{dn} are the limited signal-to-noise ratio, and the effects of the high pass filters. Both these have been discussed in Chapter 9 (Sections 9.3.4, 9.3.5, 9.5.4 and 9.5.5).

11.3.6   Practical limitations

There are, in addition to the errors already discussed, two major limitations of the multigate technique. These are that the method assumes that there is partial symmetry in the velocity profile, and that in practice the sample volumes are of a finite size in both length and width. The former means that measurements should only be made remote from branches, bifurcations, curves, disease sites, and any other geometry changes that might disturb the symmetry of the flow profile (see Chapter 2), whilst the latter limits the technique to use on large vessels where the sample volumes will be small compared with the vessel diameter. The effect of the finite sample volume is to distort the shape of the measured velocity profile, and hence the measured value of flow. This has been discussed in some depth by Jorgensen et al (1973) and Baker et al (1978) who showed that it is, in principle, possible to apply a deconvolution process to determine the real velocity profile. Such a process has been used in the MAVIS device to improve its accuracy (Fish 1981), but still the method may not be used with small blood vessels.

A further practical problem of the multigate method is ensuring that the narrow beam of ultrasound passes through the centre of the vessel. If this is not achieved precisely, the flow may be seriously underestimated since both arterial diameter and mean velocity are underestimated.

11.4   FLOW MEASUREMENT – ATTENUATION-COMPENSATED METHOD

The attenuation-compensated (AC) volume flowmeter is an ingenious variation of the uniform insonation technique first described by Hottinger and Meindl (1979). It differs from all flow techniques so far described in that it makes use not only of the Doppler shift frequencies, but also of the total power of the Doppler signal.

A Doppler transducer is arranged to transmit two concentric pulsed beams. The sample volume of one totally encompasses the vessel (Fig. 11.7a) and is used to make both Doppler shift and power measurements. The sample volume of the other lies totally within the vessel and is used to make power measurements for the purpose of calibrating the power measurements from the first (Fig. 11.7b).

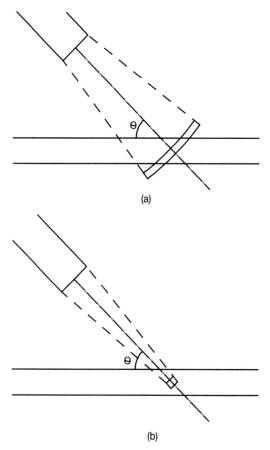

(a)

(b)

Figure 11.7 The sample volumes interrogated by the two ultrasound beams necessary for the attenuation compensated flowmeter. (a) The entire cross-section of the vessel is insonated. (b) The sample volume lies entirely within the vessel lumen. (Redrawn with permission from Evans et al (1986). © The Institute of Physical Sciences in Medicine)

11.4.1 Theoretical aspects

Provided the vessel is uniformly insonated, the total Doppler power is proportional to the vessel cross-section and inversely proportional to the cosine of θ, and therefore we may write the area as:

$$A(t) = P(t) \cos \theta / M(z) \qquad 11.10$$

where $P(t)$ is the total power in the Doppler signal and $M(z)$ is a depth-dependent constant. If eqns 11.2 and 11.10 are now substituted into eqn 11.1, the cosine term vanishes and we are left with:

$$\bar{Q} = (c/2M(z)f_t) \int_{t=0}^{T} (P(t)\bar{f}_d(t)/T)\, dt \qquad 11.11$$

Therefore the volumetric flow can be calculated from the integrated product of power and Doppler shift frequency, provided the constant $M(z)$ can be evaluated.

Using the notation of Hottinger and Meindl (1979), $M(z)$ may be shown to be given by:

$$M(z) = T(z)\, \eta\, I_1(z)\, \Delta z \qquad 11.12$$

where $T(z)$ is the round trip transmission efficiency representing the effects of attenuation caused by the tissue between the transducer and vessel at range z, η the volumetric scattering coefficient representing the scattering of the blood, $I_1(z)$ a parameter indicating the transducer sensitivity, and Δz the Doppler sample volume length. Both $T(z)$ and η are dependent on the measurement configuration and must therefore be evaluated for each flow determination, and it is the method of so doing that is the crux of the AC method. A second smaller ultrasound beam, concentric with the first, is used to make power measurements from entirely within the lumen (Fig. 11.7b). This power, P_2, may be written:

$$P_2 = T(z)\, \eta\, I_2(z)\, A_2(z)\, \Delta z \qquad 11.13$$

where $I_2(z)$ indicates the sensitivity of transducer 2, and $A_2(z)$ is the cross-sectional area of the sample volume. Substituting eqn 11.13 into eqn 11.12 leads to:

$$M(z) = P_2[I_1(z)/A_2(z)\, I_2(z)] \qquad 11.14$$

The expression in the square bracket is, for a particular transducer design, dependent only upon the range z, and can be either calculated or better measured experimentally, and therefore $M(z)$ is easily calculated from P_2 given z.

11.4.2 Practical limitations

The greatest practical problem to be overcome with the AC method is that of producing two ultrasound beams, one of which is uniform over the vessel lumen, and one of which is smaller than the lumen. The former is technically the more difficult to overcome, and any non-uniformity will affect both the mean Doppler shift (in the same way as for the Duplex scanner) and the total power received. At present the most promising solution to producing a large uniform beam seems to be to use an annular array transducer (Fu and Gerzberg 1983, Evans et al 1986).

The production of a small sample volume is not very demanding and may be achieved with the same transducer used to produce the large beam, but the AC method may not be used on small vessels where difficulty arises in keeping the sample volume entirely within the blood vessel lumen.

A major advantage of the AC method is that both the cross-sectional area of the vessel and the angle θ are automatically taken into account, but it is susceptible to the same problems with high pass filters and limited signal-to-noise ratios that all other Doppler flowmeters have. Potentially the method may be of some considerable utility, but it is as yet at an early stage of its development.

11.5   FLOW MEASUREMENT – ASSUMED VELOCITY PROFILE METHOD

The assumed velocity profile method is a variation of the Duplex method. The cross-sectional area of the vessel and the angle θ are determined as for that method, but the mean velocity, $\bar{v}$, is determined from the time-averaged maximum Doppler shift. The equation evaluated is a slight modification of eqn 11.4, i.e.:

$$\bar{Q} = (KcA/2f_t \cos \theta) \int_{t=0}^{T} (\hat{f}_d(t)/T) \, dt \qquad 11.15$$

where $\hat{f}_d(t)$ is the instantaneous maximum Doppler shift, and K a constant which depends on the time-averaged velocity profile. The instantaneous maximum Doppler shift frequency may be found using a maximum frequency follower and is generally more reliable than the estimated mean frequency because of its immunity to the effects of attenuation, noise and high pass filters (see Section 9.4).

The assumed velocity profile technique is of value in two distinct circumstances, when the time-averaged velocity profile is flat, and when it is parabolic. The former is at least approximately true of the flow in the ascending aorta and the arch of the aorta, and a number of workers (e.g. Light and Cross 1972, Mackay 1972, Light 1974, Huntsman et al 1983) have used the assumed velocity profile method to measure and to monitor changes in cardiac output. Since the velocity profile is almost flat, the maximum velocity and the mean velocity are very similar and therefore the constant K in eqn 11.15 is approximately equal to unity. In practice, skewing of the velocity profile will cause a slight spread in the velocities across the vessel lumen and the mean velocity is therefore slightly overestimated.

In vessels in which the flow is fully established (Sections 2.3.2 and 2.4.4) the time-averaged velocity profile is parabolic, and in this case the time-averaged maximum velocity is twice the mean velocity, provided the lamina with the maximum velocity is always in the centre of the vessel (Evans 1985). Under these circumstances K in eqn 11.15 has a value of 0.5. It is as yet uncertain which arterial sites meet the appropriate criteria sufficiently closely for this method to be of value, but the common carotid artery is one likely candidate, and the method seems to be of value for measuring mean velocities in the neonatal cerebral arteries. (Unfortunately flow is unobtainable because the vessel diameters are too small to be measured.)

11.6   MISCELLANEOUS TECHNIQUES

11.6.1   First moment followers

The first moment follower is not a method of measuring flow, but is mentioned here because its output is, at least theoretically, more closely related to flow in a highly pulsatile vessel than the output of a mean frequency follower. Essentially it works in the same way as the attenuation-compensated flowmeter (Section 11.4) except that no absolute calibration is performed, and therefore its output is only proportional to flow, and the constant of proportionality changes with the measurement configuration. The reader is referred to Section 9.6.1 for more details.

11.6.2   Angle estimation techniques

A wide variety of methods have been used to measure the angle, θ, between the direction of the ultrasound beam and the flow axis. Some of them rely on imaging the vessel, and some on combining or comparing the output from two or more Doppler transducers.

11.6.2.1   Imaging techniques

The most widely used method of measuring θ is the imaging one described in Section 11.2.4 and illustrated in Fig. 11.1. In addition it is possible to define the axis of the vessel, and hence θ, by making cross-sectional images of the vessel some distance apart but close to the Doppler measurement site, using either a Doppler imaging technique (Fish 1978), or a B-scan technique (Gill 1979). The orientation of the line joining the centre of these two images is an estimate of the vessel axis, and θ can thus be calculated by simple three-dimensional geometry.

11.6.2.2   Definition of normal

Another approach to measuring θ is to start by defining a normal to the vessel, either by minimizing the Doppler shift frequency from a Doppler transducer (Histand et al 1973) or by maximizing the size of the vessel wall echoes using an A-mode transducer (Doriot et al 1975, Uematsu 1981) or a combination of both (Marquis et al 1983).

There are two variations of this technique. In one the ultrasound probe is constructed from two transducers held at a fixed angle, ϕ, to each other, so that when the angle-finding transducer is perpendicular to the vessel the angle of the other with respect to the vessel axis is known to be $\pi/2-\phi$ (Fig. 11.8a). In the second variation the same Doppler transducer is rotated through a known angle from the known perpendicular, and the angle of insonation derived in the same way (Fig. 11.8b).

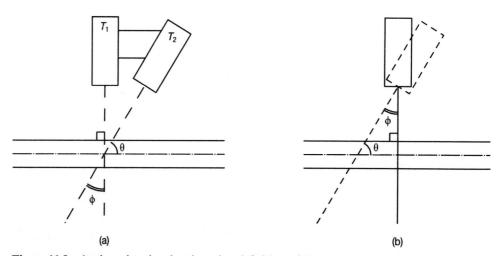

(a)   (b)

Figure 11.8   Angle estimation by the prior definition of the normal (a) using a fixed pair of transducers, and (b) by rotating a single transducer through a known angle. Note that the two ultrasound beams in (a) may not necessarily intersect with the same part of the vessel

11.6.2.3   Dual and triple transducer systems

There are a number of methods of measuring θ by comparing the signals from two transducers set at a known angle to each other. The simplest, which is only applicable to vessels that are parallel or nearly parallel to the skin, is to arrange two crystals symmetrically about a perpendicular so that they intersect within or near the vessel lumen (Fig. 11.9a). The orientation of the transducer pair is then adjusted until the Doppler shifts detected by each crystal are equal and opposite, in which case θ is equal to $\alpha/2$, where α is the angle between the two crystals (Safar et al 1981).

A more sophisticated method is to use a similar transducer arrangement set at an arbitrary angle to the vessel (Fig. 11.9b) and to use the relationship between the two Doppler shift frequencies to determine the angles (Peronneau et al 1972, 1977, Wang and Shao 1986).

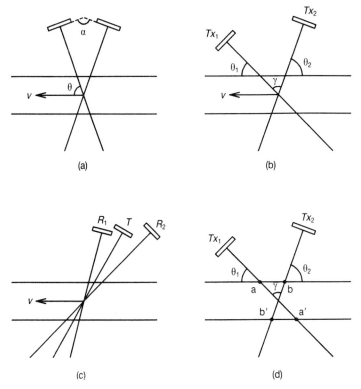

Figure 11.9   Angle estimation by (a) equalizing forward and reverse Doppler shifts, (b,c) comparing the Doppler shift frequencies from two transducers, and (d) measuring two apparent diameters with a pulse Doppler system

Referring to Fig. 11.9b, the Doppler shift measured by transducer $1 (f_{d1})$ may be written $Kv \cos \theta_1$ where K is a constant equal to $2f_t/c$. Similarly the shift measured by transducer 2 (f_{d2}) will be $-Kv \cos \theta_2$. Combining these two expressions to eliminate Kv gives:

$$f_{d2} \cos \theta_1 = -f_{d1} \cos \theta_2 \qquad\qquad 11.16$$

Writing θ_2 in terms of θ_1 and γ, the angle between the two ultrasound beams, gives:

$$f_{d2} \cos \theta_1 = f_{d1} \cos(\theta_1 + \gamma) \qquad 11.17$$

Finally, expanding the term in the bracket and rearranging the equation leads to:

$$\theta_1 = \arctan \frac{f_{d1} \cos \gamma - f_{d2}}{f_{d1} \sin \gamma} \qquad 11.18$$

A similar approach is adopted in the system described by Uematsu (1981) except that a single transmitting crystal is flanked by two receiving crystals (Fig. 11.9c).

A variation of the two transducer approach, described by Borodzinski et al (1976), is to measure the apparent diameters of the vessel as viewed by each transducer in a pulsed Doppler mode (Fig. 11.9d); θ is then calculated from the two diameters and the angle between the two beams.

Referring to Fig. 11.9d the distance measured by transducer 1 is aa' and may be written $D/\sin \theta_1$, where D is the true diameter of the blood vessel. Similarly distance bb' measured by transducer 2 may be written $D/\sin \theta_2$. Combining these expressions and eliminating D gives:

$$aa' \sin \theta_1 = bb' \sin \theta_2 \qquad 11.19$$

Writing θ_2 in terms of θ_1 and γ gives:

$$aa' \sin \theta_1 = bb' \sin(\theta_1 + \gamma) \qquad 11.20$$

Finally expanding the terms in the bracket and rearranging the equation leads to:

$$\theta_1 = \arctan \frac{bb' \sin \gamma}{aa' - bb' \cos \gamma} \qquad 11.21$$

11.6.3   Diameter estimation techniques

The majority of diameter and area measuring techniques have already been described earlier in this chapter. These comprise direct measurement from a B-scan image (Section 11.2.3), measurement using an A-scan transducer orientated at 90° to the vessel axis (Section 11.3.3) and the method of measuring the ratio of two powers used in the attenuation-compensated method (Section 11.4).

One additional method is a pulsed Doppler one where a range gate is moved across the vessel to establish the points at which the Doppler signal first appears and then eventually disappears. The angle θ must be other than 90° since a finite Doppler shift is required, and therefore θ must be found at the same time in order to calculate the true diameter. This method tends to overestimate the size of the vessel because of the finite sample volume size (Jorgensen et al 1973, Borodzinski et al 1976).

A number of methods have also been described for tracking the motion of the vessel wall throughout the cardiac cycle (Hokanson et al 1970, Olsen 1977, Groves et al 1982, Hoeks et al 1985, Struyk et al 1985) but at present few attempts to combine instantaneous vessel diameter and mean velocity measurements have been successful.

11.7   FLOW MEASUREMENT ERRORS

Each flow measurement method is subject to a number of errors which limits its use under some conditions. The exact types of error are dependent on the method in use, but all ultimately stem from the same sources.

It is important to distinguish between systematic and random errors since the former may be of little import if they affect all measurements to a similar degree so that comparisons of flow between patients and within the same patient are valid. Random errors on the other hand may invalidate such comparisons, although many of them may be reduced by repeated measurement. It is worth noting that one of the largest contributions to the overall error of measuring volumetric flow is usually that of vessel size measurement, and therefore mean velocity may be evaluated much more accurately than volumetric flow, and may sometimes be a useful alternative measurement.

For a more detailed discussion of errors, their sources and their interactions the reader is referred to the articles by Gill (1985) and Evans (1986).

11.8   SUMMARY

The problem of measuring volumetric flow may be split into three parts: the measurement of the component of velocity parallel to the axis of the Doppler transducer, the measurement of the angle between the flow axis and the transducer axis, and the measurement of the cross-sectional area of the blood vessel. There are a variety of approaches to each of these measurements. Velocity may be measured using a uniform insonation method, or by measuring or assuming the shape of the velocity profile. The angle θ may be found using imaging techniques, by first finding a normal to the blood vessel using a Doppler or A-scan technique, or by comparing the Doppler shift frequencies or vessel diameters measured by different transducers. The cross-sectional area of the blood vessel may be measured from a cross-sectional image, or calculated from a diameter found either by imaging, by an A-scan technique, or using pulsed Doppler. One method, the attenuation compensated method, makes use of power measurements to eliminate θ from the Doppler equation and to measure the vessel cross-section.

Each method of measurement is subject to errors and it is important that they are appreciated by the user if the best possible accuracy is to be achieved, and if realistic confidence limits are to be placed on the measurements; with care reasonably accurate measurements of blood flow can be made in a variety of vessels.

As with other ultrasound techniques, flow measurement by Doppler ultrasound has the great virtue of not interfering with the parameters being measured, and flows can therefore be measured under normal physiological conditions, and this benefit may far outweigh the cost of any inaccuracies when compared with other methods.

11.9   REFERENCES

Avasthi PS, Greene ER, Voyles WF, Eldridge MW (1984) A comparison of echo-Doppler and electromagnetic renal blood flow measurements. J Ultrasound Med 3, 213–218.
Baker DW (1970) Pulsed ultrasonic Doppler blood-flow sensing. IEEE Trans Sonics Ultrasonics SU-17, 170–185.
Baker DW, Forster FK, Daigle RE (1978) Doppler principles and techniques. In: Ultrasound: its applications in medicine and biology, Part I (Ed. FJ Fry), pp 219–254, Elsevier, Amsterdam.

Borodzinski K, Filipczynski I, Nowicki A, Powalowski T (1976) Quantitative transcutaneous measurements of blood flow in carotid arteries by means of pulse and continuous wave Doppler methods. Ultrasound Med Biol 2, 189–193.

Brandestini M (1978) Topoflow – a digital full range Doppler velocity meter. IEEE Trans Sonics Ultrasonics SU-25, 287–293.

Doriot PA, Casty M, Milakara B, Anliker M, Bollinger A, Siegenthaler W (1975) Quantitative analysis of flow conditions in simulated vessels and large human arteries and veins by means of ultrasound. Excerpta Medica International Congress Series No 363, Proc Second Eur Congr Ultrasonics Med, Amsterdam, pp 160–168.

Eik-Nes SH, Marsal K, Brubakk AO, Kristofferson K, Ulstein M (1982) Ultrasonic measurement of human fetal blood flow. J Biomed Eng 4, 28–36.

Eik-Nes SH, Marsal K, Kristoffersen K (1984) Methodology and basic problems related to blood flow studies in the human fetus. Ultrasound Med Biol 10, 329–337.

Evans DH (1985) On the measurement of the mean velocity of blood flow over the cardiac cycle using Doppler ultrasound. Ultrasound Med Biol 11, 735–741.

Evans DH (1986) Can ultrasonic Duplex scanners really measure volumetric flow? In: Physics in medical ultrasound (Ed. JA Evans), Chapter 19, pp 145–154, IPSM, 47 Belgrave Square, London.

Evans JM, Skidmore R, Wells PNT (1986) A new technique to measure blood flow using Doppler ultrasound. In: Physics in medical ultrasound (Ed. JA Evans), Chapter 18, pp 141–144, IPSM, 47 Belgrave Square, London.

Fish PJ (1978) Doppler vessel imaging and its aid to flow measurement. In: Doppler ultrasound in the study of the central and peripheral circulation (Eds JP Woodcock, RF Sequeira), Chapter 6, pp 50–54, Bristol University Press.

Fish PJ (1981) A method of transcutaneous blood flow measurement – accuracy considerations. In: Recent advances in ultrasonic diagnosis, vol 3 (Eds A Kurjak, A Kratochwil), pp 110–115, Elsevier, Amsterdam.

Fu C-C, Gerzberg L (1983) Annular arrays for quantitative pulsed Doppler ultrasonic flowmeters. Ultrasonic Imaging 5, 1–16.

Furuhata H, Kanno R, Kodaira K, Aoyagi T, Hayashi J, Matsumoto H, Yoshimura S (1978) An ultrasonic blood flow measuring system to detect the absolute volume blood flow. Jap J Med Electron Biol Eng 16, Suppl, 334.

Gill RW (1979) Pulsed Doppler with B-mode imaging for quantitative blood flow measurement. Ultrasound Med Biol 5, 223–235.

Gill RW (1982) Accuracy calculations for ultrasonic pulsed Doppler blood flow measurements. Australasian Phys Eng Sci Med 5, 51–57.

Gill RW (1985) Measurement of blood flow by ultrasound: accuracy and sources of error. Ultrasound Med Biol 11, 625–641.

Gill RW, Kossoff G, Warren PS, Garrett WJ (1984) Umbilical venous flow in normal and complicated pregnancy. Ultrasound Med Biol 10, 349–363.

Griffin DR, Teague MJ, Tallet P, Willson K. Bilardo C, Massini L, Campbell S (1985) A combined ultrasonic linear array scanner and pulsed Doppler velocimeter for the estimation of blood flow in the fetus and adult abdomen. II: Clinical evaluation. Ultrasound Med Biol 11, 37–41.

Groves DH, Powalowski T, White DN (1982) A digital technique for tracking moving interfaces. Ultrasound Med Biol 8, 185–190.

Hartley CJ, Hanley HG, Lewis RM, Cole JS (1978) Synchronized pulsed Doppler blood flow and ultrasonic dimension measurement in conscious dogs. Ultrasound Med Biol 4, 99–110.

Histand MB, Miller CW, Mcleod FD (1973) Transcutaneous measurement of blood velocity profiles and flow. Cardiovasc Res 7, 703–712.

Hoeks APG, Reneman RS, Peronneau PA (1981) A multi-gate pulsed Doppler system with serial data processing. IEEE Trans Sonics Ultrasonics SU-28, 242–247.

Hoeks APG, Ruissen CJ, Hick P, Reneman RS (1985) Transcutaneous detection of relative changes in artery dimensions. Ultrasound Med Biol 11, 51–59.

Hokanson DE, Strandness DE, Miller CW (1970) An echo-tracking system for recording arterial-wall motion. IEEE Trans Sonics Ultrasonics SU-17, 130–132.

Hottinger CF, Meindl JD (1979) Blood flow measurement using the attenuation-compensated volume flowmeter. Ultrasonic Imaging 1, 1–15.

Huntsman LL, Stewart DK, Barnes SR, Franklin SB, Colocousis JS, Hessel EA (1983) Non-invasive Doppler determination of cardiac output in man: clinical validation. Circulation 67, 593–602.

Jorgensen JE, Campau DN, Baker DW (1973) Physical characteristics and mathematical modelling of the pulsed ultrasonic flowmeter. Med Biol Eng 11, 404–421.

Keller HM, Meier WE, Anliker M, Kumpe DA (1976) Non-invasive measurement of velocity profiles and blood flow in the common carotid artery by pulsed Doppler ultrasound. Stroke 7, 370–377.

Lewis P, Psaila JV, Davies WT, McCarty K, Woodcock JP (1986) Measurement of volume flow in the human common femoral artery using a duplex ultrasound system. Ultrasound Med Biol 12, 777–784.

Light H (1974) Initial evaluation of transcutaneous aortovelography – a new non-invasive technique for haemodynamic measurements in the major thoracic vessels. In: Cardiovascular applications of ultrasound (Ed. RS Reneman), Chapter 27, pp 325–360, Elsevier, New York.

Light H, Cross G (1972) Cardiovascular data by transcutaneous aortovelography. In: Blood flow measurement (Ed. VC Roberts), Chapter 11, pp 60–63, Sector, London.

Mackay RS (1972) Non-invasive cardiac output measurement. Microvasc Res 4, 438–452.

Marquis C, Meister JJ, Mirkovitch V, Depeursinge C, Mooser E, Mosimann R (1983) Femoral blood flow determination with a multichannel digital pulsed Doppler: an experimental study on anesthetized dogs. Vasc Surg 17, 95–103.

Olsen CF (1977) Doppler ultrasound: a technique for obtaining arterial wall motion parameters. IEEE Trans Sonics Ultrasonics SU-24, 354–358.

Peronneau P, Xhaard M, Nowicki A, Pellet M, Delouche P, Hinglais J (1972) Pulsed Doppler ultrasonic flowmeter and flow pattern analysis. In: Blood flow measurement (Ed. VC Roberts), Chapter 2, pp 24–28, Sector, London.

Peronneau P, Sandman W, Xhaard M (1977) Blood flow patterns in large arteries. In: Ultrasound in medicine 3B (Eds DN White, RE Brown), pp 1193–1208, Plenum Press, New York.

Qamar MI, Read AE, Skidmore R, Evans JM, Williamson RCN (1985) Transcutaneous Doppler ultrasound measurement of coeliac axis blood flow in man. Br J Surg 72, 391–393.

Safar ME, Peronneau PA, Levenson JA, Toto-Moukouo JA, Simon AC (1981) Pulsed Doppler: diameter, blood flow velocity and volumetric flow of the brachial artery in sustained essential hypertension. Circulation 63, 393–400.

Struyk PC, Pijpers L, Wladimiroff JW, Lotgering FK, Tonge M, Bom N (1985) The time–distance recorder as a means of improving the accuracy of fetal blood flow measurements. Ultrasound Med Biol 11, 71–77.

Teague MJ, Willson K, Battye CK, Taylor MG, Griffin DR, Campbell S, Roberts VC (1985) A combined ultrasonic linear array scanner and pulsed Doppler velocimeter for the estimation of blood flow in the fetus and adult abdomen. I: Technical aspects. Ultrasound Med Biol 11, 27–36.

Uematsu S (1981) Determination of volume of arterial blood flow by an ultrasonic device. J Clin Ultrasound 9, 209–216.

Wang W, Shao Q (1986) Reduced error in double beam Doppler ultrasound flow velocity measurement. Ultrasound Med Biol 12, L413–L414.

Chapter 12

DOPPLER TEST PHANTOMS AND QUALITY CONTROL

12.1 INTRODUCTION

Doppler instruments depend for their operation on small changes in electronic signals, for example the phase changes in the detected echo relative to the internal reference signal from which the transmitted signal is generated. The equipment employed to analyse the Doppler signal to give sonograms or traces for chart recorders is complex. The ultrasonic equipment used for blood flow studies will produce erroneous results if it is only slightly out of calibration. A common example of this is error in the indicated direction of flow.

There are four common ways of checking the performance of a Doppler unit:

1. Recording flow signals from a blood vessel in which the flow pattern is considered to be known
2. Detection of signals from simple moving objects
3. The use of flow test phantoms
4. Electronic checks of signal levels and phases.

The measurement of ultrasonic output power, intensity and pressure amplitude is described in Chapters 5 and 13.

A check of flow in a blood vessel is quick and practical but it suffers from two drawbacks; the flow pattern is not accurately known and it cannot be varied over a wide range to increase the range of the calibration. From a safety point of view regular checks on the same individual are not recommended.

Simple moving objects permit useful tests of constancy of operation but the signal from them is rarely truly representative of that from many blood cells or from variable flow patterns. Flow phantoms are the most attractive option for checking performance. Although they are still at an early stage of development they are sufficiently good for some tests and for checking the constancy of performance at different times. The flow pattern can be known if their structure is kept simple but they rarely simulate flow in elastic or complex arteries. Electronic checks are not convenient in the hospital situation and are the province of the service engineer.

Since the function of Doppler equipment is dependent on subtle changes in electronic signals which can be affected by drifts in the equipment, quality control procedures should be carried out fairly frequently, say once per day. This is more often than is practised in most centres at present, but the situation may change as test objects become more readily available and easier to use. The results that can be obtained by the use of each test phantom should be clearly ascertained since the simulation of flow as encountered *in vivo* may be quite crude. The limitations of each test must also be clearly understood.

12.2   SIMPLE TEST OBJECTS

12.2.1   Rotating disc or cylinder

In their simplest form these consist of a disc or cylinder, of diameter 5–15 cm, whose curved surface has been made rough to act as an acoustic scatterer. The maximum rate of rotation of a 10 cm cylinder is typically 3 rev s^{-1}, giving a rim speed of 94 cm s^{-1}. Rotation is performed in an oil bath which attenuates multiple reflection echoes. A refinement of this test-object is to construct the disc or cylinder from tissue-mimicking material as shown in Fig. 12.1 (McDicken et al 1983). Tissue-mimicking material has been described for test phantoms designed to evaluate ultrasonic pulse echo imaging instruments and takes the form of graphite particles in gelatin or of reticulated foam (Madsen et al 1978, McCarty and Stewart 1982, Lerski et al 1982). Using a test object made of such material, simulated flow signals can be obtained from within the body of the

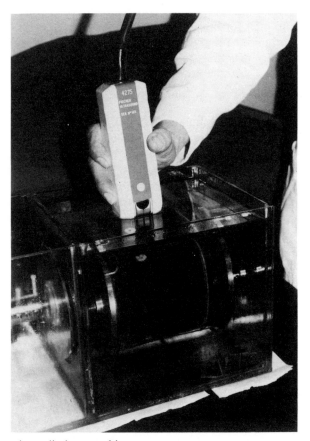

Figure 12.1   A rotating cylinder test object

disc or cylinder. Another advantage of these materials is that the ultrasound is reflected from small scatterers moving through the sample volume as is the case with blood. However, in the design of such phantoms, described to date, the reflectivity of the material is closer to that of soft tissue than blood. Since the tissue-mimicking material attenuates ultrasound, rotation can be performed in oil or water without serious problems arising from multiple reflections.

With this type of test object a known speed can be related to that measured by a Doppler instrument and shown in a sonogram. The accuracy of sample volume registration at several places in the field of view of a duplex imager can be checked by moving the sample volume across the rim of the object and noting the onset of a Doppler sound. The accuracy of the direction-sensing circuitry can be tested by recording signals from different parts of the rotating material. Finally, the moving structure can be used to check the performance of a flow mapping Doppler system and the adequacy of its colour-coded display.

12.2.2 Oscillating piston

Oscillating pistons which have been briefly mentioned in the literature (Reid 1983) present a flat moving surface to the ultrasonic beam. They are used for quick checks. Since at high speeds unwanted vibrations are usually produced in oscillating mechanisms, this approach has little to offer for blood flow testing. It has been described as a means of testing systems that examine cardiac motions and of observing the types of Doppler signal resulting from oscillating structures such as fluttering valves.

12.3 FLOW PHANTOMS

Development of phantoms that simulate flow is still progressing (McCarty and Locke 1986). It has reached a stage where real or artificial blood can be pumped in a continuous or pulsed fashion through a tube surrounded by tissue-mimicking material. Such phantoms represent simplified vascular vessels since the tube diameters and the flow patterns cannot be varied over the ranges encountered in clinical practice. Further advances can be expected relating to the materials used, the variability of the impedance to flow and computer control of the pump. However, even as they stand, flow phantoms are useful for calibrating Doppler systems, i.e. for velocity or quantitative flow checks and for testing flow pattern classification techniques. Details of their structure are discussed in the following sections.

12.3.1 Artificial blood

Real blood, human and animal, has been used in phantoms but, given the problems of infection and difficulty of supply, a blood substitute is normally preferred. Typically, artificial blood is made from small particles suspended in a mix of degassed water and glycerol. A mixture of four parts glycerol to five parts water gives a viscosity of 5 cP at 25°C which is similar to that of blood. The values quoted for the viscosity of blood in large vessels cover a surprising range (McDonald 1974). They depend strongly on the method used for the measurement since the viscosity of blood alters as it flows through different structures, i.e. it is non-Newtonian (Section 2.2.1). Sephadex particles have been widely used as the scattering centres (Fish 1981). A supply of Sephadex, type G25, is available from Pharmacia, Uppsala, Sweden. It has particles of size 20–80 μm which are rather large but which give an acceptable signal provided the tests to be implemented do not depend on the power level of the reflected sound. The diameters of these particles are less than the ultrasound wavelengths encountered in soft tissue for frequencies in the range 1–10 MHz. Ultrasound therefore scatters from these particles in a similar manner to blood.

The optimum type of particle to simulate cells in blood has still to be identified, so test-object designers have used a varied selection. An ideal particle would be 5 μm in diameter, chemically stable, remain in suspension and be damage-resistant during circulation. It would also have acoustic scattering properties similar to those of blood cells. The formation of rouleaux in regions of low velocity or the non-Newtonian properties of real blood are unlikely to be obtained with artificial blood. An example of a scattering particle which has been recommended as a standard for sensitivity calibration is Cellulose Pulver MN300 of size 2–20 μm (Newhouse et al 1982). Other examples of substitutes that have been tried are corn starch (Vera et al 1978, Lefebvre 1981), latex spheres (Arts and Roevros 1972), albumin spheres (De Jong et al 1975), granular polythene (Shirasaka et al 1981, Seo et al 1982). Another approach to artificial blood is the use of an emulsion, e.g. automobile transmission oil in water (Michie and Fried 1973), silicon oil in water (Hassler 1982) and acetate/ethylene copolymer emulsion (Teague et al 1984). An attraction of emulsion is that the particle size can be adjusted by the method of preparation and the suspension can be made stable.

12.3.2 Structure and dimensions

The basic structure of a flow phantom is shown in Fig. 12.2. Latex rubber tubing is often used as a vessel since it has an acoustic impedance similar to that of soft tissue. This property ensures that the ultrasound beam is not badly distorted at the vessel causing

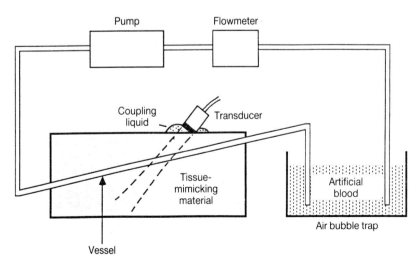

Figure 12.2 A basic flow phantom. Reproduced from McDicken WN (1986), by permission of Pergamon Press Ltd

parts of the lumen to be in shadow. A problem of thin-walled latex tubing is that it does not retain a fixed shape. Heat-shrink sleeving, which is used in electronics to cover cables, has also been used since it can be tailored to different diameters, between 0.5 and 20 mm, and its shape can be varied along its length (McDicken 1986). Changes in shape have been used to make stenoses and exact alterations of diameter. Further work will be required on materials for artificial vessels if they are accurately to mimic blood vessels. To avoid disturbed flow at the point of measurement the ultrasound beam is usually arranged to intersect the tube at a distance greater than 50 diameters from the flow entrance. The tubing is surrounded by tissue-mimicking material such as gelatin loaded with graphite particles (Madsen et al 1978, McCarty and Stewart 1982) or reticulated polyurethane foam immersed in water (Lerski et al 1982). Reticulated foam is versatile in that different thicknesses may be selected to give the desired penetration range. A disadvantage is that, with the need to immerse it in degassed water, the phantoms are difficult to transport.

A desirable feature of a flow phantom is a pump that can deliver up to $5 \, l \, min^{-1}$ through the vessel, and which is driven by a stepping motor. This type of motor can be readily controlled by a microcomputer and can therefore generate pulsatile flow patterns. A magnetic coupling between the motor and the pump ensures that bubbles are not introduced to the liquid at this part of the system. Part of the value of a test object is to be able to check reproducibility of results over a period of time. Simplicity of design and digital control of the pump are therefore highly desirable. Reproducibility is made easier by the fact that the scattering power of the particles in the liquid is not critically dependent on their concentration (Newhouse et al 1982).

12.4   PERFORMANCE OF FLOW PHANTOMS

12.4.1   Signal quality

The Doppler signals from a flow phantom based on the above design sound like flow in arteries and veins and produce realistic sonograms. Figures 12.3a and 12.3b show continuous and pulsed flow sonograms respectively from such a phantom. The flow calibration results shown below confirm the suitability of the signals for test purposes. When no scatterers are included in the circulating liquid a weak fluctuating signal is heard, presumably due to a few particles or bubbles which are not removed by the trap in the system. The addition of scatterers greatly increases the signal level rendering the background level insignificant.

12.4.2   Test procedures

Flow phantoms have several uses:

1. To check accuracy of quantities calculated automatically by Doppler instrumentation, e.g. mean forward or maximum velocity, pulsatility index and quantitative flow
2. To check reproducibility of results from equipment over a period of time
3. To check reproducibility of operator performance over a period of time
4. To educate new operators

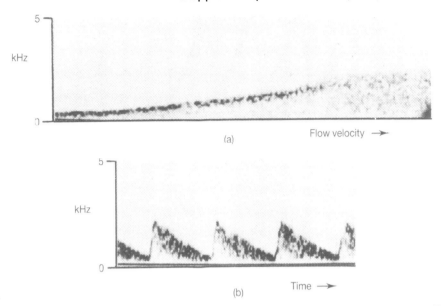

kHz

(a)   Flow velocity →

kHz

(b)   Time →

Figure 12.3 (a) Continuous flow sonogram in which the velocity is steadily increased. The effect of turbulence is seen at high velocities. (b) Pulsed flow sonogram. Both these recordings were made using the phantom illustrated in Fig. 12.2. Reproduced from McDicken WN (1986), by permission of Pergamon Press Ltd

Figure 12.4 shows sonograms on which traces representing the maximum velocity and the intensity-weighted mean forward velocity have been plotted. Similar arterial sonograms can be produced by flow test systems and give a check of the entire Doppler instrumentation, i.e. Doppler unit, recorder and analyser. The value of the calculated indices can also be checked (Law et al 1987).

Quantitative flow values in ml s^{-1} are most easily calibrated by collecting the artificial blood in a measuring jar and timing the duration with a stopwatch. Although superficially crude, this method is in fact accurate to around 1%. Figure 12.5 shows a comparison of ultrasonically and directly measured pulsed flow. These results were obtained using a 3.5 MHz pulsed Doppler device; the mean flow was calculated by noting the maximum velocity from the sonogram and assuming a parabolic velocity profile in the vessel. Similar good agreement between direct and ultrasonically measured flow has been obtained by several investigators (Greene et al 1981, McDicken 1986). The flow rate calculated automatically by a number of machines and based on the intensity-weighted mean velocity has also been shown to be reasonably accurate for several machines. It is worth emphasizing that this measurement is highly dependent on uniform insonation of the vessel (Evans 1982), a state of affairs which is more likely to exist in a test phantom than in genuine anatomy.

12.4.3 Limitations of phantoms

Flow phantoms will probably always be much idealized representations of flow in blood vessels. Their limitations are worth bearing in mind since they reinforce the point that the

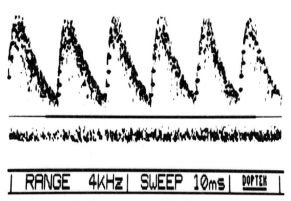

Figure 12.4   Umbilical artery and vein sonograms. Maximum velocity and mean forward velocity in the arterial sonogram are denoted by superimposed dots

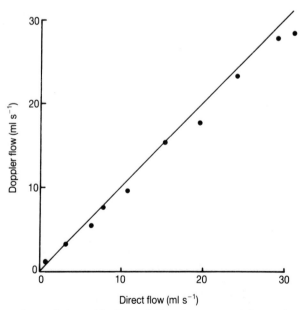

Figure 12.5   Comparison of ultrasonically and directly measured flow using a 3.5 MHz pulsed wave Doppler unit. Reproduced from McDicken WN (1986), by permission of Pergamon Press Ltd

results relate to the acoustic, mechanical and fluid dynamic conditions that exist in the phantom:

1. At present the changing diameter of arteries is not simulated.
2. The influence of proximal and distal impedances or of collateral flow are rarely simulated.
3. Physiological variation between cardiac cycles is not normally considered.
4. The artificial blood does not have identical properties to real blood either in terms of acoustic scattering or viscosity.
5. The surrounding tissue-mimicking material is usually much more uniform than real tissue.

12.5   SAMPLE VOLUME TEST OBJECTS

To interpret and to make some calculations from a Doppler signal it is necessary to know the size and shape of the sample volume. In some applications it may be sufficient to know that the sample volume straddles the vessel in question. It is important to note that the sample volume depends on the effective beam width at the depth of the flowing blood. Assuming that the beam has not been greatly distorted, the effective beam width still depends on the sensitivity settings of the instrument. Increasing the gain, for example, will result in blood cells at the side of the beam contributing to the Doppler signal. The machine settings for which measurements are made should therefore be quoted with the sample volume size.

For a continuous wave unit the sample volume is usually large (Evans and Parton 1980), e.g. 15 cm in length and 2 cm in width for a 2 MHz instrument operating at maximum sensitivity. On the other hand a 2 MHz pulsed unit may have a sample volume of length 1 cm and width 1 cm. The sample volumes of high-frequency devices can be made quite small with dimensions down to around 0.5 mm. The above figures are quoted to give an appreciation of the range of sample volumes to be measured. Sample volumes do not have simple cylindrical shapes. Consideration of the volume of blood cells which will contribute to the echo signal accepted by the instrument during a specific range gate time, shows that the sample volume will be teardrop-shaped if the insonating pulse is a few cycles long and the beam is cylindrical (Baker and Yates 1973). A number of test objects have been employed to measure sample volumes. No one technique has been recognized as the best approach. Indeed the problem of plotting the shape of a small sample volume to obtain a result which would apply to blood cells has still to be fully solved. For this case the test object is required to consist of moving scattering centres in a volume of dimensions around 0.1 mm.

12.5.1   Miniature flow probe

The flow phantom described earlier has been reduced in diameter to 0.5 mm (McDicken 1986) and still provides a readily detected signal even when the flow is perpendicular to the ultrasound beam axis. This signal arises from spectral broadening effects (Section 8.3.5). With a thin-walled tube, distortion of the ultrasonic beam does not appear to be a serious problem. The sample volume is explored by moving the fine tube across the ultrasonic beam. This is most easily undertaken by moving the transducer rather than the tube. The scattering element in this test object comprises particles in liquid moving in a line and thus resembles blood in a vessel. The sample volume plotted is therefore appropriate for blood flowing in a straight line.

12.5.2   Jet flow probe

A fine jet of particles suspended in liquid has been used as a probe to plot sample volume size and shape (Baker and Yates 1973) (Fig. 12.6). The jet comprises a solution by volume of 0.2% silicone antifoam emulsion in distilled water. The jet forms a line of moving particles under water in the test tank. An exhaust tube collects the jet so that the particles do not contaminate the water of the test tank. Using a jet rather than a fine tube eliminates the risk of reflections in the walls of the tube affecting the results. Filtering was employed to keep the size of the particles in the emulsion to less than 20 μm. The diameter of the jet can be made 0.25 mm over a length of 2–3 cm.

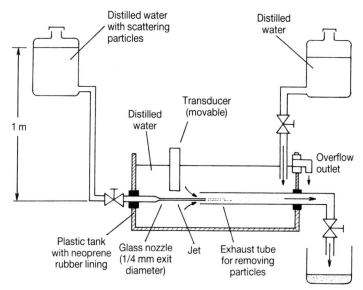

Figure 12.6   Test probe utilizing a jet of fine particles in a water tank. (Reproduced from Baker DW, Yates WG (1973), by permission of the International Federation for Medical and Biological Engineering)

12.5.3   Moving string test target

A number of investigators have made a small moving linear target from string or thread. The phantom described by Walker *et al.* (1982) is shown in Fig. 12.7. Surgical silk thread was used as a scatterer of ultrasound. Two strings could be arranged to pass through the same volume while moving in the same or opposite directions at the same or different speeds. The angle of the string to the ultrasound beam could be varied from 0° to 90°. In addition to measuring the sample volume length and width at different ranges, this device was also employed to check the accuracy of the direction-sensing circuits and the accuracy of the sample volume registration in a duplex system image. Since the target is moved at a

Figure 12.7 A moving string test object. (Reproduced from Walker AR, Phillips DJ, Powers JE (1982), by permission of John Wiley & Sons, Inc)

well defined speed, the performance of frequency spectrum analysers may also be checked. The main difficulty with this type of phantom is the elimination of vibrations in the string.

12.5.4 Moving point target

It is argued that a moving line of targets is valid for delineating a sample volume, since blood flow is often laminar in practice and the random spacing of the scatterers simulates blood cells. However, a moving point target also has attraction for the detailed plotting of a sample volume since the detected sound is not the summed contribution from several points. It is therefore easier to understand the factors that contribute to the final result. If symmetry is assumed in the shape of the sample volume the number of required measurements can be reduced to one axial and one lateral plot. A small vibrating sphere, diameter 0.8 mm, supported on a fine wire, diameter 0.01 mm, has been employed to plot sample volumes (Hoeks et al 1984) (Fig. 12.8). The sphere vibrating at a frequency of 500 Hz modulates the ultrasound which is detected in the Doppler unit. No significant signal is detected from the fine wire. Both axial and lateral shapes of the sample volume have been measured with this device. These investigators have also demonstrated that the lateral plot obtained with a vibrating sphere and a Doppler unit is very similar to that from a static sphere measured using a pulse echo plotting technique. The same type of comparison cannot be made in the axial direction since pulse echo signals are obtained all along the axis and not just in the sample volume.

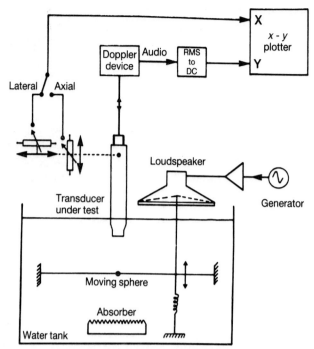

Figure 12.8   A moving point test object. (Reproduced from Hoeks APG, Ruissen CJ, Hick P, Reneman RS (1984), by permission of Pergamon Press)

12.6 QUALITY CONTROL

The result obtained from a Doppler examination is highly dependent on a number of technical as well as clinical factors. Steps are therefore necessary to limit the variability of results due to technical factors. This is best done by examining a phantom in which the flow conditions are exactly known. Given the large number of factors and the slight degree of change in the circuitry which can affect the results, a quick check of the system performance prior to each clinical session is desirable. This check should be carried out with the settings of the controls at typical working positions which can be accurately set. The lack of simple reliable test objects makes this difficult at present and most investigators only carry out crude tests. Nationally and internationally accepted standards are just being drawn up. The whole subject of reproducibility of results, from both a technical and a clinical point of view, is rather neglected. The technical factors that affect the outcome are:

1. Ultrasound frequency.
2. The sensitivity settings (gain, TGC, transmitted power, electronic thresholds, filters). Some of these controls are to be found on the analysing and recording equipment as well as the Doppler instrument.
3. Sample volume size. It should be borne in mind that this factor is highly dependent on the sensitivity settings and on the position of the sample volume along the beam.
4. The brilliance and contrast of the display, chart record or film on which the sonogram is presented.

Fortunately the determination of the maximum velocity from a sonogram and hence the shape of the corresponding waveform is not too dependent on machine settings. This is not true of spectral broadening or quantitative flow measurement which depends on mean velocity.

12.7 SUMMARY

In this chapter details of a number of test devices have been provided. These devices can be used with all types of Doppler instrument. The value and also the limitations of test phantoms have been emphasized. The need for quality control has been noted particularly since many factors can influence the output from a Doppler unit. Doppler test phantoms are now becoming commercially available.

12.8 REFERENCES

Arts MGJ, Roevros JMJG (1972) On the instantaneous measurement of bloodflow by ultrasonic means. Med Biol Eng 10, 23–34.
Baker DW, Yates WG (1973) Technique for studying the sample volume of ultrasonic Doppler devices. Med Biol Eng 11, 766–770.
De Jong DA, Megens PHA, De Vlieger M, Thon H, Holland WJP (1975) A directional quantifying Doppler system for measurement of transport velocity of blood. Ultrasonics 13, 138–141.
Evans DH (1982) Some aspects of the relationship between instantaneous volumetric blood flow and continuous wave Doppler ultrasound recordings. 1. The effect of ultrasonic beam width on the output of maximum, mean and rms frequency processors. Ultrasound Med Biol 8, 605–609.
Evans DH, Parton L (1980) The directional characteristics of some ultrasonic Doppler blood flow probes. Ultrasound Med Biol 6, 51–62.
Fish PJ (1981) A method of transcutaneous blood flow measurement – accuracy considerations. In: Recent advances in ultrasound diagnosis 3 (Eds A Kurjak, A Kratochwil), pp 110–115, Excerpta Medica, Amsterdam.
Greene ER, Venters MD, Avasthi PS, Conn RL, Jahnke RW (1981) Noninvasive characterization of renal artery blood flow. Kidney Int 20, 523–529.
Hassler D (1982) Messung der blutgeschwindigkeit, des blut-volumenstromes und der aderquerschnittsflache nach der integralen ultraschall-Dopplermethode-vergleich und synthese zweier lösungen. Ultraschall Med 3, 24–29.
Hoeks APG, Ruissen CJ, Hick P, Reneman RS (1984) Methods to evaluate the sample volume of pulsed Doppler systems. Ultrasound Med Biol 10, 427–434.
Law YF, Cobbold RSC, Johnston RW, Bascom PAJ (1987) Computer-controlled pulsatile pump system for physiological flow simulation. Med Biol Eng Comp 25, 590–595.
Lefebvre JP (1981) Mésure du profile de vitesse d'un écoulement stationnaire axisymmétrique par anémometrie Doppler ultrasonore en mode continu. J Mécanique 20, 827–848.
Lerski RA, Duggan TC, Christie J (1982) A simple tissue-like ultrasound phantom material. Br J Radiol 55, 156–157.
McCarty K, Stewart M (1982) A simple calibration and evaluation phantom for ultrasound scanners. Ultrasound Med Biol 8, 393–401.
McCarty K, Locke DJ (1986) Test objects for the assessment of the performance of Doppler shift flowmeters. In: Physics in medical ultrasound (Ed. JA Evans), pp 94–106, IPSM, 47 Belgrave Square, London.
McDicken WN (1986) A versatile test-object for the calibration of ultrasonic Doppler flow instruments. Ultrasound Med Biol 12, 245–249.
McDicken WN, Morrison DC, Smith DSA (1983) A moving tissue-equivalent phantom for ultrasonic real-time scanning and Doppler techniques. Ultrasound Med Biol 9, L455–L459.
McDonald DA (1974) Blood flow in arteries, 2nd Edn, Edward Arnold, London.

Madsen EL, Zagzebski JA, Banjavic RA, Jutila RE (1978) Tissue mimicking materials for ultrasound phantoms. Med Phys 5, 391–394.

Michie DD, Fried WI (1973) An *in vitro* test medium for evaluating clinical Doppler ultrasonic flow systems. J Clin Ultrasound 1, 130–133.

Newhouse VL, Nathan RS, Hertzler LW (1982) A proposed standard target for ultrasound Doppler gain calibration. Ultrasound Med Biol 8, 313–316.

Reid JM (1983) Methods of measuring the performance of continuous-wave ultrasonic Doppler diagnostic equipment. Draft IEC standard, Sub-committee 29D, Working Group 10.

Seo Y, Hongo H, Komatsu K, Sasaki H, Shirasaka T, Iinuma K (1982) The blood flow phantom for ultrasonic pulsed Doppler system. Ultrasound Med Biol 8, Suppl 1, 176.

Shirasaka T, Hongo H, Seo Y, Sasaki H, Iinuma K (1981) Quantitative blood volume measurement by pulsed Doppler method. Jap Soc Ultrasound Med November, 557–558.

Teague SM, von Ramm OT, Kisslo JA (1984) Pulsed Doppler spectral analysis of bounded fluid jets. Ultrasound Med Biol 10, 435–441.

Vera JC, Lefort MF, L'Huillier JP, Stoltz JF (1978) Contribution a l'étude du debit sanguin par analyse spectrale basses fréquences, des signaux Doppler ultrasonores application à un modèle d'écoulement permanent. Biorheology 15, 181–191.

Walker AR, Phillips DJ, Powers JE (1982) Evaluating Doppler devices using a moving string test target. J Clin Ultrasound 10, 25–30.

Chapter 13

SAFETY CONSIDERATIONS IN DOPPLER ULTRASOUND

13.1 INTRODUCTION

Ultrasound is a form of energy and as with any form of energy if its level is increased it will eventually produce effects in tissue, some deleterious. When developing both equipment and techniques the aim should be to keep the exposure of the patient to the minimum commensurate with obtaining a diagnosis. In the event of there being an identifiable risk from a procedure, the user then has to attempt the difficult task of weighing up benefit versus risk. For more than a decade diagnostic ultrasound has been considered safe if the time-averaged intensity in the region of the maximum in the beam, often the focus, is less than $100 \, \text{mW cm}^{-2}$. Although this is a useful upper limit for intensity, it represents a rather oversimplified view of the question of potential hazard. The whole subject is discussed in further detail below. It is worth noting that the outputs of some pulsed Doppler units now exceed the $100 \, \text{mW cm}^{-2}$ level. If high intensity levels are employed the user should certainly be aware of it and be able to justify the levels used.

The aim of this chapter is to provide a summary of the material relevant to the bioeffects of ultrasound and to give a list of references for further study. There is considerable activity in this field at present so the user of ultrasonic equipment is obliged to keep up to date with current opinion and to study reports which are put out by a number of organizations.

13.2 ULTRASONIC FIELD MEASUREMENTS

Virtually all measurements of the parameters of ultrasonic fields have been performed with water as the propagating medium. If it is desired to know the value of a parameter such as intensity at a certain depth in tissue, the corresponding value in water is used with allowance being made for attenuation in tissue. The accuracy of this is questionable since it does not take into account the often very variable degradation of an ultrasound beam by different tissue structures. In addition, dispersive attenuation and non-linear propagation are rarely included in the calculation. Nevertheless a simple extrapolation from water to tissue is a practical way of estimating field parameters in tissue.

Extensive work has been undertaken to develop instruments capable of accurate measurement of ultrasonic fields. If care is taken it is now possible to measure power, intensity and pressure amplitude accurately (Brendel 1985, Preston et al 1983, Stewart

1982). These instruments are commercially available. A more detailed discussion of ultrasonic field measurement is given in Chapter 5.

13.2.1 Power

The power of an ultrasonic beam is the rate of flow of energy through the cross-sectional area of the beam. Power at diagnotic levels is most often measured using a force (radiation pressure) balance. With a carefully designed balance, power can be measured down to levels around 1 mW with an accuracy of 0.1 mW in ideal conditions. The force balance works on the principle that when an ultrasonic beam is completely absorbed by a target it exerts a force of P/c on the target (where P is the power of the beam and c is the velocity of sound in the propagating medium). If the target completely reflects the ultrasound, the force on it is $2P/c$. A number of force balances have been described over the years (Freeman 1963, Kossoff 1965, Hill 1970). They have now been reduced in size to make them portable and therefore much more convenient for calibrating medical instruments (Farmery and Whittingham 1978, Duck et al 1985). Figure 13.1 shows a portable power balance.

Figure 13.1   A portable ultrasonic radiation balance for the measurement of power. The ultrasonic beam is directed at the target through the thin membrane window. (By courtesy of Doptek Ltd, Chichester, UK)

13.2.2 Pressure amplitude

The pressure amplitude of a CW or PW beam is of considerable interest in hazard studies since it can be measured throughout the beam and is of direct relevance in the explanation of non-thermal biological effects.

When the pressure amplitude of a CW ultrasound beam is to be measured the position in the beam should be specified. An ultrasonic beam can be represented schematically as shown in Fig. 13.2. The pressure amplitude is often measured near the transducer or at the spatial peak (p_{sp}), which is commonly the focus, though it may be the largest axial maximum of an unfocused beam. The pressure may also be spatially averaged across the beam at the specified range (p_{sa}).

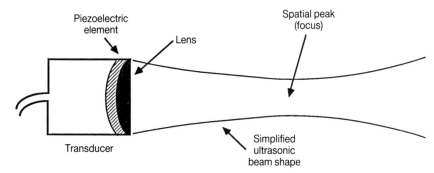

Figure 13.2   Schematic representation of an ultrasonic beam. The spatial peak intensity does not necessarily coincide with the focus

Prior to measuring the pressure amplitude of a pulsed ultrasonic wave it is necessary to define the exact quantity to be measured. The pressure amplitude may be measured at the spatial peak and at the temporal peak of the pulse, p_{sptp}, or it may be averaged over the cross-section of the beam and over the duration of the whole exposure, p_{sata}. Another common measurement is the pressure amplitude at the spatial peak averaged over the pulse length, p_{sppa}. Other combinations of peak and average quantities are possible as shown in Table 13.1. The quantities of greatest interest from the point of bioeffects are p_{sp}, p_{sptp} and p_{sppa}. When the ultrasonic waveform is not symmetrical about the ambient

Table 13.1   Some parameters used to specify an ultrasonic field. The pressure parameters may be quoted for both positive and negative amplitudes

Pressure (spatial peak)	p_{sp}
Pressure (spatial average)	p_{sa}
Pressure (spatial peak, temporal peak)	p_{sptp}
Pressure (spatial average, temporal average)	p_{sata}
Pressure (spatial peak, temporal average)	p_{spta}
Pressure (spatial average, temporal peak)	p_{satp}
Pressure (spatial peak, pulse average)	p_{sppa}
Pressure (spatial average, pulse average)	p_{sapa}
Intensity (spatial peak)	I_{sp}
Intensity (spatial average)	I_{sa}
Intensity (spatial peak, temporal peak)	I_{sptp}
Intensity (spatial average, temporal average)	I_{sata}
Intensity (spatial peak, temporal average)	I_{spta}
Intensity (spatial average, temporal peak)	I_{satp}
Intensity (spatial peak, pulse average)	I_{sppa}
Intensity (spatial average, pulse average)	I_{sapa}
Intensity averaged over largest half-cycle in pulse	I_m

pressure level, the above quantities may be measured for both positive and negative pressure amplitudes.

It was seen in Chapter 5 that hydrophones exist for the accurate measurement of pressure amplitude though great care is required particularly in the case of p_{sptp}. The values obtained for quantities that involve averaging depend on how pulse length and beam width are defined. Two or three definitions are in common usage and fortunately they produce similar results for the ultrasonic fields encountered in diagnostic ultrasound (Duck et al 1985).

13.2.3 Intensity

The intensity of a continuous or pulsed wave beam is of interest since it is directly related to thermal biological effects. It is normally calculated by inserting the square of the pressure amplitude into a simple formula:

$$I = p^2/2c \qquad\qquad 13.1$$

This formula is valid only at points in the beam where pressure and particle velocity are in phase, i.e. at the focus and in the far field but not in the near field of the transducer (Preston 1985).

Just as for pressure, the particular intensity to be measured must be precisely specified. Derived from the pressure amplitudes defined above are corresponding intensities (Table 13.1). Of most interest in the field of bioeffects are I_{sp}, I_{sata}, I_{spta} and I_{sppa}.

13.3   ULTRASONIC OUTPUT FROM DOPPLER UNITS

13.3.1 CW Doppler

Table 13.2 presents typical pressure amplitude, power and intensity values for CW instruments. It should not be assumed that a Doppler unit will have output values close to the centre of these ranges. Intensity values have been reported over a wide range (Duck et al 1987). The lower intensities correspond to fetal heart detectors, the higher to vascular units.

Table 13.2   Some typical output values for pressure amplitude, intensity and power from CW and PW units. Values outside the quoted ranges can also be encountered. Data from Duck FA, Martin K (1986) Br Med Ultrasound Soc Bull 40

CW Doppler	
Spatial peak pressure (p_{sp})	0.01–0.12 MPa
Spatial peak intensity (I_{sp})	6–455 mW cm^{-2}
Power (P)	1–84 mW
PW Doppler	
Spatial peak temporal peak pressure (p_{sptp})	0.1–6.5 MPa
Spatial peak temporal average intensity (I_{spta})	55–825 mW cm^{-2}
Power (P)	1.5–100 mW

13.3.2 PW Doppler

Table13.2 also shows pressure amplitude, power and intensity values for pulsed wave instruments. Not a great deal of data is available in the literature for these devices but again the values of intensity vary over a wide range.

13.3.3 Doppler imaging (flow mapping)

Doppler imaging instruments are being increasingly used in clinical practice. Where the CW or PW ultrasonic beam is moved by hand, i.e. relatively slowly, the intensity in a particular region of tissue may be taken to be the same as that for a static beam (Table 13.2) though the duration of the exposure will usually be less than 1 s at each point. For real-time Doppler flow mapping where the beam moves quickly the number of pulses per second transmitted along a specified beam direction is typically one-twentieth of the number transmitted in a static pulsed Doppler beam. The time-averaged intensity in any one direction is therefore also reduced to one-twentieth of that of a static beam. A contribution from neighbouring beams should be included which results in this intensity being increased by a factor in the range 2 to 10. The intensity (I_{spta}) from a Doppler flow mapping unit is therefore typically one-tenth to one-half of that from a static pulsed Doppler device. For a sector scanner the contribution from neighbouring beams is obviously dependent on the depth of the point considered. The pressure amplitude of CW and PW imaging devices is not influenced by the fact that the beam is moving and has similar values to the static beam values quoted in Table 13.2. The output power along a slow-moving beam can be taken to be the same as that of a static beam whereas the power of a real-time flow mapping beam is reduced by one-tenth to one-half compared to a static beam, just as intensity is.

13.3.4 Duplex systems

When a Doppler instrument is part of a duplex system, the patient is obviously receiving ultrasound in two ways. The interaction of the pulse from the real-time B-scanner is of most relevance to bioeffects in terms of its pulse pressure amplitude. Heating is not thought to be a problem with this type of scanner. On the other hand, with a CW or a PW Doppler unit, both pressure and intensity are of interest since both non-thermal and thermal effects are feasible. The outputs of Doppler instruments in duplex systems are similar to those of static beam devices (Table 13.2).

13.3.5 Fetal monitoring

Fetal monitoring instruments are designed with divergent beams and also low power and intensity outputs since the duration of their use may be several hours. The power output from a monitoring machine should be less than 10 mW.

13.3.6 Cuff pressure measuring instruments

The outputs from Doppler devices in cuff pressure measuring instruments are as for other CW Doppler blood flow units (Table 13.2).

13.4   PHYSICAL EFFECTS OF ULTRASOUND

13.4.1   Mechanisms

Ultrasound interacts with tissue via three well established mechanisms, namely heating, streaming and cavitation (Nyborg 1985, Williams 1983). Heating results when the orderly vibrational energy of the wave is converted into the random vibrational motions of heat by absorption processes described in Chapter 3. Streaming occurs as a result of the radiation force which is generated when the wave energy is absorbed by a liquid (Chapter 3). Micro-streaming also takes place in the vicinity of small oscillating bubbles as discussed below in relation to cavitation.

Cavitation takes two forms; stable and transient. Stable cavitation is a phenomenon whereby small bubbles in a liquid grow under the influence of the wave until they reach a particular size at which resonant oscillations occur. Transient cavitation takes place at higher intensities and is due to the creation and rapid growth of a bubble during the low-pressure half of the wave cycle followed by a violent collapse in the high-pressure half.

There is some evidence that other mechanisms may be involved though they have not been precisely described and are grouped together under the collective heading of 'direct' mechanisms. If an effect is seen under conditions where heating, streaming or cavitation is not considered to be significant the effect is attributed to a direct mechanism. Examples of this type of mechanism are the muscle stimulation reported by Forester et al (1982) and the functional changes in the cochlea described by Barnett (1980).

13.4.2   Therapeutic levels

The effects of heating, streaming and cavitation have been extensively studied at therapeutic intensity and the results can be reproduced (National Council on Radiation Protection and Measurement 1983, British Journal of Radiology 1986). Therapeutic levels are typically 3 W cm^{-2} (I_{sata}) of 3 MHz ultrasound applied for 5 minutes. This could result in a temperature rise of several degrees Celsius in tissue (Lehmann and Guy 1972). In practice the rise is highly dependent on the vascularity of the tissue. Haemolysis *in vivo* due to heating has been demonstrated above 3 W cm^{-2} at 3.4 MHz (Williams et al 1986). Evidence for the formation of cavitation bubbles *in vivo* with therapy ultrasound has been reported (ter Haar and Daniels 1981). Cavitation can affect biological material by temperature elevation, mechanical stress and free radical generation. Streaming of liquid in the vicinity of oscillating bubbles or needles has been shown to disrupt cells (Gershoy and Nyborg 1973). Studies of bioeffects at therapy levels emphasize that outputs of Doppler instruments should be kept below 1 W cm^{-2}

13.4.3   Diagnostic levels

At the levels of ultrasound normally encountered in diagnostic echo imaging, heating is not usually thought to be of concern in mammalian tissue. Temperature rises of up to 1°C can be accommodated by the body since they are experienced in normal daily variations. However, recently some echo imaging machines have been shown to generate average intensities in excess of 100 mW cm^{-2} (I_{spta}) (Duck et al 1985) which has generally been regarded as a maximum acceptable level (AIUM 1984). Also the outputs of several pulsed Doppler units are in the neighbourhood of 500–1000 mW cm^{-2} (I_{spta}). These levels are approaching those of therapy machines which can produce temperature rises of a few degrees Celsius in 2–3 minutes.

Streaming is not thought to be a likely occurrence in tissue with diagnostic ultrasound. It is most easily set up in *in vitro* preparations, for instance where stable cavitation is generated.

The growing awareness of non-linear propagation of diagnostic pulses and the improved accuracy of hydrophones have caused the values of the pulse pressure amplitudes from imaging equipment to be revised upwards recently (Bacon 1984, Duck and Starritt 1984, Muir and Cartensen 1980). It is therefore not possible to say categorically that cavitation does not occur with imaging or Doppler beams in tissue. One theoretical study has concluded that small bubble nuclei suspended in liquid could be made to generate cavitation when irradiated by short ultrasonic pulses as utilized in diagnosis (Flynn 1982). This has been confirmed experimentally in water (Crum and Foulkes 1986, Carmichael et al 1986).

13.5   BIOEFFECTS OF ULTRASOUND

A large number of publications have been produced on the biological effects of ultrasound on molecules, subcellular structures, cells, individual tissues and whole animals. The subject has been treated from the point of view of almost all biological subjects, i.e. biochemistry, physiology, genetics and epidemiology. Several textbooks have appeared on the subject (Nyborg and Ziskin 1985, Williams 1983). Most of the confirmed results are related to therapy levels of ultrasound and there is a great lack of reliable data for the diagnostic situation. It is not possible to extrapolate from the therapy results to lower intensity levels since many ultrasonic processes are likely to have thresholds. The best known example is the transient cavitation threshold which is highly dependent on the conditions in the propagating medium but is well defined. References to selected papers are listed at the end of the chapter. The subjects that receive most attention are briefly outlined below.

13.5.1   Chromosome damage

In the early 1970s the main topic of study was the possibility of chromosome damage. A few positive results were reported but these could not be confirmed by other investigators (Macintosh and Davey 1970, Boyd et al 1971, Hill et al 1972, Brock et al 1973). In retrospect it is felt that too much attention was paid to this work since the high level of damage reported would have led to cell death and not development defects (Thacker 1973).

13.5.2   Sister chromatid exchanges

In the early 1980s sister chromatid exchanges (SCEs) were used to look for an effect of ultrasound at diagnostic levels. An SCE is the interchange of equivalent parts of the chromosome (Fig. 13.3). It can be detected using fluorescent labelling techniques. The process can occur spontaneously without the influence of any agent and its significance is not fully understood. Certain chemicals are well established agents of this effect and are used as controls in experimental practice, e.g. mytomycin. The SCE test is reported to be

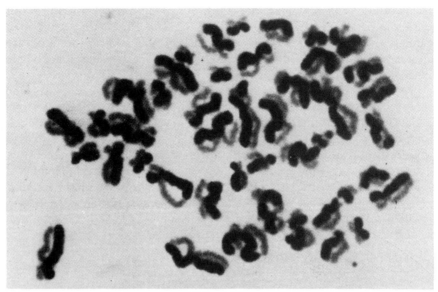

Figure 13.3 Chromatid staining and demonstration of sister chromatid exchanges. (Reproduced from Jacobson-Kram D (1984) J Clin Ultrasound 12, 5–10, by permission of John Wiley & Sons, Inc)

about 100 times more sensitive than chromosome damage but this depends on the agent causing the exchange; surprisingly ionizing radiation is not particularly effective.

Both continuous and pulsed beams have been employed in the investigation of SCEs. The dosimetry and biology have been of variable quality. To date about four studies have reported positive results whereas more than 20 have reported negative ones (Miller 1985, Goss 1984).

13.5.3 Growth retardation

Growth retardation has been reported in several studies for fetal mice and rats irradiated by therapeutic levels of ultrasound, e.g. CW, 1 MHz, I_{sa} 0.5–5.5 W cm^{-2}, exposure time 10–300 s (O'Brien 1983, 1984). A radiation regimen such as this can produce quite large temperature rises in the mother and fetus, e.g. 5°C in a few minutes. A dependence of fetal weight on exposure was identified in this work. Extrapolation of this dependence to diagnostic conditions showed that fetal weight reduction would be expected to be quite small and possibly biologically insignificant. Further work is required to see if the extrapolation to diagnostic levels is valid.

13.5.4 Damage to blood vessels

Dyson et al (1974) have demonstrated damage to the inside wall of blood vessels by transmitting therapeutic levels of CW and PW ultrasound (1–5 MHz, up to 12 W cm^{-2} peak intensity, 15 min irradiation time) in an arrangement that created standing waves.

These are obviously conditions different from those found in clinical examinations but they do serve to remind designers and users of an upper limit to acceptable ultrasonic field strengths.

13.5.5 Cavitation in moving blood

There has been a suggestion that moving blood was a medium in which micrometre-sized bubbles could be spontaneously produced and that these bubbles might cavitate under the influence of an ultrasonic wave (Williams 1985). Similar fears have been expressed with regard to contrast agents employing microbubbles. However in subsequent work it has been proven that microbubbles do not exist in blood *in vivo* and that even artificially injected bubbles do not generate stable cavitation when irradiated by CW and PW ultrasound of spatial peak intensities up to 16 W cm^{-2} and frequencies in the range 0.5–1.6 MHz (Williams et al 1985).

13.5.6 Epidemiological surveys

The most convincing evidence for the safety of ultrasound would be from large-scale population surveys. There are many obstacles to this approach, e.g. the difficulty of obtaining a control population and in guaranteeing accurate dosimetry. The latter is virtually impossible in retrospective studies. Rapidly changing techniques present another major hurdle. A number of small surveys have been reported, typically on around 200–2000 patients (Lyons 1986, Hellman et al 1970, Kinnier Wilson and Waterhouse 1984, Cartwright et al 1984). They are limited from a statitical point of view and all that can be said from them is that diagnostic ultrasound does not cause gross effects. A few large-scale studies are being attempted and preliminary results indicate that no deleterious effects can be attributed to diagnostic ultrasound (Lyons et al 1980). It remains to be seen if they can avoid the major difficulties and pitfalls in this field.

13.6 SAFETY SURVEYS

13.6.1 Summary of survey conclusions

Several organizations have set up working parties to search and evaluate the world literature on bioeffects and safety. These organizations include the World Health Organization, The American Institute of Ultrasound in Medicine, the European Federation of Societies for Ultrasound in Medicine and Biology, and the Japanese Society of Ultrasonics. The available reports are referenced at the end of the chapter. These organizations have all come to the conclusion that there have been no confirmed deleterious effects of diagnostic ultrasound. Most surveys also highlight the severe lack of reliable and confirmed results on the effects of low-intensity ultrasound.

The AIUM issued a new statement on safety in December 1987:

> In the low megahertz frequency range there have been (as of this date) no independently confirmed significant biological effects in mammalian tissues exposed *in vivo* to

unfocused ultrasound with intensities[a] below 100 mW cm^{-2}, or to focused ultrasound[b] with intensities below 1 W cm^{-2}. Furthermore, for exposure times[c] greater than 1 second and less than 500 seconds (for unfocused ultrasound) or 50 seconds (for focused ultrasound) such effects have not been demonstrated even at higher intensities, when the product of intensity and exposure time is less than 50 J cm^{-2}.

[a]Free-field spatial peak, temporal average (SPTA) for continuous wave exposures and for pulsed mode exposures with pulses repeated at a frequency of greater than 100 Hz.
[b]Quarter-power (-6 dB) beam width smaller than four wavelengths or 4 mm, whichever is less at the exposure frequency.
[c]Total time includes off-time as well as on-time for repeated pulse exposures.

Higher intensity is quoted for focused fields since heat is lost more readily from small focal zones. The majority of Doppler fields fall into the unfocused category.

It should be remembered that the statement is based on limited information (Hill and ter Haar 1982). For example, results are scarce on short pulse exposures and also those of long duration at low intensity. Most of the information is related to small mammals rather than to man. It was not intended that the 100 mW cm^{-2} level should be regarded as a threshold level for bioeffects below which safety is guaranteed for every application. It is more a guideline, which may be altered as new knowledge is accumulated. Techniques using ultrasound above this level should be used in the full knowledge of the intensity being transmitted. It can be seen from Table 13.2 that this situation often arises with pulsed Doppler units.

A number of instrument designers state that imaging units and pulsed Doppler devices can be made satisfactorily with output intensities (I_{spta}) below 100 mW cm^{-2}. However, a number of manufacturers are unhappy with this restriction. As a result, the Food and Drug Administration of America has permitted the supply in the USA of instruments that comply with new regulations which allow higher outputs. Different upper limits are permitted for different applications. These new regulations specify the intensity limit a machine may generate at a typical tissue depth for each application – the *in situ* intensity. Another factor which determines the new allowable *in situ* intensity levels is that no new instrument should generate higher intensities than were in use prior to 28 May 1976, the 'pre-enactment date'. Table 13.3 lists the *in situ* intensities allowable for different applications and the corresponding levels in water where attenuation is not significant. When an output intensity is measured in water, the *in situ* value may be calculated from it to see if it is below the permitted level. These regulations are still under discussion.

Table 13.3 Maximum *in situ* values of ultrasonic intensity (I_{spta}) recommended by the Food and Drug Administration (U.S.A.)

Use	I_{spta} **in situ** (mW cm^{-2})	I_{spta} **in water** (mW cm^{-2})
Cardiac	430	730
Peripheral vessel	720	1500
Ophthalmic	17	68
Fetal imaging and other*	94	180

*Abdominal, intraoperative, paediatric, small organ (breast, thyroid, testes), neonatal cephalic, adult cephalic.

13.6.2 Intensity versus exposure time diagrams

In hazard assessments two quantities are of primary importance, the magnitude of the quantity describing the radiation, e.g. intensity or pressure amplitude, and the duration of the exposure. A number of investigators have drawn up diagrams of intensity versus exposure time indicating combinations of these two quantities for which bioeffects have been reported. Figure 13.4 has been derived from the work of these investigators. The boundary between the bioeffect zone and the no-effect zone is shown as a shaded region to emphasize that for most effects thresholds have not been established. The diagnostic ultrasound zone into which most CW and PW Doppler examinations fall is also indicated. The box indicated by the dashed lines shows the region *above* the 100 mW cm^{-2} level in which the outputs of some diagnostic instruments now fall.

A similar diagram should be drawn for wave pressure amplitude versus number of transmitted pulses but the required information is not available at present to do this.

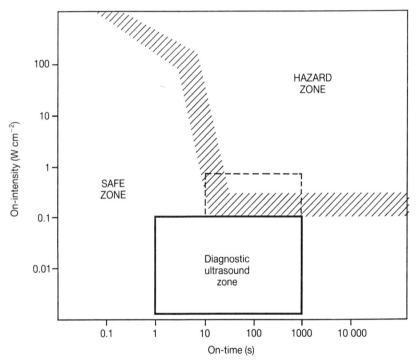

Figure 13.4 Intensity versus exposure time diagram. The box indicated by the dashed line shows the region above the 100 mW cm^{-2} level in which some diagnostic intensities now fall

13.6.3 Minimizing patient exposure

Since the factors relevant to biological effects in patients have not been identified for low-intensity ultrasound, it is not possible to calculate the dose to the patient as might be done for ionizing radiation. However, the prudent use of ultrasound dictates that the exposure of the patient should be minimized. A number of simple points add up to good practice:

1. Use the lowest transmitted power that will give a result.

2. Use the minimum duration of examination.
3. Use the lowest PRF of the pulsed Doppler unit that will allow the highest velocity to be measured.
4. Use CW rather than PW Doppler if it will give a result.
5. Do not leave the Doppler beam irradiating a particular region for longer than is necessary. With a duplex system, switch back to the imaging mode as soon as the Doppler recording is completed.

These measures can easily reduce the total amount of ultrasonic energy delivered to the patient by as much as a factor of 1000.

13.7   SUMMARY

In this chapter the output parameters that specify the transmitted ultrasonic field were defined. Values of these parameters as found in practice were listed. To acquaint the user of Doppler techniques with biological effects, a brief survey of interaction mechanisms and fields of study relevant to safety was given. Finally current opinion on the safety of diagnostic ultrasound was summarized and procedures for minimizing the dose to the patient were presented. The reference list provides further in-depth reading material for this extensive field.

13.8   REFERENCES

AIUM (1984) Safety considerations for diagnostic ultrasound, American Institute of Ultrasound in Medicine, Bethesda, Maryland.

Bacon DR (1984) Finite amplitude distortion of the pulsed field used in diagnostic ultrasound. Ultrasound Med Biol 10, 189–195.

Barnett SB (1980) Structural and functional changes in the cochlea following ultrasonic irradiation. Ultrasound Med Biol 6, 25–32.

Boyd E, Abdulla U, Donald I, Fleming JEE, Hall A, Ferguson-Smith MA (1971) Chromosome breakage and ultrasound. Br Med J 2, 501–502.

Brendel K (1985) Ultrasonic power measurement in liquids in the frequency range 0.5 MHz to 25 MHz. International Electrotechnical Commission Draft TC29 29D WG8.

British Journal of Radiology (1987) Supplement 20, The safety of diagnostic ultrasound (Ed. PNT Wells), The British Institute of Radiology, London.

Brock RD, Peacock WJ, Geard CR, Kossoff G, Robinson DE (1973) Ultrasound and chromosome aberrations. Med J Austral 2, 533–536.

Carmichael AJ, Mossoba MM, Riesz P, Christman CL (1986) Free radical production in aqueous solutions exposed to simulated ultrasonic diagnostic conditions. IEEE Trans Ultrasonics Ferroelectrics Frequency Control UFFC-33, 148–155.

Cartwright RA, McKinney PA, Hopton PA, Birch JM, Hartley AL, Mann JR, Waterhouse JAH, Johnston HE, Draper GJ, Stiller C (1984) Ultrasound examinations in pregnancy and childhood. Lancet ii, 999–1000.

Crum LA, Foulkes JB (1986) Acoustic cavitation generated by microsecond pulses of ultrasound. Nature 319, 52–54.

Duck FA, Starritt HC (1984) Acoustic shock generation by ultrasonic imaging equipment. Br J Radiol 57, 231–240.

Duck FA, Starritt HC, Aindow JD, Perkins MA, Hawkins AJ (1985) The output of pulse-echo ultrasound equipment: a survey of powers, pressures and intensities. Br J Radiol 58, 989–1001.

Duck FA, Starritt HC, Anderson SP (1987) A survey of the acoustic output of ultrasonic Doppler equipment. Clin Phys Physiol Meas 8, 39–49.

Dyson M, Pond JB, Woodward B, Broadbent J (1974) The production of blood cell stasis and endothelial damage in the blood vessels of chick embryos treated with ultrasound in a stationary wave field. Ultrasound Med Biol 1, 133–148.

Farmery MJ, Whittingham TA (1978) A portable radiation-force balance for use with diagnostic ultrasonic equipment. Ultrasound Med Biol 3, 373–379.

Flynn HG (1982) Generation of transient cavities in liquids by microsecond pulses of ultrasound. J Acoust Soc Am 72, 1926–1932.

Forester GV, Roy OZ, Mortimer AJ (1982) Enhancement of contractility in rat isolated papillary muscle with therapeutic ultrasound. J Mol Cell Cardiol 14, 475–477.

Freeman FE (1963) Measurement of liquid-borne ultrasonic power. Ultrasonics 1, 27–34.

Gershoy A, Nyborg WL (1973) Perturbation of plant-cell contents by ultrasonic micro-irradiation. J Acoust Soc Am 54, 1356–1367.

Goss SA (1984) Sister chromatid exchange and ultrasound. J Ultrasound Med 3, 463–470.

Hellman LM, Duffus GM, Donald I, Sunden R (1970) Safety of diagnostic ultrasound in obstetrics. Lancet 1, 1133–1135.

Hill CR (1970) Calibration of ultrasonic beams for bio-medical applications. Phys Med Biol 15, 241–248.

Hill CR, Joshi GP, Revell SH (1972) A search for chromosome damage following exposure of Chinese hamster cells to high intensity, pulsed ultrasound. Br J Radiol 45, 333–334.

Hill CR, ter Haar G (1982) Ultrasound. In: Nonionising radiation protection (Ed. MJ Suess) WHO Regional Publication, European Series, No. 10, pp 199–228, World Health Organization, Copenhagen.

Kinnier Wilson LM, Waterhouse JAH (1984) Obstetric ultrasound and childhood malignancies. Lancet i, 1130–1134.

Kossoff G (1965) Balance technique for the measurement of very low ultrasonic power outputs. J Acoust Soc Am 38, 880–881.

Lehmann JF, Guy AW (1972) Ultrasound therapy: interaction of ultrasound and biological tissue (Eds JM Reid, MR Sikov), pp 141–151, DHEW Publication (FDA) 73-8008, US Government Printing Office, Washington, DC.

Lyons EA (1986) Human epidemiological studies. Ultrasound Med Biol 12, 689–691.

Lyons EA, Coggrave M, Brown RE (1980) Follow-up study in children exposed to ultrasound *in utero* – analysis of height and weight in the first six years of life. In: Proc 25th Annual Meeting AIUM, New Orleans, p 49.

Macintosh IJC, Davey DA (1970) Chromosome aberrations induced by an ultrasonic foetal pulse detector. Br J Radiol 48, 230–232.

Miller MW (1985) In vitro studies: single cells and multicell spheroids. In: Biological effects of ultrasound (Eds WL Nyborg, MC Ziskin), pp 35–48, Churchill Livingstone, New York.

Muir TG, Cartensen EL (1980) Prediction of nonlinear acoustic effects at biomedical frequencies and intensities. Ultrasound Med Biol 6, 345–357.

National Council on Radiation Protection and Measurements (1983) Report No 74, Biological effects of ultrasound: mechanisms and clinical implications, NCRP, Bethesda, Maryland.

Nyborg WL (1985) Biophysical mechanisms of ultrasound. In: Essentials of medical ultrasound (Eds MH Repacholi, DA Benwell), pp 35–72, Humana, Clifton, New Jersey.

Nyborg WL, Ziskin MC (Eds) (1985) Biological effects of ultrasound, Churchill Livingstone, New York.

O'Brien WD Jr (1983) Dose-dependent effect of ultrasound on fetal weight in mice. J Ultrasound Med 2, 1–8.

O'Brien WD Jr (1984) Ultrasonic bioeffects: a view of experimental studies. Birth 11, 149–157.

Preston RC (1985) Measurement and characterisation of ultrasonic fields using hydrophones in the frequency range 0.5 MHz to 15 MHz. International Electrotechnical Commission Draft TC29 29D WG8.

Preston RC, Bacon DR, Livett AJ, Rajendran K (1983) PVDF membrane hydrophone performance properties and their relevance to the measurement of the acoustic output of medical ultrasonic equipment. J Phys E: Sci Instrum 16, 786–796.

Stewart HF (1982) Ultrasonic measurement techniques and equipment output levels. In: Essentials of medical ultrasound (Eds MH Repacholi, DA Benwell), pp 77–116, Humana, Clifton, New Jersey.

ter Haar G, Daniels S (1981) Evidence for ultrasonically induced cavitation *in vivo*. Phys Med Biol 26, 1145–1149.

Thacker J (1973) The possibility of genetic hazard from ultrasonic radiation. Curr Topics Radiation Res Q 8, 235–258.

Williams AR (1983) Ultrasound: biological effects and potential hazards, Academic Press, London.

Williams AR (1985) Effects of ultrasound on blood and the circulation. In: Biological effects of ultrasound (Eds WL Nyborg, MC Ziskin), pp 49–65, Churchill Livingstone, New York.

Williams AR, Gross DR, Miller DL (1985) Cavitation in mammalian blood: an in vivo search. In: Proc 4th Mtg World Fed Ultrasound Med Biol (Eds RW Gill, MJ Dadd), p 484, Pergamon, Sydney.

Williams AR, Miller DL, Gross DR (1986) Haemolysis *in vivo* by therapeutic intensities of ultrasound. Ultrasound Med Biol 12, 501–509.

Chapter 14

DOPPLER APPLICATIONS IN THE LOWER LIMB

14.1 INTRODUCTION

Blood vessels were imaged using ultrasound as early as 1957 (Howry), but it was not until the development of Doppler methods that ultrasound offered any advance in the diagnosis and management of patients with peripheral vascular diseases. Three types of Doppler ultrasound equipment are commonly used nowadays in the vascular laboratory. These are the continuous wave velocimeter with a facility for spectral analysis, Doppler imaging systems, and duplex scanners. The information obtained with this equipment might consist of blood pressures at various positions along the supine lower limb, Doppler spectra from the pulse sites on the lower limb and either Doppler flow maps or pulse-echo real-time images of the arterial system. The aim of these measurements is to produce both anatomical and functional assessments of the severity of arterial disease.

14.2 MEASUREMENT OF BLOOD PRESSURE

The first reported attempt to measure systolic blood pressure using a Doppler velocimeter appears to be that of Yao et al (1968). In this technique the stethoscope of the conventional Korotkoff method is replaced by the velocimeter. In studying the lower limb, inflation cuffs are typically placed on the upper thigh, calf and arm. The flowmeter is placed over the posterior tibial or dorsalis pedis artery and the cuff inflated until the flow signal disappears, and then deflated until the signal appears again. The pressure in the cuff at this time is taken to be the systolic pressure of the artery under the cuff. Blood pressures measured in this way have been shown to correlate with walking time (Carter 1979), patient symptoms (Yao 1970) and angiography (Carter 1969). Pressure measurements have been used to assess the severity of occlusive arterial disease (Carter 1969), and its progression (Lewis 1974), to determine the correct level of amputation (Yao 1979), to predict graft failure (Dean et al 1975), and to evaluate the prognosis for skin healing in severe ischaemia (Carter 1973). In all of these studies the pressure index (defined as the ratio of systolic pressure in the thigh or ankle to the systolic pressure in the arm) is preferred to the absolute value at any particular site. For a more detailed account of the use of pressure measurements in the study of lower limb arterial disease the reader is referred to Carter (1979) and Baker et al (1987).

Table 14.1   Typical blood pressure values in normal subjects and in patients with occlusive arterial disease of the lower limb. (Pearce et al 1983)

Site and measurement	Blood pressure
Ankle systolic pressure	> arm systolic pressure
	< 40 mmHg in threatening ischaemia
Ankle pressure index	> 1.0
Thigh systolic pressure	
Upper thigh	30–40 mmHg > arm
Lower thigh	20–30 mmHg > arm
Thigh pressure index	> 1.1
Pressure gradients	< 30 mmHg between adjacent sites
Toe systolic pressure index	0.7 ± 0.19
	0.35 ± 0.15 claudication
	0.11 ± 0.10 rest pain
Finger systolic pressure index	> 0.95

Many authors have published normal data resulting from blood pressure measurements; typical values are those due to Pearce et al (1983), which are reproduced in Table 14.1.

14.3   VELOCITY WAVEFORMS FROM THE LOWER LIMB

The shape of the velocity–time waveform over the cardiac cycle was originally demonstrated to contain useful clinical information by Strandness et al (1967). It is now routine procedure to examine the Doppler spectra from the pulse sites in the lower limb, i.e. at the common femoral, popliteal and posterial tibial arteries. Typical normal spectra from these sites are shown in Fig. 14.1.

The presence of arterial disease modifies the appearance of these spectra, as exemplified in Fig. 14.2. Proximal stenosis initially produces a reduction of the peak frequency in the second and third phases of the Doppler spectrum, then as the stenosis increases further the third component disappears and the second is markedly reduced. Eventually flow becomes continuous over the whole of the cardiac cycle.

If the proximal arterial disease is accompanied by superficial femoral occlusion, a characteristic shoulder often appears on the downstroke of systole, as illustrated in Fig. 14.3.

14.4   VELOCITY WAVEFORM ANALYSIS

Numerous attempts have been made to quantify the maximum frequency envelope of the Doppler spectrum to give an objective, quantifiable indication of the severity of arterial disease. Several of these methods are described in Chapter 10.

14.4.1   Pulsatility index

The first quantitative index of waveform shape to be described was the pulsatility index (Gosling et al 1969), later modified by Gosling and King (1974) – see Section 10.4.3.1. The modified *PI* is defined as the peak-to-peak excursion of the maximum frequency

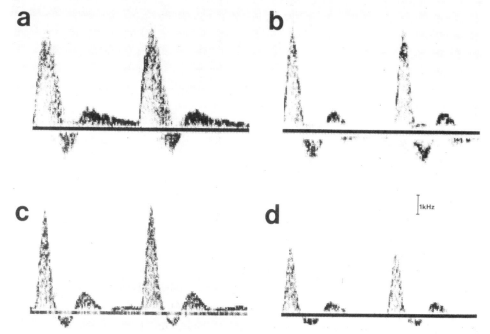

Figure 14.1 Typical normal spectra from (a) the common femoral, (b) the popliteal, (c) the posterior tibial and (d) the dorsalis pedis arteries. Note the triphasic shape normally found in Doppler signals from normal lower limb vessels. All these recordings were made using the same analyser frequency range

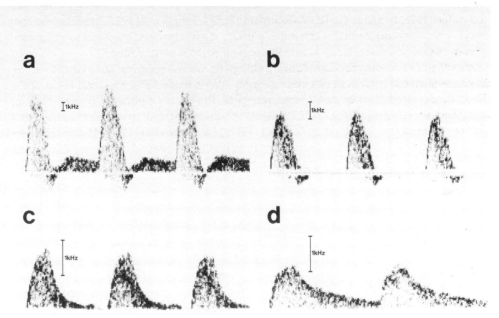

Figure 14.2 Doppler spectra recorded from common femoral arteries with various degrees of common iliac stenosis. See text for explanation. Three different analyser ranges have been used as indicated by the frequency calibration markers

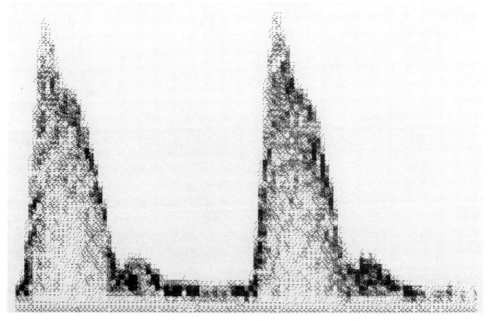

Figure 14.3 Doppler spectra recorded from the common femoral artery in a patient with mixed iliac and superficial femoral artery disease. Note the shoulder on the downstroke of systole

envelope divided by the mean value over the cardiac cycle. There are a large number of published studies on the effect of arterial disease on *PI* (see for example Johnston and Taraschuk 1976, Evans et al 1980, Johnston et al 1983, Junger et al 1984, Macpherson et al 1984). There is no doubt that it can reliably detect major stenoses, but it is also influenced to some degree by distal stenosis and changes in peripheral resistance.

A method of assessing the disease in a single given arterial segment, based on *PI* values, has been described by Woodcock and colleagues (Woodcock 1970, Fitzgerald et al 1971). The *PI* values from two adjacent pulse points on the leg are compared by finding their ratio (Section 10.4.10.1). This ratio, known as the damping factor, has been evaluated by a number of investigators including Gosling (1976), Johnston et al (1978), Humphries et al (1980), Campbell et al (1983), and Baker et al (1986). In general, however, it has been found necessary to combine damping factor with 'transit time' (Section 10.4.10.1) to extract diagnostically useful information from the Doppler signals.

14.4.2   Laplace transform analysis

A more sophisticated method of waveform shape analysis, first described by Skidmore and Woodcock (1978), allows a physiological interpretation of changes in the velocity waveform shape (Section 10.4.7). In this method the waveform is represented by three coefficients (ω_0, γ and δ) derived from the Laplace transform of the waveform. It has been demonstrated that ω_0 is related to the elastic properties of the major conduit arteries, γ to the distal impedance, and δ to the diameter of the iliofemoral arterial segment (Skidmore and Woodcock 1980a, 1980b, Skidmore et al 1980). The term δ is called the Laplace transform damping factor (LTD), and is sensitive to the presence of occlusive arterial disease.

Baird et al (1980) have compared the sensitivity and specificity of PI and LTD. LTD values correlated reasonably well with the severity of iliac artery stenosis ($r = -0.75$). There was a significant difference between minor stenosis of $<50\%$ diameter and major stenosis of $>50\%$ diameter. For minor stenosis $LTD = 0.5 \pm 0.15$, and for major stenosis $LTD = 0.78 \pm 0.16$, $P < 0.02$. There was a highly significant difference between normal volunteers and patients with minor iliac disease, $P < 0.001$. LTD was not affected by an increased distal impedance and there was a fair correlation ($r = 0.73$) with iliac stenosis in the presence of occlusion of the superficial femoral artery. In the same study there was a good correlation between PI and iliac stenosis if the distal vessels were patent ($r = 0.75$) but not so good when the distal vessels were occluded together with iliac stenosis ($r = 0.51$).

Several other authors have subsequently evaluated LTD as a method of assessing aorto-iliac disease (Johnston et al 1984, Junger et al 1984, Law et al 1984, Macpherson et al 1984) and have found it to give similar results to the much simpler PI. One reason for these disappointing results may be that a third-order Laplace equation is not sufficient to model the diseased arterial system. This possibility is supported by the finding of some authors that, in contrast to theory, LTD is in practice affected by peripheral resistance (Evans et al 1981, Junger et al 1984, Law et al 1984).

Campbell et al (1983) described the ω_0 gradient, which is the ratio of the femoral to ankle ω_0 values, and found it to be superior to PI damping factor over this segment. However, as previously indicated, PI damping factor is of little diagnostic value when used in isolation, and should be combined with transit time. In addition, as pointed out by Baker et al (1986), ω_0 at any site can only be calculated if there is a pair of complex poles in the solution of the Laplace transform equation, and therefore if all three poles at the common femoral are real, the technique can give no information about the femoropopliteal segment.

14.4.3 Principal component analysis

Principal component analysis (PCA) has been discussed in detail in Section 10.4.8. It is a method of describing the shape of the whole waveform using the coefficients of an orthogonal series. This is similar to describing a waveform in terms of its Fourier components, except that principal components are not sinusoidal, they are specifically chosen to best model the population of waveforms being studied. Because the method is so efficient, a waveform can be described to a good degree of accuracy using the coefficients of only two components, and therefore the entire waveform shape may be represented by a point on a two-dimensional graph. Waveforms recorded from different disease conditions occupy different areas of 'feature space' and may therefore be separated.

Macpherson et al (1984) compared PI, LTD and PCA with direct femoral artery pressure measurements in 50 patients. They showed that even the coefficient of the first PC alone produced significantly better agreement with intra-arterial pressure measurements than LTD ($\chi^2 = 5.8$, $P < 0.02$). In addition, PCA of the femoral waveform was able to predict the presence or absence of disease in the superficial femoral artery. In the same study the coefficient of the first PC alone just failed to perform significantly better than PI.

A larger study (116 patients, 221 limbs) was reported by Campbell et al (1984a) in which

LTD, PCA, *PI*, thigh pressure ratios, and other non-invasive tests were compared with clinical judgement and arteriography. Multiple regression analyses were performed for the prediction of the severity of aorto-iliac disease greater than 50%. It was found that *LTD* was worse than *PI* which in turn was worse than PCA for the detection of stenoses of greater than 50%, but that clinical assessment alone was better than all these methods. For stenoses less than 50% of the vessel diameter, *LTD* was the most sensitive technique, and was better than clinical judgement alone. Results from this study are summarized in Table 14.2. PCA could have been expected to perform better than reported if more sophisticated algorithms had been used to combine the coefficients of the first two components (Section 10.5.2). It is interesting to see that the more non-invasive tests that are carried out the better are the results.

Table 14.2   Multiple regression analysis for the prediction of the severity of aorto-iliac stenosis of less than 50% by clinical assessment and waveform analysis methods (Campbell et al 1984a)

	Value of R^2
Clinical assessment	0.14
Clinical/*LTD*	0.19
Clinical/*PI*	0.14
Clinical/*LTD*/*PI*	0.20
Clinical/*PCI*	0.16
Clinical/*PC1*/*PC2*	0.17
Clinical/*LTD*/*PC1*/*PC2*	0.19
Clinical/*LTD*/*PI*/*PC1*/*PC2*	0.22

14.5   VARIABILITY AND REPRODUCIBILITY OF DOPPLER SHIFT WAVEFORMS

Clifford et al (1981) investigated the variability of *PI* in patients in normal sinus rhythm (NSR) and in atrial fibrillation (AF). *PI* values showed a mean variability of 17% in subjects in NSR and of 26% in patients in AF but with no clinical arterial disease. The variability was reduced using a ten-beat average, to 7.5% in NSR and 9% in AF.

Campbell et al (1984b) studied the variability and reproducibility of Doppler shift waveforms in the calculation of Laplace indices. They postulated that they are influenced by operator experience, physiological changes in the patient, and the type of Doppler processing used. Two groups of patients were studied 9 months apart, the first group being early on in the investigators' experience. The variability of femoral *LTD* decreased from 16.9% to 12.5% over the 9-month period, and that of the posterior tibial *LTD* from 23.3% to 17.9% ($P < 0.05$ in each case). Normal subjects examined in the longer term with a mean interval of 28 days between recordings did not show significantly greater variability than those examined over 1 hour.

14.6   MEASUREMENT OF ABSOLUTE VELOCITY

Dilley and Fronek (1979) have discussed quantitative velocity measurements in lower limb arterial disease. Patients were divided into three groups according to the site of the disease. Group 1 consisted of patients with isolated aorto-iliac occlusions, group 2 of those with isolated femoropopliteal occlusion and a normal aorto-iliac segment, and group 3 of those with aorto-iliac occlusion with significant femoral, popliteal or distal occlusive disease. Their results are summarized in Table 14.3. The quantitative velocity measurements were not able to distinguish the level of arterial obstruction, but peak reverse velocity and deceleration accurately differentiated patients with multi-level disease from those with single-level disease.

Table 14.3 Measurements of absolute velocity in the femoral and posterior tibial arteries of normal subjects, and in patients with occlusive arterial disease of the lower limb. Groups 1, 2 and 3 are defined in the text. (Dilley and Fronek 1979)

	Peak forward velocity (cm s^{-1})	Peak reverse velocity (cm s^{-1})	Mean velocity (cm s^{-1})	Acceleration (cm s^{-2})	Deceleration (cm s^{-2})
Femoral					
Normal	40.7 ± 10.9	6.5 ± 3.6	9.8 ± 5.3	353 ± 113	250 ± 60
Group 1	25.8 ± 9.4*	3.5 ± 3.5	8.9 ± 2.9	260 ± 176	123 ± 75*
Group 2	30.3 ± 15.4*	4.2 ± 4.4	8.9 ± 4.2	352 ± 193	181 ± 117*
Group 3	20.9 ± 11.2*	0.8 ± 1.9*†	7.9 ± 4.2	208 ± 166*	91 ± 71*†
Posterior tibial					
Normal	16.0 ± 10.0	2.0 ± 2.3	4.0 ± 3.5	145 ± 74	130 ± 76
Group 1	13.4 ± 11.5	2.2 ± 2.9	4.4 ± 3.3	166 ± 192	79 ± 62
Group 2	13.3 ± 6.6	1.2 ± 1.5	7.4 ± 7.0*	122 ± 60	77 ± 83*
Group 3	11.7 ± 8.2	0.4 ± 1.1*†	5.2 ± 4.2	89 ± 65*	43 ± 40*†

*Control group versus group 1, 2 or 3, significant at $P < 0.01$.
†Group 3 versus groups 1 and 2, significant at $P < 0.01$.

14.7 DOPPLER IMAGING

Mozerskey et al (1972) have produced Doppler images of the iliac, common, superficial and profunda femoral arteries, and saphenous vein bypass grafts. Baird et al (1979) produced images of the femoral bifurcation and arterial prostheses in three orthogonal planes. The internal diameter of 21 arterial grafts implanted into the femoral artery (14 aorto-iliofemoral and 7 axillo-bifemoral) were compared with 14 ilio-popliteal grafts. All grafts measured 10 mm internal diameter (ID) at implantation. At the time of the study the ilio-popliteal grafts were reduced to 7.6 ± 0.4 mm ID at the inguinal level and to 6.3 ± 0.4 mm ID just above the popliteal anastomosis. The inguinal IDs of the axillo-bifemoral grafts were not reduced, but some localized stenoses were identified elsewhere in the grafts. No distal narrowing was found in any of the 14 aortofemoral grafts.

Sequential measurements on 14 iliopopliteal grafts were made up to five years after implantation. Popliteal narrowing was found to be progressive with time ($r = 0.54$, $P < 0.01$).

14.8 DUPLEX SCANNING

Jager et al (1985) have reported how duplex scanning compares with arteriography in the assessment of lower limb arterial disease. The overall accuracy for predicting five degrees of severity of arterial disease was 76%. The sensitivity was 96% and the specificity 81%.

Clifford et al (1980) reported the use of duplex scanning in combination with Doppler imaging to study neo-intimal hyperplasia, and Clifford et al (1982) used duplex scanning to investigate the behaviour of arterial grafts at the hip and knee joints.

14.9   COLOUR FLOW MAPPING

Colour flow mapping allows the simultaneous acquisition of anatomical and functional information from arteries. Some examples of colour flow images of the arteries supplying the lower limb are shown in Plates 14.1–14.6. It is too early as yet to assess the impact of colour flow mapping on the assessment of peripheral vascular disease.

14.10   SUMMARY

In the investigation of lower limb arterial disease the measurement of segmental blood pressure and the calculation of pressure indices is an important first step. Lower limb blood pressure measurements correlate with patient symptoms, walking time and angiography, and help in the assessment of the severity and site of the disease, and in disease progression. Investigation of the Doppler spectra and indices derived from them provide further information in the minor to major arterial disease categories where non-invasive pressure measurements alone are insensitive. Duplex scanning and colour flow mapping allow the investigation of minor disease and also the simultaneous collection of anatomical and functional information.

14.11   REFERENCES

Baird RN, Lusby RJ, Bird DR, Giddings AEB, Skidmore R, Woodcock JP, Horton RE, Peacock JH (1979) Pulsed Doppler angiography in lower limb arterial ischaemia. Surgery 86, 818–825.

Baird RN, Bird DR, Clifford PC, Lusby RJ, Skidmore R, Woodcock JP (1980) Upstream stenosis diagnosed by Doppler signals from the femoral artery. Arch Surg 115, 1316–1322.

Baker AR, Prytherch DR, Evans DH, Bell PRF (1986) Doppler ultrasound assessment of the femoro-popliteal segment: comparison of different methods using ROC curve analysis. Ultrasound Med Biol 12, 473–482.

Baker AR, Macpherson DS, Evans DH, Bell PRF (1987) Pressure studies in arterial surgery. Eur J Vasc Surg 1, 273–283.

Campbell WB, Baird RN, Cole SEA, Evans JM, Skidmore R, Woodcock JP (1983) Physiological interpretation of Doppler shift waveforms: the femorodistal segment in combined disease. Ultrasound Med Biol 9, 265–269.

Campbell WB, Cole SEA, Skidmore R, Baird RN (1984a) The clinician and the vascular laboratory in the diagnosis of aortoiliac stenosis. Br J Surg 71, 302–306.

Campbell WB, Skidmore R, Baird RN (1984b) Variability and reproducibility of arterial Doppler waveforms. Ultrasound Med Biol 10, 601–606.

Carter SA (1969) Clinical measurement of systolic pressures in limbs with arterial occlusive disease. J Am Med Assoc 207, 1869–1873.

Carter SA (1973) The relationship of distal systolic pressures to healing of skin lesions in limbs with arterial occlusive disease, with special reference to diabetes mellitus. Scand J Clin Lab Invest 31, Suppl 128, 239–247.

Carter SA (1979) Role of pressure measurements in vascular disease. In: Noninvasive diagnostic techniques in vascular disease (Ed. EF Bernstein), pp 261–287, CV Mosby, St Louis, Missouri.

Clifford PC, Skidmore R, Bird DR, Lusby RJ, Baird RN, Woodcock JP, Wells PNT (1980) Pulsed Doppler and real-time duplex imaging of Dacron arterial grafts. Ultrasonic Imaging 2, 381–390.

Clifford PC, Skidmore R, Bird DR, Woodcock JP, Baird RN (1981) The role of pulsatility index in the clinical assessment of lower limb ischaemia. J Med Eng Technol 5, 237–241.

Clifford PC, Skidmore R, Woodcock JP, Baird RN (1982) Behaviour of arterial grafts at the hip and knee joints. Ultrasonic Imaging 44, 83–91.

Dean RH, Yao ST, Thompson RG, Bergan JJ (1975) Predictive value of ultrasonically derived arterial pressure in determination of amputation level. Ann Surg 41, 731–736.

Dilley RB, Fronek A (1979) Quantitative velocity measurements in arterial disease of the lower extremity. In: Non-invasive diagnostic techniques in vascular disease (Ed. EF Bernstein), pp 294–303, CV Mosby, St Louis, Missouri.

Evans DH, Barrie WW, Asher MJ, Bentley S, Bell PRF (1980) The relationship between ultrasonic pulsatility index and proximal arterial stenosis in a canine model. Circ Res 46, 470–475.

Evans DH, Macpherson DS, Bentley S, Asher MJ, Bell PRF (1981) The effect of proximal stenosis on Doppler waveforms: a comparison of three methods of waveform analysis in an animal model. Clin Phys Physiol Meas 2, 17–25.

Fitzgerald DE, Gosling RG, Woodcock JP (1971) Grading dynamic capability of arterial collateral circulation. Lancet i, 66–67.

Gosling RG (1976) Extraction of physiological information from spectrum-analyzed Doppler-shifted continuous-wave ultrasound signals obtained non-invasively from the arterial system. In: IEE Medical electronics monographs 18–22 (Ed. DW Hill, BW Watson), Chapter 4, pp 73–125, Peter Peregrinus, Stevenage, Hertfordshire.

Gosling RG, King DH (1974) Continuous wave ultrasound as an alternative and complement to X-rays in vascular examination. In: Cardiovascular applications of ultrasound (Ed. RS Reneman), Chapter 22, pp 266–282, North-Holland, Amsterdam.

Gosling RG, King DH, Newman DL, Woodcock JP (1969) Transcutaneous measurement of arterial blood velocity by ultrasound. In: Ultrasonics for Industry Conference Papers, pp 16–32, IPC Press, Guildford.

Howry DH (1957) Techniques used in ultrasonic visualization of soft tissues. In: Ultrasound in biology and medicine (Ed. E Kelly), p 49, AIBS, Washington.

Humphries KN, Hames TK, Smith SWJ, Cannon VA, Chant ADB (1980) Quantitative assessment of the common femoral to popliteal arterial segment using continuous wave Doppler ultrasound. Ultrasound Med Biol 6, 99–105.

Jager K, Ricketts HJ, Strandness DE (1985) Duplex scanning for the evaluation of lower limb arterial disease. In: Non-invasive techniques in vascular disease (Ed. EF Bernstein), pp 619–631, CV Mosby, St Louis, Missouri.

Johnston KW, Taraschuk I (1976) Validation of the role of pulsatility index in quantitation of the severity of peripheral arterial occlusive disease. Am J Surg 131, 295–297.

Johnston KW, Maruzzo B, Cobbold RS (1978) Doppler methods for quantitative measurement and localization of peripheral arterial occlusive disease, by analysis of the blood flow velocity waveform. Ultrasound Med Biol 4, 209–223.

Johnston KW, Kassam M, Cobbold RSC (1983) Relationship between Doppler pulsatility index and direct femoral pressure measurements in the diagnosis of aortoiliac occlusive disease. Ultrasound Med Biol 9, 271–281.

Johnston KW, Kassam M, Koers J, Cobbold RSC, MacHattie D (1984) Comparative study of four methods for quantifying Doppler ultrasound waveforms from the femoral artery. Ultrasound Med Biol 10, 1–12.

Junger M, Chapman BLW, Underwood CJ, Charlesworth D (1984) A comparison between two types of waveform analysis in patients with multisegmental arterial disease. Br J Surg 71, 345–348.

Law YF, Graham JC, Cotton LT, Roberts VC (1984) Validity of the transfer function model of the human arterial system of the lower limb in man. Med Biol Eng Comp 22, 537–542.

Lewis JD (1974) Pressure measurements in the long-term follow-up of peripheral vascular disease. Proc R Soc Med 67, 443.

Macpherson DS, Evans DH, Bell PRF (1984) Common femoral artery Doppler waveforms: a comparison of three methods of objective analysis with direct pressure measurements. Br J Surg 71, 46–49.

Mozerskey DJ, Hokanson DE, Sumner DS, Strandness DE (1972) Ultrasonic visualization of the arterial lumen. Surgery 72, 253–259.

Pearce WH, Yao ST, Bergan JJ (1983) Noninvasive vascular diagnostic testing. Curr Problems Surg 20, 460–465.

Skidmore R, Woodcock JP (1978) Physiological interpretation of arterial models derived using transcutaneous ultrasonic flowmeters. Proc Physiol Soc 277, 29–30P.

Skidmore R, Woodcock JP (1980a) Physiological interpretation of Doppler shift waveforms. I. Theoretical considerations. Ultrasound Med Biol 6, 7–10.

Skidmore R, Woodcock JP (1980b) Physiological interpretation of Doppler shift waveforms. II. Validation of the Laplace transform method for characterization of the common femoral blood-velocity/time waveforms. Ultrasound Med Biol 6, 219–225.

Skidmore R, Woodcock JP, Wells PNT, Bird DR, Baird RN (1980) Physiological interpretation of Doppler shift waveforms. III. Clinical results. Ultrasound Med Biol 6, 227–231.

Strandness DE, Schultz RD, Sumner DS, Rushmer RF (1967) Ultrasonic flow detection: a useful technique in the evaluation of peripheral vascular disease. Am J Surg 113, 311–319.

Woodcock JP (1970) The transcutaneous ultrasonic flowmeter and the significance of changes in the velocity–time waveform in occlusive arterial disease of the leg. PhD Thesis, University of London.

Yao ST (1970) Haemodynamic studies in peripheral arterial disease. Br J Surg 57, 761–766.

Yao ST (1979) Surgical use of pressure studies in peripheral arterial disease. In: Noninvasive diagnostic techniques in vascular disease (Ed. EF Bernstein), pp 288–293, CV Mosby, St Louis, Missouri.

Yao ST, Hobbs J, Irvine NT (1968) Pulse examination by an ultrasonic method. Br Med J 4, 555–557.

Chapter 15

DOPPLER APPLICATIONS IN THE CEREBRAL CIRCULATION

15.1 INTRODUCTION

Doppler ultrasound methods and instruments are available to investigate both the extracranial and the intracranial circulation. Initially these were developed to detect the presence of atheroma in the extracranial carotid arteries, but now information is readily available relating to the changes in blood flow in the neonatal brain and also in the adult brain through the intact skull.

15.2 THE EXTRACRANIAL CIRCULATION

The carotid arteries and vertebral arteries are readily accessible to Doppler flowmeters and duplex scanners. Both CW and PW Dopplers are routinely used and the present state of the art in their clinical applications is discussed below.

15.2.1 Carotid arteries

15.2.1.1 Waveform analysis

Typical blood-velocity spectra for the common, internal and external carotids in a normal subject are shown in Fig. 15.1. There are obvious differences in the spectra; for example, the internal carotid has a markedly higher diastolic velocity than the external carotid, where the diastolic velocity may in some cases be almost zero. Based on the hypothesis that the shape of these spectra is due in part to the pathophysiology of the circulation, attempts have been made to characterize these waveforms and relate them to the known pathological state of the vessels.

Planiol and Pourcelot (1971) introduced the Pourcelot resistance index, defined as $(S-D)/S$, where S is the systolic height and D the end-diastolic height of the common carotid waveform (Section 10.4.3.2). The normal range, regardless of age, is between 0.55 and 0.75. Higher values imply that there is an increased peripheral resistance.

Baskett et al (1977) used the A/B ratio illustrated in Fig. 15.2 (Section 10.4.3.5). In a total of 800 asymptomatic volunteers aged from 5 to 90 years, the ratio A/B decreased with age until the sixth decade, after which it remained at the value of approximately 1.2.

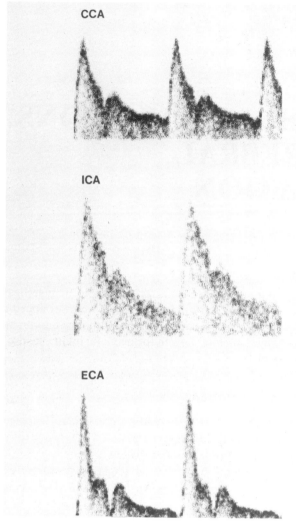

Figure 15.1 Typical Doppler shift spectra from the normal common carotid artery (CCA), internal carotid artery (ICA) and external carotid artery (ECA)

Baskett et al also recorded the A/B ratios from the common carotid and ipsilateral supraorbital arteries in 101 patients, and compared their results with arteriography. They showed that if the A/B ratio in either vessel is less than 1.05 there is an 88% probability of disease at the carotid bifurcation. When the A/B ratio in both vessels is greater than 1.05, there is an 80% probability of a normal bifurcation.

Archie (1981) suggested comparing waveforms from the two common carotid arteries by taking the ratio of the A/B ratios from each vessel.

Spencer and Reid (1979) developed an index in which the ratio of the maximum frequency recorded in the internal carotid artery at the angle of the jaw is divided by the maximum frequency recorded in the stenosis (Section 10.4.10.3). For stenoses of greater than 70% the sensitivity of the method was estimated to be 63%, the specificity 85% and the overall accuracy 85%.

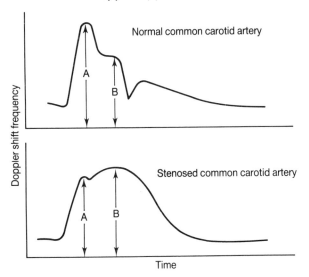

Figure 15.2 Measurement of the A/B ratio of maximum velocity waveforms from the common carotid artery

These same authors investigated the relationship between the maximum Doppler shift frequency in 77 stenoses and the minimum arterial diameter shown on arteriography (Section 10.4.2). For an angle of insonation of 60° and a transmitted frequency of 5 MHz, and when vessels with a diameter of less than 1.5 mm were excluded, the residual diameter D was related to the maximum Doppler shift frequency f_{max} by:

$$D = 8.77 f_{max}^{-0.57}$$

with a correlation coefficient of 0.74.

Rutherford et al (1977) identified five waveform parameters (Section 10.4.6) that contribute significantly to an expression for the severity of the disease. The Rutherford classification is given by:

$$0.07(PSV) - 0.5(EDV) - 0.1(PDV) - 7.1(w/PSV) + 0.1(dV_s/dt) - 6.8$$

where PSV is the peak systolic velocity, EDV the end diastolic velocity, PDV the peak diastolic velocity, w the width of the systolic peak at mid-height, and dV_s/dt the systolic upstroke rate. All these variables are calculated or measured from the common carotid artery. Internal carotid artery disease can be classified into normal, less than 50% diameter stenosis, greater than 50% diameter stenosis, and occlusion. Young volunteers have a score of $+6$, whereas patients with occlusions have a score of approximately -8.

Martin et al (1980) applied principal component analysis (Section 10.4.8) to the waveform from the common carotid artery and found it to be better at grading disease than the more commonly used A/B ratio. Sherriff et al (1982) later applied PCA to the entire sonogram from the common carotid artery and obtained similar results.

15.2.1.2   Peri-orbital Doppler investigation

Further information can be gained from the peri-orbital circulation concerning, particularly, the internal carotid. The principle of the peri-orbital methods is to establish the flow direction in vessels such as the supratrochlear and supraorbital arteries, and to study changes in flow direction resulting from manual compression of supply arteries. Brockenborough (1969) first showed that if flow is monitored over one of these terminal branches of the internal carotid, manual compression of the ipsilateral superficial temporal artery produces an enhanced signal, in a direction out of the orbit, in the normal circulation. If the pressure in the internal carotid artery is reduced because of occlusive disease, there will be a flow from the higher-pressure temporal artery through the collaterals to the supraorbital/frontal arteries, and the flow direction will be into the orbit. If manual compression is applied to the temporal artery, flow will be decreased. This test is known as the temporal artery occlusion test (TAOT), and a positive response, in conjunction with a flow reversal, is highly suggestive of severe arterial disease present in the internal carotid. Barnes et al (1977), using this simple TAOT, found that for stenosis of the internal carotid of greater than 50% of the diameter the result was positive in 64% of cases.

A more extensive examination procedure was developed by Barnes et al, which included manual compression of all branches of the external carotid artery, and common carotid compression. For stenoses of the internal carotid of greater than 50%, this new method was accurate in detecting the disease in 98% of cases. For stenoses of less than 50% the test was positive in only 9% of cases. The examination was only 56% sensitive in detecting carotid disease of potential clinical significance. The predictive value of these techniques for carotid endarterectomy was 39%, which means that only 39% of those presenting with abnormal Doppler signals are candidates for carotid endarterectomy. Conversely, more than one-third of the vessels associated with a normal Doppler examination were suitable for endarterectomy.

Padayachee et al (1982) combined A/B ratio with compression manoeuvres to markedly improve the success rate of these non-invasive tests. Keller et al (1976a) produced a set of eight criteria based on flow direction, compression pulse rise times, etc., which were also successful in detecting haemodynamically significant disease.

15.2.1.3   Doppler spectral broadening

Reinertson and Barnes (1977) showed that in order reliably to detect carotid artery stenoses of less than 50% in diameter, the whole Doppler spectrum has to be visualized. Typical, normal Doppler spectra are shown in Fig. 15.1. It can be seen that in systole there is a definite 'window' in peak systole in which no Doppler shift frequencies are present. It is found that this window fills in when the Doppler spectrum is examined just distal to a stenosis.

Two simple qualitative systems of assessing internal carotid disease by means of spectral broadening in the internal carotid artery have been described by Langlois et al (1983) and Sumner et al (1985).

In the Langlois method the Doppler spectra are classified into one of five groups. These are:

A. Systolic peak frequency up to 4 kHz, with no spectral broadening
B. Systolic peak frequency up to 4 kHz, spectral broadening in deceleration phase of systole

C. Systolic peak frequency up to 4 kHz, spectral broadening throughout systole
D. Systolic peak frequency above 4 kHz, spectral broadening throughout systole, increased frequencies in diastole
E. No signal from internal carotid artery, zero diastolic velocity in the common carotid

All these definitions assume a transmitted frequency of 5 MHz and an angle of insonation of 60°. Category A was used to describe a normal artery, B one with a 1–15% diameter reduction, C a 16–49% diameter reduction, D a 50–99% diameter reduction, E an occlusion. The agreement of this classification system against arteriography is shown in Table 15.1.

Table 15.1 Results of spectral classification used by Langlois et al (1983) for categorizing degrees of stenosis of the internal carotid artery

Angiography	Spectral classification					Total
	A	B	C	D	E	
Normal	4	4				8
1–15%	1	9	3			13
16–49%		4	16	3		23
50–99%		1	2	25		28
Occlusion					5	5

Sumner et al (1985) classified spectra into four categories based on the presence or absence of the systolic window. These categories are:

1. Spectra with windows and peak frequencies less than 4 kHz;
2. Spectra with reduced or absent windows and peak frequencies of less than 4 kHz;
3. Spectra with or without windows but with frequencies exceeding 4 kHz;
4. Pulsatile spectra typical of an external carotid artery signal.

As with the Langlois classification, all these definitions assume a transmitted frequency of 5 MHz and an angle of insonation of 60°. Table 15.2 shows the relationship found between the above classification and the angiographic classification, and Table 15.3 shows the sensitivity, specificity, positive predictive value, negative predictive value, and accuracy reported for the classification system.

Table 15.2 Classification of Doppler spectra based on peak frequency and on the presence or absence of a systolic window. (Sumner et al 1985)

Category	Angiographic diagnosis			
	Normal	1–49% stenosis	50–99% stenosis	Occluded
1	94	28	0	1
2	23	23	4	0
3	6	5	40	1
4	3	0	1	2

Table 15.3 Sensitivity, specificity, accuracy, positive predictive and negative predictive values for detecting any degree of stenosis of the spectral classification system described by Sumner et al (1985)

	Sensitivity (%)	Specificity (%)	PPV (%)	NPV (%)	Accuracy (%)
Spectral classification	75	75	73	76	75

15.2.1.4 Quantitative analysis of spectral broadening

Many attempts have been made to quantify spectral broadening (see Section 10.4.5 and Table 10.2). The simplest is to calculate the ratio f_{max}/f_{mean} in systole, where f_{max} is the maximum Doppler shift frequency and f_{mean} is the mean Doppler shift frequency at peak systole. Woodcock et al (1983) described the frequency spread function (*FSF*), defined as $(f_{max}-f_{min})/f_{mean}$, where f_{max} and f_{mean} are as before, and f_{min} is the minimum frequency present during systole. Sheldon (1985) has compared these indices of spectral broadening, together with others such as the spectral width at the half-power level, and the ratio f_{max}/f_{mode} in systole, and concluded that the simple f_{max}/f_{mean} is the most reliable in objectively detecting stenoses of greater than 40% in diameter. Aldoori (1986) in a similar study came to the same conclusion.

Rittgers et al (1983) used a method in which five parameters were extracted from the Doppler spectrum recorded from the proximal and distal internal carotid artery during the 100–200 ms following peak systole. These were the mode frequency, the highest and lowest frequencies whose amplitudes exceeded one-quarter of the amplitude of the mode frequency (i.e. the -12 dB frequencies), the highest frequency whose amplitude exceeded one-eighth of the amplitude of the mode frequency (i.e. the -18 dB frequency), and the systolic window defined in terms of the difference between the -12 dB frequencies. Three indices proved to be useful in predicting disease. These were:

1. $\dfrac{\text{Peak } (-18 \text{ dB) proximal ICA frequency}}{\text{Peak } (-18 \text{ dB) distal ICA frequency}} > 1.1$

2. $\dfrac{\text{Mode proximal ICA frequency}}{\text{Mode distal ICA frequency}} > 1.3$

3. Distal systolic window $< 0.38\%$

Greene et al (1982) have described a computer-based pattern recognition system (Section 10.4.6) in which the disease states can be separated with accuracies of upwards of 94%. The method was correct in 97.5% of cases in differentiating normal from diseased, 98.2% in distinguishing greater than 50% diameter stenoses from less than 50%, and 85.9% in distinguishing greater than or less than 20% stenoses.

15.2.1.5 Doppler imaging

Doppler imaging systems can be based on either pulsed or continuous wave Doppler flowmeters (see Sections 4.5.1 and 4.5.2). CW imaging systems are quite successful in distinguishing greater than or less than 50% diameter stenoses. Shoumaker and Bloch

(1978) were correct in 92% of cases when the stenosis was greater than 50%, and similar results were reported by Spencer et al (1977). White and Curry (1978), using a colour-coded Doppler image, reported similar results for stenoses greater than 50% but only 25% agreement between their system and angiography for stenoses of less than 25%.

With pulsed Doppler systems the results are rather better. Sumner et al (1985) have reviewed the accuracy of pulsed Doppler imaging systems in detecting all levels of disease of the internal carotid artery (Table 15.4). For all levels of disease the negative predictive value was only 64% which means that a negative pulsed Doppler image cannot rule out the presence of some disease (Sumner et al 1985). A positive predictive value of 88%, however, means that if the image shows a stenosis there is a high probability that it is correct.

Table 15.4 Sensitivity, specificity, accuracy, and positive and negative predictive values of pulsed Doppler imaging systems in the study of non-haemodynamically significant disease, and in distinguishing occluded from non-occluded vessels. (Sumner et al 1985)

	Sensitivity (%)	Specificity (%)	PPV (%)	NPV (%)	Accuracy (%)
Any disease	76	80	88	64	78
Non-haemodynamically significant disease	90	87	80	94	88
Distinguishing occluded from non-occluded vessels	82	97	80	98	95

In a previous section concerned with qualitative analysis of spectral broadening, the results from Sumner et al were quoted to show the success of a simple classification system. Table 15.5 shows how the situation is improved if the classification system is combined with the pulsed Doppler image.

Table 15.5 The success rate of spectrum analysis alone against the pulsed Doppler image plus spectrum analysis for detecting any degree of stenosis (Sumner et al 1985)

	Sensitivity (%)	Specificity (%)	PPV (%)	NPV (%)	Accuracy (%)
Spectrum	75	75	73	76	75
Image and spectrum	88	93	91	90	91

15.2.1.6 Duplex scanning

Duplex scanners combine a real-time imaging capability with a Doppler flowmeter. This allows the accurate placement of the sample volume and the subsequent analysis of the spectrum. The real-time imaging capability allows the classification of atherosclerotic plaques into categories characterized by their ultrasonic appearance. Reilly et al (1983) described two major echo patterns, homogeneous and heterogeneous. Homogeneous lesions are characterized by uniformly high or medium level echoes, an example of which is shown in Fig. 15.3. Heterogeneous lesions are characterized by mixed, high, medium and low level echoes, and often have an area within the lesion of the same echogenicity as blood. An example of a heterogeneous plaque is shown in Fig. 15.4. Homogeneous lesions correlate highly with histologically demonstrated fibrous lesions with no intraplaque haemorrhage or ulceration. Heterogeneous lesions, however, correlate highly with intraplaque haemorrhage and ulceration (Reilly et al 1983, Bluth et al 1986).

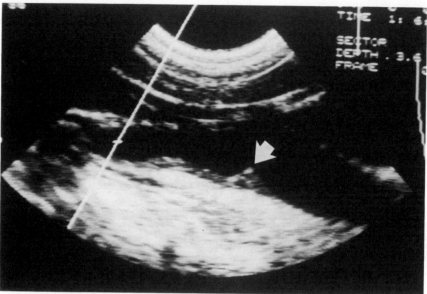

Figure 15.3   Homogenous plaque at the origin of the internal carotid artery

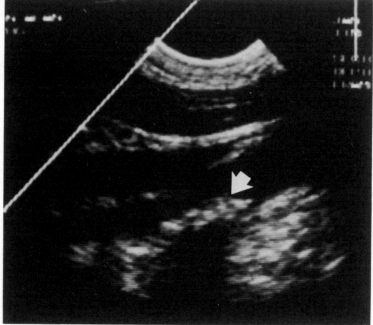

Figure 15.4   Heterogeneous plaque in the proximal internal carotid. The plaque surface is irregular and an acoustic shadow is seen behind the plaque

The ability to study plaque structure and the consequent flow disturbance, using spectral broadening methods, results in instrumentation capable of following the progression or regression of atherosclerosis in an objective way. Hames et al (1985) have compared the duplex scanner and Doppler spectral analysis methods with angiography, for stenosis of less than 50%. These authors show that a combination of image, maximum frequency envelope changes and spectral broadening, are useful in the study of minor stenoses. In particular, for stenoses in the 1–24% diameter reduction category, the most accurate method for the detection and measurement of plaque is a combination of the real-time image and flow disturbance in the downstroke of systole. In the 25–49% category, the best combination is the image plus full spectral broadening. These results are shown in Tables 15.6 and 15.7. More recently Rubin et al (1987) have compared both angiography and duplex scanning with surgical and pathology findings, and have shown both to be accurate for determining diameter reduction, but that duplex scanning is more accurate for detecting important plaque characteristics.

Table 15.6 Comparison of Doppler spectral analysis estimation of the diameter stenosis compared with biplanar angiography. (Hames et al 1985)

Diameter of stenosis by biplanar angiography	Diameter of stenosis by Doppler spectral analysis			
	Normal	1–24%	25–49%	Total
Normal	13	8	1	22
1–24%	8	20	2	30
25–49%	0	6	11	17

Table 15.7 Comparison of duplex scanning against biplanar angiography in the measurement of diameter stenosis in the internal carotid artery. (Hames et al 1985)

Diameter of stenosis by biplanar angiography	Diameter of stenosis by duplex scanning			
	Normal	1–24%	25–49%	Total
Normal	11	10	1	22
1–24%	1	25	4	30
25–49%	0	4	13	17

An example of a duplex scan of a normal carotid bifurcation is shown in Fig. 15.5. Figures 15.6 and 15.7 show the duplex information from a patient with a moderate stenosis (approx. 40%) and a patient with a major stenosis of the internal carotid (>80%).

15.2.1.7 Colour flow mapping

Colour flow mapping (CFM) systems produce a two-dimensional colour image of the Doppler shift frequency information from a predetermined scan plane within the body (see Sections 4.5.3 and 4.6). This colour flow map is superimposed on the real-time gray scale image and allows the visualization of flow disturbances in relation to the anatomical

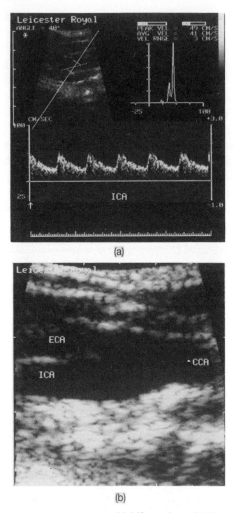

Figure 15.5   (a) Duplex scan of a normal carotid bifurcation. (b) Enlarged image from (a)

features present in the blood vessel. A number of examples of colour flow maps from the neck vessels are shown in Plates 15.1–15.6. CFM devices represent a major step forward in the simultaneous visualization of flow disturbance and plaque morphology, and in the study of their interrelationship.

15.2.2 Vertebral arteries

Miyazaki (1966) showed it is possible to study the vertebral arteries using Doppler shift velocimeters. Kaneda et al (1977) produced a method of assessment based on the measurement of the systolic and diastolic heights of the sonogram. Although this index was not normalized, the diagnostic reliability of the percutaneous vertebral Doppler

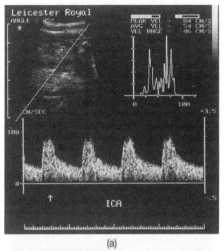

(a)

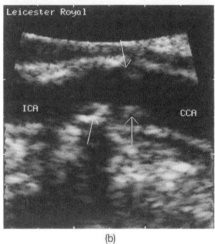

(b)

Figure 15.6 (a) Duplex scan of an internal carotid artery with a diameter stenosis of approximately 40%. (b) Enlarged image from (a)

technique was 67% where no flow was detected, 94% for poor flow, 97% for normal vessels, and 92% for all vessels examined.

Keller et al (1976b) used a peroral route and achieved a correct diagnosis in 82% of cases, with a 90% reliability in those cases of subclavian steal syndrome, missing vertebral artery, and normal findings.

Ackerstaff (1985) gave an excellent review of Doppler ultrasound in investigation of the vertebrobasilar arterial system. The results for the use of duplex ultrasound in the detection of stenoses of greater than 50% diameter in the innominate, subclavian and vertebral arteries are shown in Table 15.8.

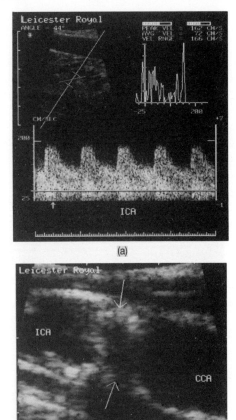

Figure 15.7   (a) Duplex scan of an internal carotid artery with a major diameter stenosis of greater than 80% diameter reduction. (b) Enlarged image from (a). Although the image appears to show a complete occlusion, the Doppler signal shows it to be a high grade stenosis. Note also that because of the very high velocities aliasing is present on the sonogram

Table 15.8   The success of duplex ultrasound in the study of stenosis of diameter greater than 50% of the innominate, subclavian and vertebral arteries. (Ackerstaff 1985)

Vessel	Sensitivity (%)	Specificity (%)	Positive predictive value (%)	Negative predictive value (%)	Accuracy (%)
Innominate	100	99	75	100	99
Right subclavian	43	99	75	96	95
Left subclavian	88	98	78	99	97
Vertebral	81	88	67	94	87

15.3   THE INTRACRANIAL CIRCULATION

15.3.1   Adults

Aaslid et al (1982) first reported the use of Doppler ultrasound to examine the intracerebral circulation through the intact skull. Ultrasound at 2 MHz was transmitted through the temporal bone just above the zygomatic arch. By pulsing the ultrasound the middle cerebral artery (MCA), anterior cerebral artery (ACA) and posterior cerebral artery (PCA) can be examined. With the transcranial approach all the arteries lie approximately in line with the direction of the ultrasound beam, so that the cosine term in the Doppler equation is approximately unity. This means that the absolute velocity can be calculated. The normal range of the time-averaged peak velocity in these arteries is 62 ± 12 cm s^{-1} for the MCA, 51 ± 12 cm s^{-1} for the ACA, and 44 ± 11 cm s^{-1} for the PCA.

Aaslid (1986a) described in detail the techniques used in the transcranial Doppler examination. There are three main approaches to gain access to the intracranial arteries, namely:

1. The transcranial approach to the basal arteries including the circle of Willis
2. The transorbital route to the carotid siphon, first described by Spencer (1983)
3. The suboccipital or transforamenal route to the basilar artery and the intracranial segments of the vertebral arteries.

15.3.1.1   Terminal internal carotid artery

In order to identify the terminal internal carotid artery, Aaslid recommends a depth setting on the pulsed Doppler of between 55 and 65 mm. The terminal ICA is defined as the position where the ACA branches off. The sample volume of the flowmeter is moved in a superior/inferior direction to track along the ICA. The signals from the most superior part will be of lower frequency than those from the MCA or ACA because of the less favourable angle of inclination, and originate from approximately the plane of the circle of Willis. Aaslid also points out that manual compression of the ipsilateral common carotid artery usually results in cessation of flow in the ICA.

15.3.1.2   Middle cerebral artery

The middle cerebral artery runs laterally and slightly anteriorly as a continuation of the intracranial ICA. It is normally the only artery seen between 25 and 50 mm from the temporal window, and has the highest velocities towards the probe. The signal diminishes with ipsilateral compression of the CCA.

15.3.1.3   Anterior cerebral artery

The anterior cerebral artery is the last branch of the ICA. It initially runs medially and then more anteriorly until it reaches the anterior communicating artery close to the midline of the brain. It then ascends around the genu of the corpus callosum. Aaslid shows that the ACA is found at the termination of the ICA and can be followed down to the brain midline. Flow is away from the probe.

15.3.2 Role of transcranial Doppler

Transcranial Doppler in the study of the adult intracerebral circulation has increased enormously in the last four years. Aaslid (1986b) has produced an excellent book devoted to this subject, and it is recommended to all those interested in the technique. Transcranial Doppler has been used to evaluate intracranial stenoses (Aaslid and Lundegaard 1986) and cerebral arteriovenous malformations (Lundegaard et al 1986), in the evaluation of cerebral vasospasm (Seiles and Aaslid 1986), and in the study of cerebral haemodynamics in general.

15.3.3 Neonatal cerebral circulation

Doppler ultrasound offers the neonatologist the opportunity to monitor cerebral haemodynamics non-invasively. Initially measurements were made from the anterior cerebral artery through the anterior fontanelle (Bada et al 1979), but it has since been realized that good images and Doppler signals may also be obtained from the middle cerebral artery by placing a transducer over the temporal bone. A typical duplex scan of the normal neonatal brain showing flow in the anterior cerebral artery is shown in Fig. 15.8.

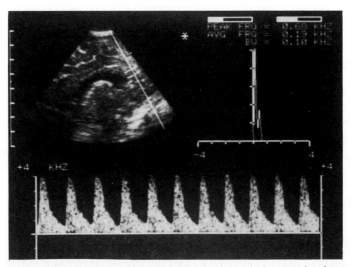

Figure 15.8 Duplex scan of the normal neonatal brain through the anterior fontanelle showing flow in the anterior cerebral artery

The use of Doppler has been reported in a wide variety of pathological conditions including birth asphyxia (Bada et al 1979, Archer et al 1986), intraventricular haemorrhage (Bada et al 1979), patent ductus arteriosus (Perlman et al 1981, Lipman et al 1982), pneumothorax (Hill et al 1982), hydrocephalus (Hill and Volpe 1982), respiratory distress syndrome (Perlman et al 1983) and brain death (McMenamin and Volpe 1983).

Drayton and Skidmore (1986) and Archer and Evans (1988) have recently reviewed the value of Doppler ultrasound in the neonatal cerebral circulation. Much of its value may lie in helping to elucidate the pathophysiology of various intracranial abnormalities rather than as a clinical tool.

One potentially valuable clinical application is for predicting the outcome of perinatal asphyxia, and in a recent study (Archer et al 1986) of a group of 43 term infants with clinical neurological manifestations of intrapartum asphyxia, the Doppler waveform, analysed in terms of Pourcelot's resistance index, predicted adverse outcome with a sensitivity of 100% and a specificity of 81%. A preliminary study of principal component analysis indicates that this type of analysis may be even better at identifying babies with abnormal cerebral haemodynamics (Evans et al 1985).

15.4 SUMMARY

Continuous wave velocimeters, quantitative waveform analysis methods and spectral analysis still have a part to play in the detection of extracranial carotid disease. If the major question to be answered is 'Is the stenosis greater or less than 50% of the diameter?', a combination of these techniques will distinguish these two categories of disease.

For a more detailed description of plaque morphology and the associated flow disturbances, and also as a means of monitoring progression/regression of atheromatous plaques, the duplex scanner is the instrument of choice. The addition of colour flow mapping allows a more rapid assessment of the severity of the disease, and also a means of studying the interaction of flow disturbance with the underlying plaque morphology.

The study of the intracranial circulation in adults is a rapidly expanding field and the use of dedicated transcranial Doppler instruments is helping in the general study of cerebral haemodynamics.

The use of duplex scanners in the neonatal cerebral circulation is increasing and much of the value of the method may be in helping to elucidate the pathophysiology of various abnormalities.

15.5 REFERENCES

Aaslid R (1986a) Transcranial Doppler examination techniques. In: Transcranial Doppler sonography (Ed. R Aaslid), pp 39–59, Springer-Verlag, Vienna, New York.

Aaslid R (1986b) Transcranial Doppler sonography. Springer-Verlag, Vienna, New York.

Aaslid R, Lundegaard KF (1986) Cerebral haemodynamics. In: Transcranial Doppler sonography (Ed. R Aaslid), pp 60–85, Springer-Verlag, Vienna, New York.

Aaslid R, Markwalder T, Nornes H (1982) Noninvasive transcranial Doppler ultrasound recording of flow-velocity in basal cerebral arteries. J Neurosurg 57, 769–774.

Ackerstaff RGA (1985) Ultrasonic duplex scanning in atherosclerotic disease of the vertebrobasilar arterial system, Drukkerij Elinkwijk, Utrecht.

Aldoori MIK (1986) Ultrasound and related studies in carotid disease. PhD Thesis, University of Bristol.

Archer LNJ, Evans DH (1988) Doppler assessment of the neonatal cerebral circulation. In: Fetal and neonatal neurology and neurosurgery (Eds MI Levene, MJ Bennett, J Punt), Chapter 10, Churchill Livingstone, Edinburgh.

Archer LNJ, Levene MI, Evans DH (1986) Cerebral artery Doppler ultrasonography for prediction of the outcome after perinatal asphyxia. Lancet, ii, 1116–1118.

Archie JP (1981) A simple, non-dimensional, normalized common carotid Doppler velocity waveform index that identifies patients with carotid stenosis. Stroke 12, 322–329.

Bada HS, Hajjar W, Chua C, Sumner DS (1979) Non-invasive diagnosis of neonatal asphyxia and intraventricular haemorrhage by Doppler ultrasound. J Paediatr 71, 298–299.

Barnes RW, Russell HE, Bone GE, Slaymaker EE (1977) Doppler cerebrovascular examination: improved results with refinements in technique. Stroke 8, 468–471.

Baskett JJ, Beasley MG, Murphy GJ, Hyams DE, Gosling RG (1977) Screening for carotid junction disease by spectral analysis of Doppler signals. Cardiovasc Res XI, 147–155.

Bluth EI, Kay D, Merritt CRB, Sullivan M et al (1986) Sonographic characteristics of carotid plaque: detection of hemorrhage. Am J Roent 146, 1061–1065.

Brockenborough EC (1969) Screening for the prevention of stroke: use of a Doppler flowmeter. Washington/Alaska Regional Medical Program.

Drayton MR, Skidmore R (1986) Doppler ultrasound in the neonate. Ultrasound Med Biol 12, 761–772.

Evans DH, Archer LNJ, Levene MI (1985) The detection of abnormal neonatal cerebral haemodynamics using principal component analysis of the Doppler ultrasound waveform. Ultrasound Med Biol 11, 441–449.

Greene FM, Beach K, Strandness DE Jr, Fell G, Phillips DJ (1982) Computer based pattern recognition of carotid arterial disease using pulsed Doppler ultrasound. Ultrasound Med Biol 8, 161–176.

Hames TK, Ratliff DA, Humphries KN, Gazzard VM, Birch SJ, Chant ADB (1985) The accuracy of duplex scanning in the evaluation of early carotid disease. Ultrasound Med Biol 11, 819–825.

Hill A, Volpe JJ (1982) Decrease in pulsatile flow in the anterior cerebral arteries in infantile hydrocephalus. Pediatrics 69, 4–7.

Hill A, Perlman JM, Volpe JJ (1982) Relationship of pneumothorax to occurrence of intraventricular haemorrhage in the premature newborn. Pediatrics 69, 144–149.

Kaneda H, Irino T, Minami T, Taneda M (1977) Diagnostic reliability of the percutaneous ultrasonic Doppler techniques for vertebral arterial occlusive disease. Stroke 8, 559–572.

Keller HM, Meier WE, Yonekawa Y, Kumpe DA (1976a) Non-invasive angiography for the diagnosis of carotid artery disease using Doppler ultrasound (carotid artery Doppler). Stroke 7, 354–363.

Keller HM, Meier WE, Kumpe DA (1976b) Noninvasive angiography for the diagnosis of vertebral artery disease using Doppler ultrasound (vertebral artery Doppler). Stroke 7, 364–369.

Langlois Y, Roederer GO, Chan A, Phillips DJ, Beach KW, Martin D, Chiko PM, Strandness DE Jr (1983) Evaluating carotid artery disease. The concordance between pulsed Doppler/spectrum analysis and angiography. Ultrasound Med Biol 9, 51–63.

Lipman B, Serwer GA, Brazy JE (1982) Abnormal cerebral hemodynamics in preterm infants with patent ductus arteriosus. Pediatrics 69, 778–781.

Lundegaard KF, Aaslid R, Nornes H (1986) Cerebral arteriovenous malformations. In: Transcranial Doppler sonography (Ed. R Aaslid), pp 86–105, Springer-Verlag, Vienna, New York.

McMenamin JB, Volpe JJ (1983) Doppler ultrasonography in the determination of neonatal brain death. Ann Neurol 14, 302–307.

Martin TRP, Barber DC, Sherriff SB, Prichard DR (1980) Objective feature extraction applied to the diagnosis of carotid artery disease using a Doppler ultrasound technique. Clin Phys Physiol Meas 1, 71–81.

Miyazaki M (1966) Measurement of cerebral blood flow by ultrasonic Doppler technique. Haemodynamic correlation of internal carotid artery and vertebral artery. Jap Circ J 30, 981–985.

Padayachee TS, Lewis RR, Gosling RG (1982) Detection of carotid bifurcation disease: comparison of ultrasound tests with angiography. Br J Surg 69, 218–224.

Perlman JM, Hill A, Volpe JJ (1981) The effect of patent ductus arteriosus on flow velocity in the anterior cerebral arteries: ductal steal in the premature newborn infant. J Pediatr 99, 767–771.

Perlman JM, McMenamin JB, Volpe JJ (1983) Fluctuating cerebral blood flow velocity in respiratory distress syndrome. New Engl J Med 309, 204–209.

Planiol T, Pourcelot L (1971) Etude de la circulation carotienne au moyen de l'effet Doppler. In: Traites de Radiodiagnostique, vol 17, Masson, Paris.

Reilly LM, Lusby RJ, Hughes L, Ferrell LD, Stoney RJ, Ehrenfeld WK (1983) Carotid plaque histology using real-time ultrasonography. Clinical and therapeutic implications. Am J Surg 146, 188–193.

Reinertson JE, Barnes RW (1977) Carotid flow velocity scattering: diagnostic value in Doppler ultrasonic arteriography. Clin Res 24, 594a.

Rittgers SE, Thornhill BM, Barnes RW (1983) Quantitative analysis of carotid artery Doppler spectral waveforms: diagnostic value of parameters. Ultrasound Med Biol 9, 255–264.

Rubin JR, Bondi JA, Rhodes RS (1987) Duplex scanning versus conventional arteriography for the evaluation of carotid artery plaque morphology. Surgery 102, 749–755.

Rutherford RB, Hiatt WR, Kreutzer EW (1977) The use of velocity waveform analysis in the diagnosis of carotid artery occlusive disease. Surgery 82, 695–701.

Seiles RW, Aaslid R (1986) Transcranial Doppler for evaluation of cerebral vasospasm. In: Transcranial Doppler sonography (Ed. R Aaslid), pp 118–131, Springer-Verlag, Vienna, New York.

Sheldon CD (1985) Doppler ultrasound in the assessment of cerebro-vascular ischaemia. PhD Thesis, University of Glasgow.

Sherriff SB, Barber DC, Martin TRP, Lakeman JM (1982) Use of principal component factor analysis in the detection of carotid artery disease from Doppler ultrasound. Med Biol Eng Comp 20, 351–356.

Shoumaker RD, Bloch S (1978) Cerebrovascular evaluation: assessment of Doppler scanning of carotid arteries. Ophthalmic Doppler flow and cervical bruits. Stroke 9, 563–566.

Spencer MP (1983) Intracranial carotid artery diagnosis with transorbital pulsed wave (PW) and continuous wave (CW) Doppler ultrasound. J Ultrasound Med Suppl 2, 61.

Spencer MP, Reid JM (1979) Quantitation of carotid stenosis with continuous wave (CW) Doppler ultrasound. Stroke 10, 326–330.

Spencer MP, Brockenborough EC, Davies DL, Reid JM (1977) Cerebrovascular evaluation using Doppler CW ultrasound. In: Ultrasound in medicine, vol 3B (Eds DN White, RE Brown), pp 1291–1310, Plenum Press, New York.

Sumner DS, Moore DJ, Miles RD (1985) Doppler ultrasonic arteriography and flow velocity analysis in carotid artery disease. In: Noninvasive diagnostic techniques in vascular disease (Ed. EF Bernstein), pp 349–366, CV Mosby, St Louis, Missouri.

White DN, Curry GR (1978) A comparison of 424 carotid bifurcations examined by angiography and the Doppler echoflow. In: Ultrasound in medicine, vol 4 (Eds DN White, EA Lyons), pp 363–376, Pergamon Press, New York.

Woodcock JP, Shedden J, Aldoori M, Skidmore R, Burns P, Evans J (1983) Doppler spectral broadening and anomalous vessel wall movement in the study of atherosclerosis of the carotid arteries. In: Ultrasound '82 (Eds RA Lerski, P Morley), pp 235–237, Pergamon Press, Oxford.

Chapter 16

DOPPLER APPLICATIONS IN CARDIOLOGY

16.1 INTRODUCTION

Over the last five years the number of cardiological applications of Doppler ultrasound has increased dramatically. This chapter is meant only as a brief introduction to these applications, and readers who require a more detailed treatment of the subject should consult one of the standard texts such as Nanda (1985) or Hatle and Angelsen (1985).

Doppler ultrasound is used in cardiology both to investigate the flow in major vessels close to the heart for the purpose of monitoring cardiac performance and output, and to study the flow patterns within the cardiac chambers themselves. Useful measurements may be made in the great vessels using relatively simple equipment, but some kind of imaging is usually required for intracardiac studies.

A major technical problem with cardiac Doppler is that very high velocities are encountered deep within the body, and with pulsed Doppler systems this leads either to aliasing if the pulse repetition frequency is too low, or alternatively to range ambiguity if the PRF is raised (see Sections 8.4.1.1 and 8.4.1.2).

16.2 TRANSCUTANEOUS AORTOVELOGRAPHY

Transcutaneous aortovelography (TAV) is the study of the blood velocity–time waveform in the aortic arch together with its relationship to cardiac function and performance (Light 1976, Sequeira et al 1976, Buchtal et al 1976, Mowat et al 1983). With the transducer placed in the suprasternal notch and angled towards the aortic arch the flow in the arch is almost directly away from the transducer, so that the angle of inclination between the direction of flow and the ultrasound beam is approximately zero. This means that the absolute velocity can be measured. Because of the cosine dependence of the Doppler shift and the 'in-line' incidence there is a wide tolerance on the angle of inclination, for example a variation of $18°$ from exactly in-line incidence produces an error of only 5% in the calculated velocity.

TAV can be used to monitor serial changes in central blood flow, to measure blood flow patterns in various cardiac lesions such as aortic incompetence and left-to-right shunts, as a guide to therapy in intensive care, and to study normal physiological responses. Table 16.1 is a summary of results obtained using TAV compiled by Light (1979).

Light (1986) has recently discussed how reliable information can be gained from

Table 16.1   Summary of results obtained using the TAV method (Light 1979)

Trial	Deviation from exact agreement or proportionality (SD as % of mean)	Range
Reproducibility		
within observer	6.6	42 subjects, 3–67 years,
between observers	7.3	14 observers
Cardiac output	11.0	Cardiac output
(acetylene)		7–22 l/min (13 normals)
Intra-aortic blood velocity	6.0	Vpk $=48$–120 cm s^{-1}
		(20 IHD patients)
Stroke volume	13.0	SV $= 20$–160 ml
(indocyanine green)		(20 IHD patients)
Cardiac output	9.7	CO $= 2.5$–10 l/min
(thermal dilution)		(14 patients)
Regurgitant fraction	s.e.e. $= 0.06$	Regurgitant fraction $= 0.2$–0.8
(Fick)		(10 patients with AR)

Vpk, peak velocity; IHD, ischaemic heart disease; AR, aortic regurgitation; SV, stroke volume; CO, cardiac output.

Doppler measurements in the aorta, and in particular which velocity-derived measures are most useful clinically. The concept of stroke distance, i.e. the distance along the aorta that the blood moves during each cardiac cycle, is important (Haites et al 1984). Abnormally low stroke distance, marked respiratory variation in velocity without obvious cause, and abnormally rounded waveforms all indicate forms of circulatory embarrassment. Acceleration responds to both myocardial condition and to inotropic state and should be used in conjunction with other variables such as peak velocity and stroke distance to distinguish the two. In the study of myocardial impairment, responses to stress provide the best indicators.

16.3   DOPPLER ECHOCARDIOGRAPHY

Johnson et al (1973) first reported the use of a range-gated Doppler flowmeter linked to an M-mode scan to study blood flow in the heart. Since then enormous progress has been made in probe technology and pulsed/CW Doppler flowmeter design to produce duplex scanners with high-resolution real-time images and sensitive flowmeters. The use of Bernoulli's equation with velocity information derived from such systems allows the calculation of pressure drops across valves and septal defects. The simplest use, however, is the detection and study of abnormal flow velocities.

16.3.1   Intracardiac flow patterns

16.3.1.1   Mitral valve

From a qualitative examination of Doppler-shift spectra it is possible to detect flow disturbance due to jets and regurgitation. A normal mitral valve flow pattern recorded from the ventricular side of the valve orifice is shown in Fig. 16.1. The flow spectrum is very similar to the normal M-mode scan of the mitral valve.

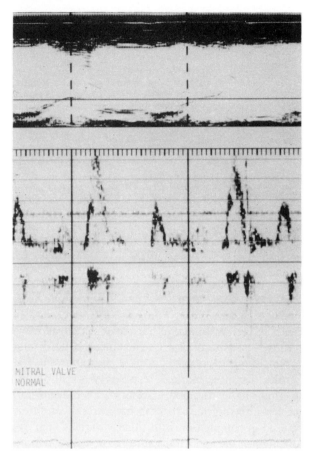

Figure 16.1   Normal mitral valve flow pattern

An example of mitral stenosis is shown in Fig. 16.2. The spectrum shows high velocities declining slowly; again, very similar to the M-mode appearance in mitral stenosis. Using this flow signal the pressure drop across the valve orifice can be calculated (see Section 16.3.2).

In cases of suspected mitral regurgitation the sample volume of the flowmeter is placed in the left atrium just behind the valve leaflet, and the presence of retrograde flow sought. Adhar et al (1985) have produced a classification system for grading the degree of regurgitation:

Grade 1 (mild): abnormal systolic flow confined to the area just above the mitral valve;
Grade 2 (moderate): flow extended to the proximal one-third of the atrium;
Grade 3 (moderately severe): regurgitant jet detected half-way up the left atrium;
Grade 4 (severe): systolic flow detected beyond mid-atrial level.

Colour flow mapping is well suited to this type of classification.

Adhar et al also reported on the use of M-mode, two-dimensional (2D) and duplex scanning in the study of mitral valve prolapse. Their results are shown in Table 16.2. It can

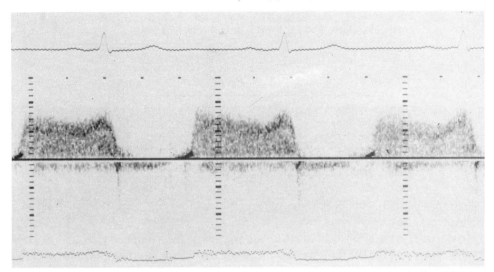

Figure 16.2   Typical flow spectrum recorded in mitral stenosis. The initial high velocity which falls slowly is very similar to the M-mode appearance

Table 16.2   The role of Doppler echocardiography in mitral valve prolapse. The addition of either two-dimensional or Doppler to M-mode increases the sensitivity of the detection. Sensitivity is increased further by the use of all three techniques. (Adhar et al 1985)

Method	Number of cases	Sensitivity (%)
M-mode	62	50
2D	85	68
Doppler	90	72
M-mode + 2D	96	77
M-mode + Doppler	100	80
2D + Doppler	111	89
M-mode, 2D, Doppler	116	93

be seen that for the highest sensitivity all three methods should be combined, but that 2D and Doppler are together almost as good.

16.3.1.2   Aortic valve

Stenosis of the aortic valve results in a high-velocity jet of blood in the ascending aorta. This jet is investigated mainly via the suprasternal notch and the right parasternal transducer position to locate the highest-pitch audio signal. Typical normal blood velocities through the various heart valves of children and adults are given in Table 16.3. A typical velocity detected in aortic stenosis is of the order of 4 m s^{-1}.

Quantitation of the pressure drop across the aortic valve can be achieved using

Table 16.3   Typical maximal blood flow velocities recorded from the hearts of children and adults. (Hatle and Angelsen 1985)

Site	Velocity (m s^{-1})			
	Children		**Adults**	
Mitral flow	1.00	(0.8–1.3)	0.90	(0.6–1.3)
Tricuspid flow	0.60	(0.5–0.8)	0.50	(0.3–0.7)
Pulmonary artery	0.90	(0.7–1.1)	0.75	(0.6–0.9)
Left ventricle	1.00	(0.7–1.2)	0.90	(0.7–1.1)
Aorta	1.50	(1.2–1.8)	1.35	(1.0–1.7)

Bernoulli's equation but it is more difficult to perform than with the other valves or prosthetic valves.

In aortic regurgitation it is usual to use the apical five-chamber view and place the sample volume in the left ventricular outflow tract. The sample volume should be tracked from side to side of the outflow tract to check whether the jet is non-parallel to the aortic root walls. Other apical views should also be used to check the three-dimensional structure of the jet. Bommer et al (1981) suggest a short axis view approach when determining the extent of the regurgitant jet. This allows for the magnitude of the regurgitant jet in all three dimensions by defining the extent of the jet in the following way. The jet is described as level 1, if the jet extends to just below the level of the aortic valve; level 2, at the level of the mitral valve leaflets; and level 3, at the papillary muscles. Gross et al (1985) state that the sensitivity and specificity for the Doppler detection of aortic regurgitation are both in excess of 90%. A typical spectrum recorded from the outflow tract is shown in Fig 16.3.

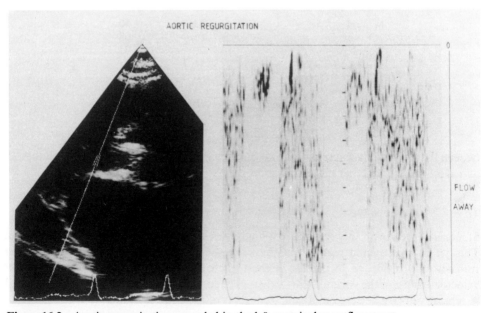

Figure 16.3   Aortic regurgitation recorded in the left ventricular outflow tract

16.3.1.3   Tricuspid valve

In order to investigate flow through the tricuspid valve it is usual to use the apical four-chamber view, with the sample volume placed in the valve orifice or in the right ventricle just distal to the valve orifice. Normal values are shown in Table 16.3. The normal Doppler spectrum has two peaks, the first due to rapid early filling and the second related to atrial contraction.

Tricuspid stenosis can be measured with the Doppler sample volume distal to the valve orifice, and tricuspid regurgitation can be detected with the sample volume in the right atrium. Quinones (1985) showed that Doppler ultrasound is a very sensitive method for detecting tricuspid regurgitation, and quoted figures for sensitivity of 94% and specificity of 100%.

16.3.1.4   Pulmonary valve

The right ventricular outflow tract is best viewed via the parasternal short axis where the sample volume can be placed parallel to the flow direction. A typical normal flow pattern is shown in Fig. 16.4.

Pulmonary stenosis can be investigated with the sample volume placed in the pulmonary artery just distal to the valve. Regurgitation is measured with the sample volume in the right ventricular outflow tract. Stevenson et al (1980) showed that pulmonary insufficiency can be detected with a sensitivity of 95% and an associated specificity of 88%.

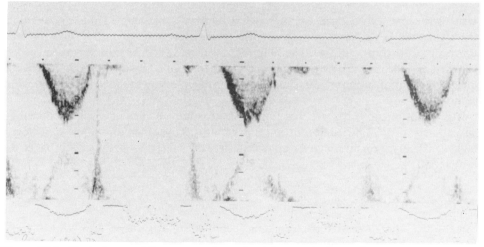

Figure 16.4 Typical normal flow pattern from the right ventricular outflow tract

16.3.1.5   Prosthetic valves

The investigation of prosthetic heart valves using Doppler ultrasound is well established. The investigations are carried out as described previously for normal valves. In most cases the flow velocities encountered in the prosthetic valve are higher than normal, and thus in 'normally' functioning prosthetic valves, high pulse repetition rates may need to be used to overcome the problems due to aliasing.

16.3.1.6 Atrial septal defect

The problem in atrial septal defect is to detect flow from one atrium to the other through the septum. Stevenson (1984) recommends the use of the subcostal four-chamber view to provide optimum imaging of the septum and Doppler orientation. The sample volume is moved along either side of the septum. Either by locating the side of the septum with maximum flow disturbance, or by locating the sample volume within the imaged defect, the direction of flow can usually be determined. Quantification of the shunt can be obtained by measuring the left and right heart cardiac output. This is achieved in the pulmonary artery by placing the sample volume just distal to the pulmonary valve. From a high precordial approach the ultrasound beam will be approximately in-line with flow so the cosine term in the basic Doppler equation is unity. Stevenson suggests that the measurement of both the aortic and the pulmonary artery diameters is carried out using the M-mode. Both long and short axis views are taken at the same time in the cardiac cycle and an average diameter used. Flow in the aorta is measured using a high right parasternal approach or a subcostal approach, and the sample volume is placed in the aortic root. The volume flow in each vessel is found by multiplying its cross-sectional area by the measured time-averaged velocity. The difference between the two gives an estimate of the size of the shunt.

16.3.1.7 Ventricular septal defect

Usually the long axis parasternal approach is used with the sample volume swept along the right side of the septum. Large defects may produce disturbed flow patterns throughout the right ventricle. The Doppler spectrum of the jet shows marked spectral broadening reflecting the flow disturbance, and the audio signal shows an increase in intensity. Quantification of the shunt can be attempted as described in the previous section.

It is important to recognize that not all ventricular septal defects will produce high velocity jets. If left ventricular function is poor, or right ventricular outflow resistance is high, ventricular pressures may be approximately the same and no flow disturbance will be detected.

16.3.1.8 Diagnostic accuracy

Stevenson (1984) reviewed the experience of the Department of Cardiology at the University of Washington in the use of Doppler echocardiography over the whole range of valvular and congenital heart disease, and his findings are shown in Table 16.4.

16.3.2 Quantification

One of the major advances in the clinical use of cardiac Doppler has been in the quantification of absolute velocities, pressure drops and valve orifice areas. Hatle and Angelsen (1985) have listed the normal velocity ranges found at various intracardiac sites

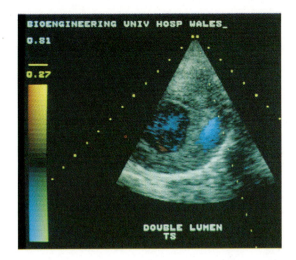

Plate 14.1 Colour flow map of an aortic cross-section showing an aneurysm with a double lumen. The dilated artery and mural thrombus can be clearly seen.

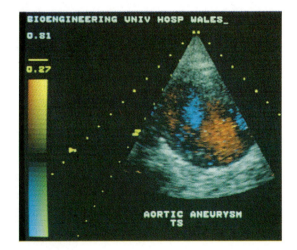

Plate 14.2 Longitudinal view of an abdominal aortic aneurysm showing marked dilation of the artery and a complex flow pattern.

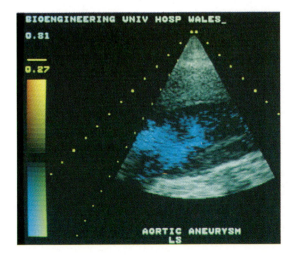

Plate 14.3 Longitudinal view of abdominal aortic aneurysm showing the dilated artery and mural thrombus.

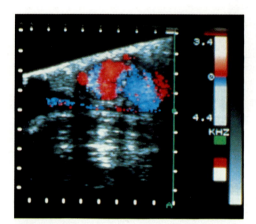

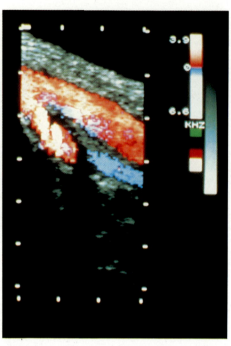

Plate 14.4 Complex flow pattern in an aneurysmal dilatation of the site of anastomosis in an axillofemoral graft. (Courtesy of Phillips Medical Systems.)

Plate 14.5 Femoral artery bifurcation with moderate wall disease. The femoral vein is seen in blue. (Courtesy of Phillips Medical Systems.)

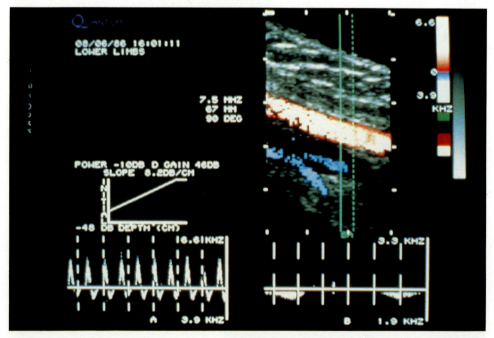

Plate 14.6 Dual Doppler spectra taken simultaneously from the femoral artery (A) and femoral vein (B). (Courtesy of Phillips Medical Systems.)

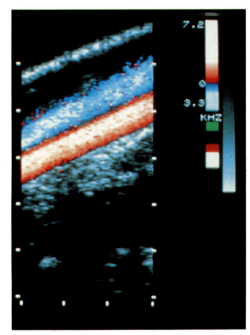

Plate 15.1 Colour flow map showing the common carotid in red and the jugular vein in blue. Under normal flow conditions the colour fills the vessels from wall to wall. The flow appears to be laminar because higher velocities (lighter red) can be seen in the centre of the vessel. (Courtesy of Phillips Medical Systems.)

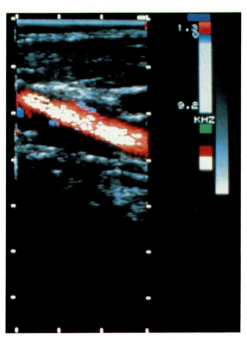

Plate 15.2 A high-velocity jet is displayed as white in this example. (Courtesy of Phillips Medical Systems.)

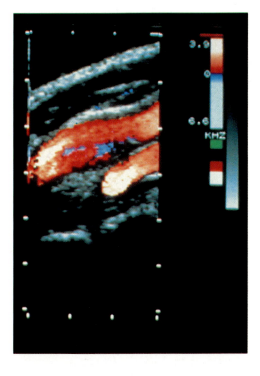

Plate 15.3 Common carotid artery with soft plaque formation at the lip of the endarterectomy site. The vertebral artery is seen posteriorly. The blue flow pattern is caused by the stenosis. (Courtesy of Phillips Medical Systems.)

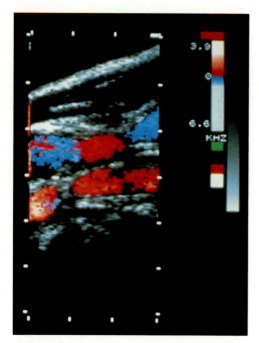

Plate 15.4 Complex flow patterns due to arterial wall roughening in a diseased carotid. Complete flow reversal is displayed in blue and plaque is evident along the posterior wall. (Courtesy of Phillips Medical Systems.)

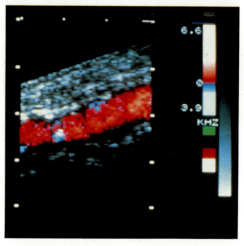

Plate 15.5 Sonolucent wall thickening and intimal roughening produces small, localized eddy current (blue). (Courtesy of Phillips Medical Systems.)

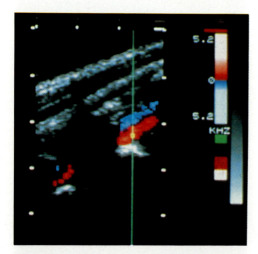

Plate 15.6 Normal vertebral artery shown in red. The vertebral vein can be seen anteriorly. (Courtesy of Phillips Medical Systems.)

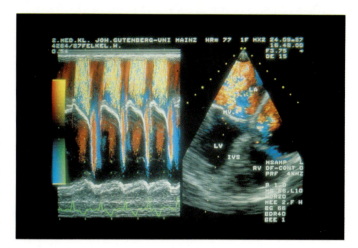

Plate 16.1 Mitral regurgitation detected via an oesophageal probe. The marked mosaic pattern in the left atrium is shown in both the M-mode and real-time display. (Courtesy of Toshiba Medical Systems.)

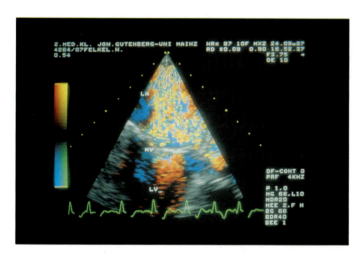

Plate 16.2 Severe mitral regurgitation imaged using an oesophageal probe. (Courtesy of Toshiba Medical Systems.)

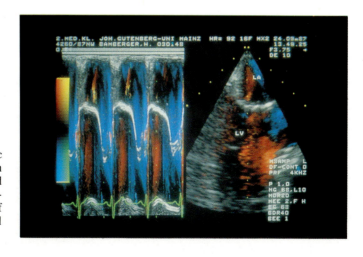

Plate 16.3 End systolic mild mitral regurgitation seen as the late systolic red regurgitant jet on the M-mode. (Courtesy of Toshiba Medical Systems.)

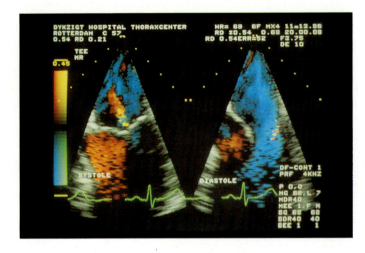

Plate 16.4 Recordings made via the oesophagus, of mild mitral regurgitation, in systole and diastole. (Courtesy of Toshiba Medical Systems.)

Plate 16.5 Transoesophageal demonstration of mitral regurgitation, showing the colour flow map, colour M-mode, and the Doppler spectrum. (Courtesy of Toshiba Medical Systems.)

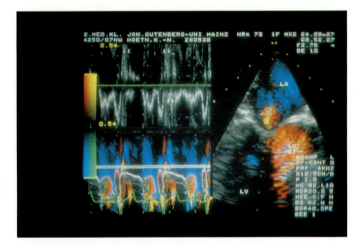

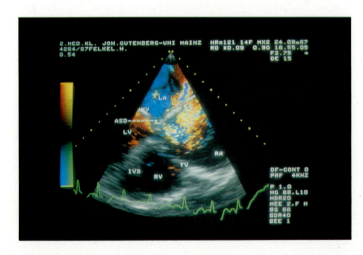

Plate 16.6 Atrial septal defect detected via the Transoesophageal route. (Courtesy of Toshiba Medical Systems.)

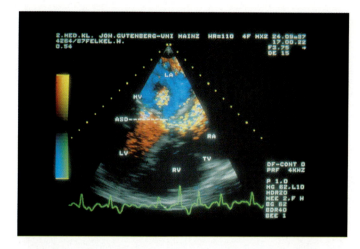

Plate 16.7 Atrial septal defect detected via the trans-oesophageal route. (Courtesy of Toshiba Medical Systems.)

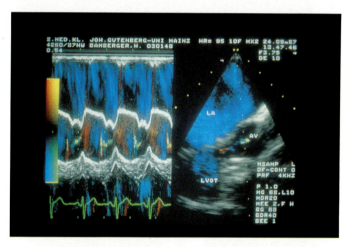

Plate 16.8 Turbulent flow at aortic valve orifice. The turbulence shows as the small area of green (via the oesophagus). (Courtesy of Toshiba Medical Systems.)

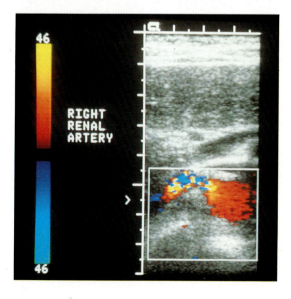

Plate 17.1 CFM of renal artery stenosis. (Courtesy of Acuson.)

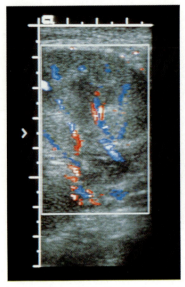

Plate 17.2 CFM of a renal transplant. (Courtesy of Acuson.)

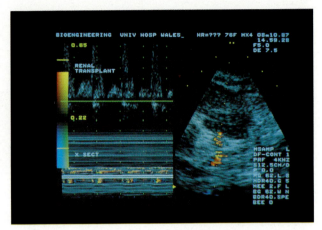

Plate 17.3 CFM of a renal transplant showing the flow spectrum from the common iliac artery, and a colour flow image of the renal artery.

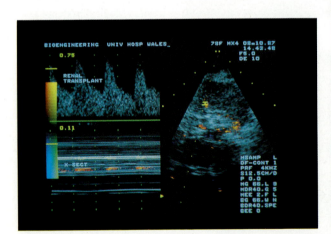

Plate 17.4 CFM showing perfusion in the cortex of a transplanted kidney.

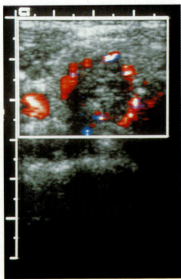

Plate 17.5 CFM of a thyroid mass with circumferential flow. (Courtesy of Acuson.)

Table 16.4 The diagnostic accuracy of Doppler echocardiography in the study of valvular and congenital heart disease. (Stevenson 1984)

Lesion	Sensitivity (%)	Specificity (%)	Positive predictive index (%)	Negative predictive index (%)
Aortic stenosis	100	83	97	—
Aortic insufficiency	100	83	88	—
Mitral regurgitation	95	90	95	90
Pulmonary insufficiency	95	93	93	96
Tricuspid regurgitation	100	100	—	—
Atrial septal defect	94	92	98	80
Ventricular septal defect	96	99	99	86

in children and adults, and these are reproduced in Table 16.3. Estimation of pressure drops and valve areas are discussed below.

16.3.2.1 Estimation of the pressure drop across stenosed valves

If P_1 and v_1 are the pressure and flow velocity at a point proximal to a constriction, and P_2 and v_2 the corresponding quantities distal to the constriction, the simplified Bernoulli equation states that:

$$P_1 - P_2 = \tfrac{1}{2}\rho(v_2^2 - v_1^2) \qquad 16.1$$

where ρ is the fluid density (see Section 2.6.1). If v_2 is very much greater than v_1, this equation reduces to:

$$P_1 - P_2 \simeq 4v_2^2 \qquad 16.2$$

where P_1 and P_2 are in mmHg, and v_2 is in m s^{-1}. This equation may be used to estimate the pressure drop across stenosed heart valves from the flow through the valve (but see Section 2.6.1 for limitations). This technique was first used by Holen et al (1976), and subsequently by Hatle et al (1978).

It is important to note that the Doppler velocimeter estimates the peak instantaneous pressure drop, whereas cardiac catheters are often used to measure the peak-to-peak pressure drop. The difference between these two measurements is shown in Fig 16.5. The Doppler estimate of the pressure drop across a valve will be higher than that calculated using the catheter.

16.3.2.2 Estimation of mitral valve orifice

The pressure drop across a valve is influenced by cardiac output and heart rate, and therefore is not totally reliable as a measure of stenosis severity. An alternative method of estimating the degree of stenosis, suggested by Libanoff and Rodbard (1966), is to measure the time taken for the pressure drop to fall to one-half of its initial value. This method has been adapted by Hatle et al (1979) for use with Doppler measurements of pressure drop. Because the pressure drop is proportional to velocity squared, a halving of pressure corresponds to a reduction of $1/\sqrt{2}$, i.e. 0.707 in velocity. Hatle et al observed a possible linear relationship between mitral valve area (MVA) and pressure half-time, $T_{\frac{1}{2}}$,

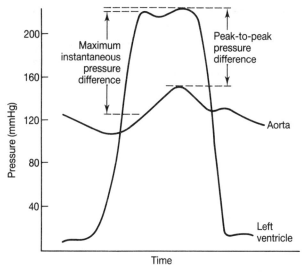

Figure 16.5 Direct pressure recordings in a patient with mitral stenosis. The Doppler flowmeter estimates peak instantaneous pressure difference whereas the cardiac catheter measures peak-to-peak pressure difference

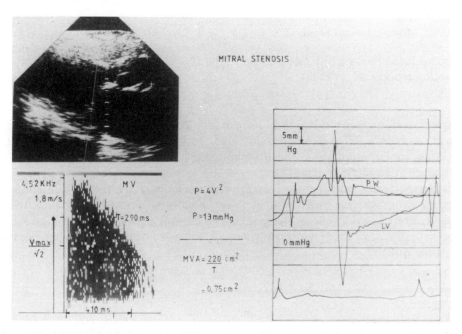

Figure 16.6 Real-time scan, Doppler spectrum and direct pressure recordings in a patient with mitral stenosis

when certain values are excluded, and Hatle and Angelsen (1982) suggested an empirical formula for estimating MVA from $T_{\frac{1}{2}}$, i.e.:

$$MVA = 220/T_{\frac{1}{2}} \qquad 16.3$$

where $T_{\frac{1}{2}}$ is measured in ms and MVA in cm^2.

In non-obstructive conditions the pressure half-time is between 20 ms and 60 ms. In the presence of obstruction it is greater than 100 ms. An example of the methods for the estimation of both pressure drop across the mitral valve and a measure of the effective valve orifice is shown in Fig. 16.6.

16.4 DOPPLER COLOUR FLOW MAPPING

It is possible, simultaneously, to produce high-resolution images of the cardiac chambers and valves, and to display velocity information. It has been shown previously that the pressure drop across stenosed valves and septal defects can also be accurately calculated using the velocity information. There are some problems, however, because the sample volume of the pulsed Doppler flowmeter can investigate only a relatively small volume at any one time. Flow disturbances such as regurgitation and stenotic jets are often eccentric and, if the conventional sample volume is placed in the wrong position, the presence and severity of the jet may be underestimated or even missed completely. This is also true for septal defects. These problems have been overcome by the development of colour flow mapping systems (Sections 4.5.3, 4.6 and 6.6.2). In the colour flow mapping system Doppler shifts are continuously sampled at multiple levels from each radial line in the sector plane. The result is a map of velocity distribution which is colour coded for flow direction and velocity. In most commercially available systems flow towards the transducer is coloured in shades of red, and away from the transducer in shades of blue.

General limitations of pulsed Dopplers have been discussed previously and it is important to realize that colour flow mapping systems suffer from the same disadvantages. Aliasing, which limits the measurement of high velocities, also occurs in the colour flow mapping system. On the colour display, a colour reversal is shown so that red areas will have a central blue zone and vice versa.

Turbulence is associated with an increased variance in the Doppler signal and this is usually indicated on the colour display by adding green. Thus, if turbulence or flow disturbance is present in a jet towards the transducer, the usual red display shifts to yellow. In conditions of flow away from the transducer the usual blue colour changes to cyan. In conditions of gross disturbance, when rotating vortices are present in front of the transducer, a mosaic of colours is produced.

Switzer and Nanda (1985) described the colour patterns seen in the normal heart using the standard scanning planes. To visualize the left ventricular inflow tract the transducer is placed in either the apical or low parasternal position and angled towards the head. Diastolic flow is seen as a diffuse red colour throughout the inflow tract. Systolic outflow is visualized as a blue colour primarily along the interventricular septum. In mitral regurgitation, when viewed from the standard apical or parasternal position, the regurgitant flow is seen as a blue area within the left atrium during systole. If turbulence is present, a mosaic of colour will also be seen. Switzer and Nanda grade the severity of the mitral incompetence as follows: Grade 1 is a blue area localized to a small area close to the valve; Grade 2 is a blue area reaching approximately 2 cm into the atrium from the plane of the mitral valve ring; Grade 3 refers to flow approaching the posterior and superior atrial walls.

Mitral stenosis, when studied from the apical position, appears as a red area within the ventricle, containing a blue central zone, due to aliasing, and a yellow peripheral zone due to turbulence.

Aortic regurgitation is best demonstrated by the parasternal and apical views. It is visualized as an abnormal area of red flow with a mosaic pattern, due to turbulence, in the left ventricular outflow tract. The degree of incompetence is assessed by the extent of the jet into the left ventricle, and also by the width of the jet. Switzer and Nanda point out that when the regurgitant jet is eccentric and towards either the interventricular septum or the anterior mitral valve leaflet, the conventional duplex scanner will not be able to grade the incompetence accurately. However, the colour flow mapping systems can visualize the full extent of the jet, and a measurement of the jet width correlates well with angiographic severity of the aortic incompetence.

In aortic stenosis a narrow blue jet will be visualized, entering the aortic root. Depending on the severity of the stenosis the central region will alias red and the mosaic appearance of turbulence may be seen.

In the normal tricuspid valve, seen in the apical long-axis view, inflow is displayed in red, extending from the orifice of the tricuspid valve into the right ventricular inflow tract. The right ventricular outflow tract and pulmonary artery most often demonstrate a uniform blue colour with small areas of turbulence around the valve cusps.

Tricuspid regurgitation is best visualized via the parasternal short-axis view. The jet is visualized during systole as a mosaic pattern in blue and yellow. Roelandt (1986) points out that significant tricuspid regurgitation is often a marker of pulmonary hypertension, and allows the estimation of right ventricular and pulmonary arterial systolic pressure from the peak velocity across the valve.

In pulmonary regurgitation, flow is visualized as a red jet extending from the centre of the pulmonary orifice backwards into the right ventricular outflow tract during diastole. Roelandt says that mild pulmonary regurgitation is seen in healthy subjects and has no clinical significance in the absence of other abnormalities and when velocities are low.

In the study of atrial septal defects the parasternal short axis or subcostal approach should be used. In the case of a left-to-right shunt, a reddish yellow flow area will be seen spreading from the left atrial cavity through the defect and towards the tricuspid orifice during systole. Colour flow mapping is particularly helpful in the detection of multiple septal defects and in single defects located atypically.

Ventricular septal defects can be examined from the apical four-chamber view. Roelandt shows that single or multiple defects are readily diagnosed. Smaller defects cause high-velocity jets with turbulence.

The advantages of colour flow mapping over conventional duplex cardiac scanning are:

1. Rapid detection and localization of single and multiple flow abnormalities
2. Visualization of flow jets such as those resulting from multiple septal defects which are difficult by conventional methods
3. Determination of the spatial orientation of jets, allowing accurate velocity and pressure calculations to be made using standard Doppler techniques
4. Study of flow patterns in prosthetic valves.

Some examples of colour flow mapping images of the heart are shown in Plates 16.1–16.8.

16.5 SUMMARY

Simple Doppler techniques can give valuable information about cardiac function. When used in conjunction with real-time 2D ultrasound and M-mode, Doppler ultrasound plays a part in both the qualitative and quantitative investigation of valvular and congenital heart disease. The addition of colour flow mapping allows the rapid detection, localization and assessment of the severity of regurgitant and stenotic lesions, and the detection of multiple septal defects.

16.6 REFERENCES

Adhar GC, Abbasi AS, Nanda NC (1985) Doppler echocardiography in the assessment of mitral regurgitation and mitral valve prolapse. In: Doppler echocardiography (Ed. NC Nanda), pp 188–210, Igaku-Shoin, New York, Tokyo.

Bommer WJ, Mapes R, Miller L, Mason DT, De Maria AN (1981) Quantitation of aortic regurgitation with two dimensional Doppler echocardiography. Am J Cardiol 47, 412 (abstract).

Buchtal A, Hanson GC, Peisach AR (1976) Transcutaneous aortovelography. Potentially useful technique in management of critically ill patients. Br Heart J 38, 451–456.

Gross CM, Wann LS, Nanda NC (1985) Doppler assessment of aortic regurgitation and dissection. In: Doppler echocardiography (Ed. NC Nanda), pp 220–261, Igaku-Shoin, New York, Tokyo.

Haites NE, McLennan FM, Mowat DHR, Rawles JM (1984) How far is the cardiac output? Lancet ii, 1025–1027.

Hatle L, Angelsen B (1982) Doppler ultrasound in cardiology: physical principles and clinical applications, 2nd edn, Lea and Febiger, Philadelphia.

Hatle L, Brubakk A, Tromsdal A, Angelsen B (1978) Non-invasive assessment of pressure drop in mitral stenosis by Doppler ultrasound. Br Heart J 40, 131–140.

Hatle L, Angelsen B, Tromsdal A (1979) Non-invasive assessment of atrio-ventricular pressure half-time by Doppler ultrasound. Circulation 60, 1096–1104.

Holen J, Aaslid R, Landmark K, Simonsen S (1976) Determination of pressure gradient in mitral stenosis with a non-invasive ultrasound Doppler technique. Acta Med Scand 199, 455–460.

Johnson WL, Baker DW, Lule RA, Dodge HT (1973) Doppler echocardiography: the localization of cardiac murmurs. Circulation 48, 810–822.

Libanoff AJ, Rodbard S (1966) Evaluation of the severity of mitral stenosis and regurgitation. Circulation 33, 218–226.

Light H (1976) Transcutaneous aortovelography: a new window on the circulation? Br Heart J 38, 433–442.

Light LH (1979) Quantitative information on central blood flow from transcutaneous aortovelography. In: Fundamentals in technical progress, vol 1 (Eds J Garson, W Gordenne, G Merchie), pp 1.4.1–1.4.8, Presse Universitaire de Liège.

Light LH (1986) Reliable information from Doppler measurements in the aorta. Abstract, 2nd World Congr Doppler Echocardiol, Kyoto, p 8–9.

Mowat DHR, Haites NE, Rawles JM (1983) Aortic blood velocity measurement in healthy adults using a simple ultrasound technique. Cardiovasc Res 17, 75–80.

Nanda NC (1985) Doppler echocardiography. Igaku-Shoin, New York, Tokyo.

Quinones MA (1985) Doppler evaluation of right-sided lesions and pulmonary hypertension. In: Doppler echocardiography (Ed. N Nanda), pp 262–292, Igaku-Shoin, New York, Tokyo.

Roelandt J (1986) Clinical experience with the Toshiba SSH-65A color Doppler flow imaging apparatus. Toshiba Medical Review No. 17, Toshiba Corporation, Tokyo.

Sequeira RF, Light LH, Cross G, Rafferty EB (1976) Transcutaneous, aortovelography: a quantitative evaluation. Br Heart J 38, 443–450.

Stevenson JG (1984) Experience with qualitative and quantitative applications of Doppler echocardiography in congenital heart disease. Ultrasound Med Biol 10, 771–796.

Stevenson JG, Kawabori I, Guntheroth WG (1980) Detection of pulmonary insufficiency by pulsed Doppler echocardiography. Validation, sensitivity, specificity and correlation with M-mode echo. Circulation 62, Suppl III, 251.

Switzer DF, Nanda NC (1985) Doppler color flow mapping. Ultrasound Med Biol 11, 403–416.

Chapter 17

DOPPLER APPLICATIONS IN OBSTETRICS, NEPHROLOGY AND ONCOLOGY

17.1 INTRODUCTION

During the last five years there has been an increase in the use of Doppler ultrasound in conditions other than cardiovascular disease. Major new areas of application include the study of flow in the placenta and fetus, the kidney, and tumours.

17.2 DOPPLER STUDIES IN OBSTETRICS

Doppler ultrasound, because of its non-invasive nature, has recently found applications in obstetrics. The three main sites of investigation are the maternal side of the placenta (uteroplacental arteries), the fetal side of the placenta (umbilical arteries) and the major fetal vessels (cerebral and aortic circulations).

Most vessels in the fetus and placenta are narrow and measurements of diameter are inaccurate, but Gill (1979) and Eik-Nes et al (1980) have succeeded in making volumetric flow measurements in the umbilical vein, and the latter group have also reported blood flow measurements from the fetal aorta. Most workers have now abandoned volume flow in favour of velocity waveform analysis.

17.2.1 Uteroplacental circulation

This subject has recently been well reviewed by Pearce (1987). The placenta is a complex network of maternal and fetal vessels. Its prime function is the transfer of oxygen and nutrients from the mother to the fetus. Many of these mechanisms of transfer are still obscure. However in pre-eclampsia and intrauterine growth retardation, which are important disorders of pregnancy where the fetus is at increased risk of death, abnormalities have been described in the maternal placental vessels (Brosens et al 1977). There is failure of cytotrophoblastic invasion of the spiral arteries which results in narrowing of these vessels and, in theory, impedes the flow of blood to the placenta. This higher vascular resistance is reflected by reduced end diastolic flow velocities in the uteroplacental vessels.

Campbell and co-workers were the first to report signals from the maternal side of the placenta (Campbell et al 1983, Griffin et al 1983). They postulated that the signals obtained were from the arcuate arteries. The typical arcuate artery signal from a normal placenta has low pulsatility and flow throughout diastole. In pregnancies complicated by intrauterine growth retardation the waveforms become more pulsatile and have a lower end diastolic velocity, and there is a dichrotic notch evident in the Doppler spectrum. The typical arcuate artery Doppler signal should be evident early in the second trimester when trophoblastic invasion of the spiral arteries is occurring. Campbell and colleagues suggested that if a high-resistance type of spectrum was seen at this stage this might be an early indicator of intrauterine growth retardation and pre-eclampsia.

Trudinger et al (1985a) studied 12 normal pregnancies from 20 weeks gestation to delivery and 91 complicated pregnancies. The 91 complicated pregnancies resulted in 25 small-for-dates fetuses, 15 of which showed reduced end diastolic flow velocities in the uteroplacental arteries. They suggested that those 15 pregnancies were a subgroup of growth-retarded fetuses in which the primary disorder was in the maternal vessels supplying the placenta. Reduced uterine artery diastolic flow velocity was associated with reduced umbilical artery diastolic flow velocity. In contrast, the other 10 small-for-dates fetuses had a normal uterine artery waveform suggesting a primary fetal cause. Trudinger et al also reported 12 patients with severe hypertensive disease in which nine had reduced uterine artery diastolic flow velocity consistent with vasospasm in the branches of the uterine artery in the placenta.

Campbell et al (1986) and Steel and Pearce (1988) have tried to develop qualitative Doppler assessment of the uteroplacental blood flow as a screening test for high-risk pregnancy. Steel and Pearce screened 200 primiparous women at 18–20 weeks and re-examined those with abnormal uteroplacental waveforms at 24 weeks gestation. They found the test to be highly specific (92%) for predicting pregnancies that did not become complicated by hypertension. However it was less sensitive (30%) at predicting patients who developed any degree of hypertension.

17.2.2 Umbilical circulation

Doppler signals from the umbilical artery were first described in 1977 by Fitzgerald and Drumm. Abnormalities in the tertiary stem villi of the placenta have been correlated with reduced end diastolic flow velocities in the umbilical artery (Giles et al 1985). As with the uteroplacental circulation, the postulate is that increased impedance in the placenta results in reduced diastolic flow in the umbilical arteries which may be of value in predicting and monitoring intrauterine growth retardation.

Trudinger et al (1985b) investigated the fetal umbilical artery Doppler signals using the ratio of systolic to diastolic velocity. In 15 normal pregnancies there was a small but significant decrease in this ratio through the last trimester. A decrease in the ratio suggests a decrease in placental resistance. In 32 of 43 fetuses subsequently shown to be small-for-dates there was an increase in placental flow resistance with reduced, absent or even reversed flow in diastole. Serial studies in patients with fetal compromise indicated increasing flow resistance.

Recently Johnstone et al (1988) have reported a study of umbilical artery waveforms in a series of 380 high-risk pregnancies and 160 patients with uncomplicated pregnancy, and concluded that absence of end diastolic flow is an indication for extremely careful surveillance, but not necessarily an indication for delivery.

Thompson et al (1986a,b) compared various objective indices of the Doppler spectrum, including the systolic/diastolic ratio, pulsatility index, Pourcelot index and acceleration time. These indices were calculated or measured from the maximum frequency and mean frequency envelope. The authors report that the maximum frequency envelope appears to be more reliable for the calculation of the various indices.

Umbilical flow velocity waveforms have been studied during labour by Stuart et al (1981) and Fairlie et al (1987). Stuart et al studied 10 patients in uncomplicated labour and found no change in the umbilical artery waveform with uterine contractions, progress in labour, or narcotic analgesia. Fairlie et al studied 103 women with normal and abnormal intrapartum fetal heart rate patterns. The umbilical pulsatility index was abnormal in 67% of small-for-dates infants whereas it was abnormal in only 13% of normal birthweight infants. They concluded that it might be of value in screening for high-risk pregnancies that require intrapartum monitoring.

Marx et al (1986) investigated umbilical Doppler signals in different maternal positions and with epidural anaesthesia. The Doppler signals were recorded in early active labour and the systolic/diastolic ratio calculated. The ratio was significantly higher in the supine than in either lateral position, suggesting that the umbilical artery vascular impedance is increased when the mother lies supine. Epidural anaesthesia lowered the placental impedance in 13 out of 16 patients and was unchanged in the remaining three.

Doppler ultrasound may also have a role in the assessment of multiple pregnancies to detect intrauterine growth retardation and twin-to-twin transfusion syndrome (Giles et al 1985).

17.2.3 The fetal circulation

The uteroplacental and umbilical circulations are relatively simple to examine with inexpensive continuous wave Doppler equipment. The purpose of examining these vessels is to produce predictive tests for detecting high-risk pregnancies. Fetal hypoxaemia results in an increase in blood flow to the cerebral and coronary arteries and a reduction in blood flow to the abdominal aorta (Peeters et al 1979). It is possible to examine the fetal cerebral and fetal aortic circulations using pulsed Doppler duplex ultrasound.

Tonge et al (1986) investigated blood flow velocity waveforms in the fetal aorta in 77 normal pregnancies and 12 cases of intrauterine growth retardation during the third trimester. In normal pregnancies the peak velocity, end diastolic velocity and pulsatility index remained unchanged with increasing gestational age. End diastolic velocity was lowered and the pulsatility index elevated in the most severe cases of intrauterine growth retardation.

Jouppila and Kirkinen (1986) studied the variation of absolute blood velocity measurements from the fetal aorta in normal and hypertensive pregnancies. In 43 normal pregnancies the velocities were measured between 30 and 43 weeks. The median value of the mean velocity was 36.5 cm s^{-1}, that of the end diastolic velocity was 10.9 cm s^{-1}. Pourcelot's resistance index was 0.84. No differences were observed in these values in uncomplicated hypertensive pregnancies. The resistance index and the mean and end diastolic velocities decreased if fetal growth retardation was associated with maternal hypertension.

Vladimiroff et al (1986) studied small-for-dates fetuses and showed an increase in impedance in the abdominal aorta and a decrease in impedance in the internal carotid artery.

Lang et al (1988) have measured the blood flow velocity in the middle cerebral arteries of fetuses between 26 and 32 weeks gestation and compared the results with neonates of similar gestation. The median velocities for the two groups were 4.5 cm s^{-1} and 8 cm s^{-1} respectively. This method has the advantages of measuring velocity directly and using a technique established in neonatal medicine.

17.3   RENAL DOPPLER STUDIES

17.3.1   The native kidney

Doppler investigation of blood flow in the renal arteries was first reported by Greene et al (1981), who studied 16 volunteer subjects and 11 patients with clinically suspected renal artery abnormality. In a further study (Avasthi et al 1984) the presence or absence of renal artery stenosis was blindly evaluated in 26 patients using a set of four criteria (a peak velocity of greater than 1 ms^{-1} in a focal area; absence of flow during diastole; total absence of flow; broadband Doppler spectra). Compared with angiography (reduction in diameter $\geq 50\%$) the sensitivity and specificity of the Doppler method were 89% and 73% respectively.

Doppler shift spectra recorded from the renal artery, the mid-region of the kidney and the arcuate vessels at the corticomedullary junction have been described by Taylor et al (1985).

Dubbins (1986) investigated the renal arteries in 40 normal patients allowing a maximum examination time of 15 minutes per patient, and was successful in recording Doppler spectra in 82.5% of vessels. He described the renal artery spectrum as having a sharp upstroke with an almost exponential fall to a steady diastolic component. Eight patients with renal artery stenosis were also investigated. Two patients had diameter stenoses of less than 50% and the Doppler spectrum exhibited a degree of spectral broadening. In five of the remaining arteries there was a marked increase in systolic velocity together with spectral broadening, consistent with stenosis of greater than 50%.

Doppler colour flow mapping is potentially of value for the identification of renal artery stenosis and Plate 17.1 is a colour flow map (CFM) of this condition.

Dubbins and Wells (1986) investigated 14 patients with renal carcinoma and described three types of Doppler signal obtained from the renal artery, namely: a normal renal artery signal; an increase in peak velocity compared with the normal kidney and spectral broadening; and, in three patients, a high-impedance signal with an occasional reverse flow phase. In these three the renal vein was occluded.

17.3.2   The transplanted kidney

Koff et al (1982) reported the use of duplex ultrasound to study renal artery Doppler signals in renal transplants. Taylor et al (1987) discussed vascular complications in renal allografts and how they may be detected with duplex ultrasound. Over a 2-year period, 334 studies were carried out on 88 patients. Severe vascular rejection resulting in vascular occlusion was seen in ten out of ten patients. The most reliable criterion for detecting stenosis of the renal artery was a high-velocity jet exceeding 7.5 kHz (using a transmitted ultrasound frequency of 5 MHz) and distal turbulence. Taylor et al also reported the diagnosis of an arteriovenous fistula by detecting an intrarenal high-velocity jet.

Rigsby et al (1987) calculated pulsatility index from signals obtained from the main renal artery, segmental, interlobar and arcuate arteries. These *PI* values from normal allografts were compared with those from cases undergoing acute rejection. Rejection produced a significantly higher *PI* value at each site. ROC curves suggested that signals obtained from the segmental arteries are the most sensitive to changes. With a threshold *PI* of 1.5, the sensitivity of this technique for the detection of acute renal allograft rejection was 75%, and the specificity 90%. In acute vascular rejection the same *PI* gave a sensitivity of 79% and a specificity of 90%.

Steinberg et al (1987) compared duplex scanning with magnetic resonance imaging (MRI) for the detection of rejection. They found that Doppler ultrasound was significantly superior to magnetic resonance imaging in identifying allograft rejection, demonstrating a higher sensitivity (95% compared with 70%), specificity (95% against 73%) and accuracy (95% against 71%). They concluded that Doppler ultrasound should become the primary modality for renal transplant screening.

Colour flow mapping in the renal transplant allows the rapid identification of the renal artery and also gives an indication of renal perfusion in real time. Examples of CFM images of transplanted kidneys are shown in Plates 17.2–17.4.

17.4   DOPPLER STUDIES OF TUMOURS

17.4.1   Breast tumours

The first report of the use of Doppler ultrasound to detect tumour circulation appears to be that of Wells et al (1977) who studied breast lumps. In a larger study carried out by the same group (Burns et al 1982), 349 patients with benign breast lumps and 55 with malignant lumps were investigated. The flow signals were assessed subjectively using a scale of 0 to 3, where 0 indicated no difference in the sound of the flow from the lump and the same site on the other breast, and 3 represented a marked difference between the two sites. A small number of the flow signals were also analysed using a spectrum analyser. Maximum systolic and minimum diastolic frequencies were compared on both sides, together with the mean frequency and the average spectral amplitude. The maximum systolic, minimum diastolic and mean frequencies were all significantly higher over the tumours, whereas the average spectral amplitude was lower. There was no significant difference between these parameters in normals.

Minasian and Bamber (1982) showed a higher mean Doppler frequency recorded over the tumour than from the normal breast, and showed that there is a correlation between the increase in tumour volume and the rise in the mean frequency.

17.4.2   Lymph nodes

Mountford and Atkinson (1979) reported the use of a simple Doppler velocimeter in the study of pathologically enlarged lymph nodes. Palpable lymph nodes were studied in 41 patients and the flow signals compared with the contralateral site. Doppler signals were generally found arising from or adjacent to the lymph gland, probably arising from abnormal feeding vessels. In a minority of cases the lymph gland itself gave rise to a characteristic Doppler signal.

17.4.3   Melanoma

The diagnosis of pigmented lesions of the skin is a common clinical problem. As most are benign there is a need for a simple non-invasive test to differentiate lesions having a high probability of malignancy. Srivastava et al (1986a) demonstrated that thin melanomas have no detectable blood supply but that vascularity develops as melanomas become thicker and nodular. Srivastava et al (1986b) showed that 96% of melanomas thicker than 0.9 mm were associated with detectable Doppler signals. Histological quantification of blood vessels using lectin staining to delineate vascular endothelium has shown a high degree of correlation between vascularity at the tumour base and tumour thickness (Srivastava et al 1986c).

If the pigmented lesion is raised above the surrounding skin and has a detectable flow signal, there is a high probability of malignancy. If, however, no flow signals are detected there is a high probability of a benign clinical diagnosis.

17.4.4   Colour flow mapping of tumours

Colour flow mapping has tremendous potential for visualizing the circulation to tumours, and in helping to decide whether a solid lesion is vascular or not. An example of a scan from a patient with a thyroid mass is shown in Plate 17.5.

17.5   SUMMARY

Doppler ultrasound is continually finding new applications in medicine; three relatively new areas of interest which appear to have some considerable potential are the study of the placental and fetal circulations, the kidney, and tumours.

17.6   REFERENCES

Avasthi PS, Voyles WF, Greene ER (1984) Noninvasive diagnosis of renal artery stenosis by echo-Doppler velocimetry. Kidney Int 25, 824–829.

Brosens I, Dixon HG, Robertson WB (1977) Fetal growth retardation and the arteries of the placental bed. Br J Obstet Gynaecol 84, 656–663.

Burns PN, Halliwell M, Wells PNT, Webb AJ (1982) Ultrasonic Doppler studies of the breast. Ultrasound Med Biol 8, 127–143.

Campbell S, Diaz-Recasens J, Griffin DR, Cohen-Overbeek TE, Pearce JM, Willson K, Teague MJ (1983) New Doppler technique for assessing uteroplacental blood flow. Lancet i, 675–677.

Campbell S, Pearce JMF, Hackett G, Cohen-Overbeek TE, Hernandez CJ (1986) Qualitative assessment of uteroplacental blood flow: an early screening test for high risk pregnancies. Obstet Gynecol 68, 648–653.

Dubbins PA (1986) Renal artery stenosis: duplex Doppler evaluation. Br J Radiol 59, 225–229.

Dubbins PA, Wells I (1986) Renal carcinoma: duplex Doppler evaluation. Br J Radiol 59, 231–236.

Eik-Nes SH, Brubakk AO, Ulstein MK (1980) Measurement of human fetal blood flow. Br Med J 280, 283–284.

Fairlie FM, Lang GD, Sheldon CD (1987) Umbilical waveforms in labour. In: Obstetric and neonatal bloodflow (Eds CD Sheldon, DH Evans, JR Salvage), Biological Engineering Society, London.

Fitzgerald DE, Drumm JE (1977) Non-invasive measurement of human fetal circulation using ultrasound: a new method. Br Med J ii, 1450–1451.

Giles WB, Trudinger BJ, Baird PJ (1985) Fetal umbilical artery flow velocity waveforms and placental resistance: pathological correlation. Br J Obstet Gynaecol 92, 31–38.

Gill RW (1979) Pulsed Doppler with B-mode imaging for quantitative blood flow measurement. Ultrasound Med Biol 5, 223–235.

Greene ER, Venters MD, Avasthi PS, Conn RL, Jahnke RW (1981) Noninvasive characterization of renal artery blood flow. Kidney Int 20, 523–529.

Griffin D, Cohen-Overbeek T, Campbell S (1983) Fetal and uteroplacental blood flow. Clin Obstet Gynecol 10, 565–579.

Johnstone FD, Haddad NG, Hoskins P, McDicken W, Chambers S, Muir B (1988) Umbilical artery Doppler flow velocity waveform: the outcome of pregnancies with absent end diastolic flow. Eur J Obstet Gynecol Reprod Biol 28, 171–178.

Jouppila P, Kirkinen P (1986) Blood velocity waveforms of the fetal aorta in normal and hypertensive pregnancies. Obstet Gynecol 67, 856–860.

Koff D, Plainfosse M, Merran S (1982) Doppler study of renal artery of transplanted kidneys. J Echog Med Ultrason 3, 111–118.

Lang GD, Levene MI, Dougall A, Shortland D, Evans DH (1988) Direct measurements of fetal cerebral blood flow velocity with duplex Doppler ultrasound. Eur J Obstet Gynecol Reprod Biol 29, 15–19.

Marx GF, Patel S, Berman JA, Farmakides G, Schulman H (1986) Umbilical blood flow velocity waveforms in different maternal positions and with epidural analgesia. Obstet Gynecol 68, 61–64.

Minasian H, Bamber JC (1982) A preliminary assessment of an ultrasonic Doppler method for the study of blood flow in human breast cancer. Ultrasound Med Biol 8, 357–364.

Mountford RA, Atkinson P (1979) Doppler ultrasound examination of pathologically enlarged lymph nodes. Br J Radiol 52, 464–467.

Pearce JFM (1987) Uteroplacental and fetal blood flow. Clin Obstet Gynaecol 1, 157–184.

Peeters LL, Sheldon RE, Jones MD, Makowski EL, Meschia G (1979) Blood flow to fetal organs as a function of arterial oxygen content. Am J Obstet Gynecol 135, 637.

Rigsby CM, Burns PN, Weltin GG, Chen B, Bia M, Taylor KJW (1987) Doppler signal quantitation in renal allografts: comparison in normal and rejecting transplants, with pathologic correlation. Radiology 162, 39–42.

Srivastava A, Hughes LE, Woodcock JP, Shedden EJ (1986a) The significance of blood flow in cutaneous malignant melanoma demonstrated by Doppler flowmetry. Eur J Surg Oncol 12, 13–18.

Srivastava A, Hughes BR, Hughes LE, Woodcock JP (1986b) Doppler ultrasound as an adjunct to the differential diagnosis of pigmented skin lesions. Br J Surg 73, 790–792.

Srivastava A, Laidler P, Hughes LE, Woodcock JP, Shedden EJ (1986c) Neovascularization in human cutaneous melanoma: a quantitative morphological and Doppler ultrasound study. Eur J Cancer Clin Oncol 22, 1205–1209.

Steel SA, Pearce JM (1988) Use of Doppler ultrasound to investigate the uteroplacental and fetal circulations. Hospital Update 14, 1094–1106.

Steinberg HV, Nelson RC, Murphy FB, Chezzmar JL, Baumgartner BR, Delaney VB, Whelchel JD, Bernardino ME (1987) Renal allograft rejection: evaluation by Doppler ultrasound and MR imaging. Radiology 162, 337–342.

Stuart B, Drumm JE, Fitzgerald DE, Duignan NM (1981) Fetal blood velocity waveforms in uncomplicated labour. Br J Obstet Gynaecol 88, 865–870.

Taylor KJW, Burns PN, Woodcock JP, Wells PNT (1985) Blood flow in deep abdominal and pelvic vessels: ultrasonic pulsed Doppler analysis. Radiology 154, 487–493.

Taylor KJW, Morse SS, Rigsby CM, Bia M, Schiff M (1987) Vascular complications in renal allografts: detection with duplex Doppler US. Radiology 162, 31–38.

Thompson RS, Trudinger BJ, Cook CM (1986a) A comparison of Doppler ultrasound waveform indices in the umbilical artery. 1. Indices derived from the maximum velocity waveform. Ultrasound Med Biol 12, 835–844.

Thompson RS, Trudinger BJ, Cook CM (1986b) A comparison of Doppler ultrasound waveform indices in the umbilical artery. II. Indices derived from the mean velocity and first moment waveforms. Ultrasound Med Biol 12, 845–854.

Tonge HM, Vladimiroff JW, Noordam MJ, VanKooten C (1986) Blood flow velocity waveforms in the descending fetal aorta: comparison between normal and growth retarded pregnancies. Obstet Gynecol 67, 851–855.

Trudinger BJ, Giles WB, Cook CM (1985a) Uteroplacental blood flow velocity–time waveforms in normal and complicated pregnancy. Br J Obstet Gynaecol 92, 39–45.

Trudinger BJ, Giles WB, Cook CM, Bombardieri J, Collins L (1985b) Fetal umbilical artery flow velocity waveforms and placental resistance: clinical significance. Br J Obstet Gynaecol 92, 23–30.

Vladimiroff JW, Tonge HM, Stewart PA (1986) Doppler ultrasound assessment of cerebral blood flow in the human fetus. Br J Obstet Gynaecol 93, 471–475.

Wells PNT, Halliwell M, Skidmore R, Webb AJ, Woodcock JP (1977) Tumour detection by ultrasound Doppler blood flow signals. Ultrasonics 15, 231–232.

Chapter 18

MISCELLANEOUS DOPPLER TECHNIQUES AND DEVICES

18.1 INTRODUCTION

Apart from Doppler techniques that involve the detection of blood flow in arteries, veins or the heart, there are a number of others that find application in medicine. Most of these additional techniques have not found widespread application. They are briefly considered in this chapter since they provide useful solutions in a diverse range of clinical problems. The more common of these techniques are discussed here; the rest are listed in the bibliography.

18.2 FETAL HEART DETECTION

One of the first applications of Doppler techniques was the detection of the fetal heart to verify the presence or absence of fetal life. Later, monitoring of fetal heart rate was developed to give information on fetal wellbeing later in pregnancy. A 2 MHz continuous wave beam is most commonly employed in both these techniques, although pulsed Doppler methods are occasionally found when range discrimination is thought to be helpful. In a basic fetal heart detector, a dual crystal transducer generates a reasonably narrow beam whereas in a monitoring system a divergent beam is produced by several crystals (Steer 1977). When the latter transducer is attached to the abdomen, some motion of the fetus can be tolerated without loss of signal. Heart walls and valves contribute most significantly to the Doppler signal. Heart rate traces based on the Doppler signal are more variable than those based on the fetal ECG but the ultrasonic technique is non-invasive. The intensity (I_{spta}) generated by fetal heart detectors and monitors is normally less than 10 mW cm^{-2}.

18.3 DOPPLER HEART AUSCULTATION

When a 2 MHz Doppler beam is directed into the heart at different sites, a variety of signals are detected which superficially resemble phonocardiogram sounds (Abelson 1968). Doppler signals are slightly more complex than heart sounds since they arise from a combination of wall, valve and blood motion. Up to nine distinct signals may be discerned. Doppler signals are recorded most distinctly in the lower part of the chest. Characteristic changes in the signals are produced by cardiac conditions such as atrial flutter and fibrillation.

18.4   PRESSURE MEASUREMENT

For the measurement of blood pressure in adults and infants, the Doppler transducer can be combined with an inflatable cuff or be distal to it on the blood vessel. The Doppler instrument detects changes in the blood flow or vessel wall movement as the pressure in the cuff is altered and hence systolic and diastolic blood pressure are measured (Kirby et al 1969). Initially the cuff is inflated above systolic pressure and is then slowly deflated. As the pressure falls below the systolic value, the commencement of wall motion and blood flow is detected and the systolic pressure is noted. On further deflation, the vessel walls cease movement and this is taken to be the diastolic pressure. The inflation and deflation cycle can be carried out automatically at regular intervals if it is desired to monitor blood pressure. The Doppler signal can be recorded on a chart recorder along with the cuff pressure.

Transducers operating at frequencies in the range 2–10 MHz are used in this technique. The error in pressure measurement is reported to be ± 2 mm. The results agree well with those obtained from direct intra-arterial measurement (Hochberg 1971).

18.5   INVASIVE DOPPLER PROBES

Doppler probes that can be inserted along a channel of an optical endoscope have been employed to identify vessels in the oesophageal, gastric and duodenal walls. Bleeding has been assessed and methods to predict the risk of rebleeding are being researched (Beckly et al 1982, Kurtz and Classen 1984). The transducer is required to be small, e.g. a cylindrical crystal of diameter 3 mm operating at 8 MHz. Both pulsed and continuous wave designs are possible. With such devices, blood flow is readily detected in the vessels associated with varices and ulcers.

Small Doppler transducers have been mounted on the ends of catheters (Nealeigh and Miller 1976). Since there are no tissues between the transducer and the blood, high ultrasonic frequencies can be used, e.g. 20 MHz. This enables higher spatial and velocity resolution to be achieved. The high ultrasonic frequency results in large Doppler shifts, for example 30 kHz, and hence pulse repetition frequencies of greater than 60 kHz are obligatory. However, since there is no attenuation in intervening tissues, the ultrasonic intensities can be maintained at low levels. The flexibility and versatility of such catheters can be increased by including mechanisms to manipulate the position of the transducer (Gichard and Auth 1975). This increase in complexity allows the orientation of the beam to the blood flow to be altered and velocity profiles across the vessel to be plotted.

Internal Doppler flow mapping or duplex scanning can be carried out by mechanical or electronic transducers placed in the oesophagus. One phased array system employs a 5 MHz, 64-element transducer mounted on the end of a flexible gastroscope. It has a 90° sector field of view which can be positioned by flexing the gastroscope in two directions as well as moving the whole device. The use of this type of device is advocated where ultrasonic access through the chest wall is difficult and in monitoring the heart during surgery. Immediate assessment of the efficacy of the surgery is also possible. Particular care is required to be taken in the design of such devices from the points of view of electrical safety and transducer temperature control. When used with a disposable sheath, endoscopic Doppler transducers can be used during electrosurgery.

18.6   FETAL MOTION AND BREATHING

Activity is of direct interest in building up a physiological profile of the fetus. Overall fetal movements can be easily detected by Doppler devices as this motion gives rise to irregular, large low-frequency signals (Lauersen and Hochberg 1982). A wide angle beam of 2 MHz ultrasound as used in fetal heart monitoring is suitable as a motion detector. Indeed fetal movements are often regarded as artefacts when the fetal heart is to be monitored. Since there is a reluctance to monitor the fetus for long periods, even with low-power CW ultrasound, fetal movements are commonly checked with a real-time pulse echo scanner.

Fetal breathing can also be detected using Doppler techniques (McHugh et al 1978, McDicken et al 1979, Gough and Poore 1979, Boyce et al 1976). Both pulsed and continuous wave Doppler instruments are employed. Apart from detecting the motion of tissues of the fetus, the influence of breathing on the fetal circulation can be observed, for example alterations in flow in the umbilical vein and arteries.

18.7   TISSUE PERFUSION

When tissue is highly vascular, strong arterial pulsatile flow can be recorded. The commonest example of this application is in the study of tumours, in particular breast tumours (Wells et al 1977, Burns et al 1982). This application of Doppler techniques is still being developed, although it can occasionally be clinically useful to ascertain the vascularity of masses in the abdomen. Detection of low-velocity perfusion has been attempted but has not been applied clinically (Kallio et al 1987). Considerable difficulties exist in separating the low Doppler shift signals due to blood flow and tissue motion. Low-velocity flow in the placenta may be demonstrated, but this is a highly vascular organ (McHugh et al 1981).

18.8   BUBBLE AND PARTICLE DETECTION

Doppler devices have been shown to be very sensitive detectors of bubbles and particulate material in blood during surgery or when extracorporeal equipment such as haemodialysis equipment is coupled to the patient (Maroon et al 1969, Badylak et al 1984). With a conventional CW blood flow unit, operating at 5 or 10 MHz, the passage of a bubble or particle distinctly increases the amplitude of the Doppler signal. Apart from macroscopic and microscopic bubbles, particles of platelet aggregates, fibrinogen–fibrin degradation products and charcoal have been shown to produce readily detected signals in both in-vitro and in-vivo studies. Particulate matter can also be identified in urethral flow (Albright and Harris 1975). When compared to other means of detecting particles and bubbles such as filtration or Coulter counters, ultrasonic instruments have the advantages of being non-invasive and sensitive. Ultrasonic pulse echo reflection and transmission techniques also find application in this field (Lubbers and van den Berg 1977).

18.9   DOPPLER TELEMETRY

The transmission of physiological signals from the patient to a recording and display system by radio waves is attractive since it allows the patient to function in a free and natural manner. This concept is particularly appealing in obstetrics since it represents little interference with the course of the pregnancy (Albert and Terry 1983). The fetal heart is sensed by a wide-angle 2 MHz beam from a transducer strapped onto the abdomen. After processing, the fetal heart rate signal is transmitted to the receiver at the central console. While the ultrasonic fetal heart rate trace is not as accurate as that of a direct fetal ECG, the technique is non-invasive and can be used before the membranes have ruptured. A pressure transducer, also on the abdomen, is usually incorporated in the telemetry system to send information on uterine contractions.

CW and PW blood flow instruments have been designed for implantation in animals for research (Allen et al 1979). Their performance is directly comparable to standard units, the complete audio Doppler signal being transmitted. These units are in direct competition with electromagnetic flowmeters but they are more stable and the PW devices can provide velocity profiles.

18.10   SUMMARY

A number of examples have been described which illustrate the versatility of Doppler techniques. The full potential of such techniques has still to be exploited. The extensive list of references gives some idea of this potential even although the applications presented are only found rarely in clinical practice at present.

18.11   REFERENCES

Abelson D (1968) Ultrasonic Doppler auscultation of the heart. J Am Med Assoc 204, 438–443.
Albert DJ, Terry HJ (1983) Non-invasive fetal monitoring using radiotelemetry. Clin Phys Physiol Meas 4, 291–298.
Albright RJ, Harris JH (1975) Diagnosis of urethral flow parameters by ultrasonic backscatter. IEEE Trans Biomed Eng BME-22, 1–11.
Allen HV, Knutti JW, Meindl JD (1979) Totally implantable directional Doppler flowmeters. Biotelem Patient Monit 6, 118–132.
Aoyagi F, Fugino T, Ohshiro T (1975) Detection of small vessels for microsurgery by a Doppler flowmeter. Plast Reconstr Surg 55, 372–373.
Badylak SF, Ash SR, Thornhill JA, Carr DJ (1984) Doppler ultrasonic detection of particulate release during hemodialysis with cellulose hollow-fiber and sorbent suspension reciprocating dialyzers. Artif Organs 8, 220–223.
Bain JA (1985) Doppler crystal fixation with double-stick discs. Can Anaesth Soc J 32, 201–202.
Beard JD, Evans JM, Skidmore R, Horrock M (1986) A Doppler flowmeter for use in theatre. Ultrasound Med Biol 12, 883–889.
Beckly DE, Casebow MP (1986) Prediction of rebleeding from peptic ulcer experience with an endoscopic Doppler device. Gut 27, 96–99.
Beckly DE, Casebow MP, Pettengell KE (1982) The use of Doppler ultrasound probe for localising arterial blood flow during upper gastrointestinal endoscopy. Endoscopy 14, 146–147.
Bingham HG, Lichti EL (1971) The doppler as an aid in predicting the behavior of congenital cutaneous hemangioma. Plast Reconstr Surg 47, 580–583.
Bishop EH (1968) Ultrasonic fetal monitoring. Clin Obstet Gynecol 11, 1154–1164.
Bitker MO, Martin B, Chatelain C, Kuss R (1983) Torsion of the spermatic cord; diagnosis by Doppler ultrasonography. Presse Med 12, 1925–1926.
Blair WF, Greene ER, Omer GE Jr (1981) A method for the calculation of blood flow in human digital arteries. J Hand Surg 6, 90–96.
Boyce ES, Dawes GS, Gough JD, Poore ER (1976) Doppler ultrasound method for detecting human fetal breathing in utero. Br Med J ii, 17–18.
Brechner TM, Brechner VL (1977) An audible alarm for monitoring air embolism during neurosurgery. J Neurosurg 47, 201–204.

Brodsky JB, Wong AL, Meyer JA (1977) Percutaneous cannulation of weakly palpable arteries. Anesth Analg (Cleve) 56, 448.

Burns PN, Halliwell M, Wells PNT, Webb AJ (1982) Ultrasonic Doppler studies of the breast. Ultrasound Med Biol 8, 127–143.

Callagan DA, Rowland TC, Goldman DE (1964) Ultrasonic Doppler observation of the fetal heart. Obstet Gynecol 23, 637.

Callahan MJ, Tajik AJ, Su-Fan Q, Bove AA (1985) Validation of instantaneous pressure gradients measured by continuous-wave Doppler in experimentally induced aortic stenosis. Am J Cardiol 56, 989–993.

Chiba Y, Utsu M, Sakakibara M, Ishida S, Kanzaki T (1983) Tracheal fluid flow according to fetal breathing movements detected by pulsed doppler ultrasound. Proc 42nd Mtg Jap Soc Ultrasonics Med, pp 545–546.

Chow PP, Horgan JG, Burns PN, Weltin G, Taylor KJ (1986) Choroid plexus papilloma: detection by real-time and doppler sonography. Am J Neuroradiol 7, 168–170.

Coppess MA, Young DF, White CW, Laughlin DE (1983) An ultrasonic pulsed Doppler balloon catheter for use in cardiovascular diagnosis. Biomed Sci Instrum 19, 9–16.

Davis WE, Lichti EL, Joseph DJ (1975) Use of the Doppler ultrasonic flowmeter for pedicle flaps. Ann Otol Thinol Laryngol 84, 213–217.

D'Luna LJ, Newhouse VL (1981) Vortex characterization and identification by ultrasound Doppler. Ultrasonic Imaging 3, 271–293.

Duck FA, Hodson CJ, Tomlin PM (1974) An esophageal Doppler probe for aortic flow velocity monitoring. Ultrasound Med Biol 1, 233–241.

Evans DS (1970) The early diagnosis of deep-vein thrombosis by ultrasound. Br J Surg 57, 726–728.

Forfang K, Otterstad JE, Ihlen H (1986) Optimal atrioventricular delay in physiological pacing determined by Doppler echocardiography. PACE 9, 17–20.

Fox MD (1978) Multiple cross-beam ultrasound Doppler velocimetry. IEEE Trans Sonics Ultrasonics SU-25, 281–286.

Freed D, Dube CM, Walker WF (1984) Quantifying microbubble populations in biologic systems. Proc 37th Ann Conf Eng Med Biol 26, 198.

Gichard FD, Auth DC (1975) Mechanically scanned catheter-tip Doppler flow measurement. IEEE Ultrasonics Symp Proc 75, CHO 994-4SU, 18–21.

Giddens DP, Khalifa AMA (1982) Turbulence measurements with pulsed Doppler ultrasound employing a frequency tracking method. Ultrasound Med Biol 8, 427–437.

Goldberg SJ, Valdes-Cruz LM, Feldman L, Sahn DJ, Allen HD (1981) Range-gated Doppler ultrasound detection of contrast echographic microbubbles for cardiac and great vessel blood flow patterns. Am Heart J 101, 793–796.

Gough JD, Poore ER (1979) A continuous wave Doppler ultrasound method of recording fetal breathing in utero. Ultrasound Med Biol 5, 249–256.

Grinomont B (1981) Improved approach of plexus brachialis by ultrasonic Doppler and by nerve stimulator. Acta Anaesthesiol Belg 32, 327–329.

Haines J (1974) An ultrasonic system for measuring activity. Med Biol Eng 12, 378–381.

Hara M, Kadowaki C, Konishi Y, Ogashiwa M, Numoto M, Takeuchi K (1983) A new method for measuring cerebrospinal fluid flow in shunts. J Neurosurg 58, 557–561.

Hartley CJ, Lewis RM, Latson LA, Entman ML (1981) Pulsed Doppler evaluation of regional myocardial function. Proc 26th Ann Conf Am Inst Ultrasound Med, 111.

Herndon JH, Bechtol CO, Crickenberger DP (1975) Use of ultrasound to detect fat emboli during total hip replacement. Acta Orthop Scand 46, 108–118.

Histand MB, Miller CW (1972) A comparison of velocity profiles measured in unexposed and exposed arteries. Biomed Sci Instrum 9, 121–124.

Hochberg HM (1971) Automatic ultrasonic blood pressure measurement in children – a feasibility study. Curr Ther Res 13, 473–481.

Hoeks APG, Ruissen CJ, Hick P, Reneman RS (1985) Transcutaneous detection of relative changes in artery diameter. Ultrasound Med Biol 11, 51–59.

Holen J, Aaslid R, Landmark K, Simonsesn S (1976) Determination of pressure gradient in mitral stenosis with a non-invasive ultrasound Doppler technique. Acta Med Scand 199, 455–460.

Holmvang G, McConville B, Tomlinson CW (1986) Noninvasive determination of valve area in adults with aortic stenosis using Doppler echocardiography. Cathet Cardiovasc Diagn 12, 9–17.

How TV, Ashmore MP, Rolfe P, Lucas A, Lucas PJ, Baum JD (1979) A Doppler ultrasound technique for measuring human milk flow. J Med Eng Technol 3, 66–71.

James PB, Galloway RW (1971) The ultrasonic blood velocity detector as an aid to arteriography. Br J Radiol 44, 743–746.

Johnson SL, Baker DW, Lute RA, Dodge HT (1973) Doppler echocardiography. The localization of cardiac murmurs. Circulation 48, 810–822.

Johnson WC, Patten DH (1977) Predictability of healing of ischemic leg ulcers by radioisotopic and doppler ultrasonic examination. Am J Surg 133, 485–489.

Kallio T, Alanen A, Kormano M (1987) Detection of slow flow velocities by ultrasound in vitro. Proc 6th Congr Eur Fed Soc Ultrasound Med Biol, 298.

Kececioglu-Draelos Z, Goldberg SJ, Areias J, Sahn DJ (1981) Verification and clinical demonstration of the echo Doppler series effect and vortex shed distance. Circulation 63, 1422–1428.

Kelly JS (1978) Doppler ultrasound flow detector used in temporal artery biopsy. Arch Ophthalmol 96, 845–846.

Klepper JR (1982) Ultrasonic methods for detection and sizing of venous gas emboli. Ultrason Symp Proc IEEE Ch 1823-4 SU, 668–672.

Kirby RR, Kemmerer WT, Morgan JL (1969) Transcutaneous Doppler measurement of blood pressure. Anesthesiology 31, 86–89.

Kopczynski HD (1974) Airborne blood pressure measurement using ultrasonics. Aerospace Med 45, 1307–1309.

Kurtz W, Classen M (1984) Blood flow measurement in esophageal varices using an endoscopic Doppler ultrasonic probe. Dtsch Med Wochenschr 109, 821–824.

Kwong PK (1982) Applications of Doppler ultrasound to foot and ankle care. Foot Ankle 2, 220–223.

Landini L, Righi E, Zacutti A (1977) A simple method for continuous recording of fetal systolic time interval. J Nucl Med Allied Sci 21, 48–50.

Lang-Jensen T (1981) Systolic time intervals measured by pulsed ultrasound-Doppler. Acta Anaesthesiol Scand 25, 461–462.

Lauersen NH, Hochberg HM (1982) Automatic detection of fetal movement by Doppler ultrasound during non-stress testing. Int J Gynaecol Obstet 20, 219–222.

Lewis RR, Padayachee TS, Beasley MG, Keen H, Gosling RC (1983) Investigation of brain death with Doppler-shift ultrasound. J R Soc Med 76, 308–310.

Lubbers J, van den Berg Jw (1976) An ultrasonic detector for microgas emboli in a bloodflow line. Ultrasound Med Biol 2, 301–310.

Lye CR, Sumner DS, Hokanson DE, Strandness D (1975) The transcutaneous measurement of the elastic properties of the human saphenous vein femoropopliteal bypass graft. Surg Gynecol Obstet 141, 891–895.

Lysikiewicz A, Chacinski A, Sternadel Z (1977) Doppler ultrasonocardiography of twins. Ginekol Pol 48, 385–388.

McDicken WN, Anderson T, McHugh R, Bow CR, Boddy K, Cole R (1979) An ultrasonic real-time scanner with pulsed Doppler and T-M facilities for foetal breathing and other obstetrical studies. Ultrasound Med Biol 5, 333–339.

McHugh R, McDicken WN, Bow CR, Anderson T, Boddy K (1978) An ultrasonic pulsed Doppler instrument for monitoring human fetal breathing in utero. Ultrasound Med Biol 3, 381–384.

McHugh R, McDicken WN, Thompson P, Boddy K (1981) Blood flow detection by an intersecting zone ultrasonic Doppler unit. Ultrasound Med Biol 7, 371–375.

Maeda K, Tatsumura M, Yoneda T, Nakamura Y (1981) Fetal mechanocardiography recorded with processing of ultrasonic Doppler fetal heart valve signals, Jap J Med Ultrason 8, 159–165.

Maroon JC, Edmonds-Seal J, Campbell RL (1969) An ultrasonic method for detecting air bubbles. J Neurosurg 31, 196–201.

Martin RW, Gilbert DA, Silverstein FE, Deltenre M, Tygat G, Gange RK, Myers J (1985) An endoscopic Doppler probe for assessing intestinal vasculature. Ultrasound Med Biol 11, 61–69.

Mianowicz J (1983) Value of combined A, B and Doppler ultrasonic methods in the diagnosis of secondary retinal detachment. Klin Oczna 85, 201–202.

Millerett R, Ng Chang Hin P, Villard J, Marion P (1978) Hemodynamic assessment of the functioning of the natural and prosthetic aortic valves by Doppler effect. Ann Cardiol Angeiol (Paris) 27, 267–275.

Minifie FD, Kelsey CA, Hixon TJ (1968) Measurement of vocal fold motion using an ultrasonic Doppler monitor. J Acoust Soc Am 43, 1165–1169.

Mountford RA, Atkinson P (1979) Doppler ultrasound examination of pathologically enlarged lymph nodes. Br J Radiol 52, 464–467.

Nealeigh RC, Miller CW (1976) A venous pulse Doppler catheter tip flowmeter for measuring arterial blood velocity, flow, and diameter in deep arteries. ISA Trans 15, 84–87.

Niederle P, Fridl P, Koudelkova E, Jabavy P, Haco M (1984) Contrast Doppler echocardiography. Arch Acoust 9, 215–221.

Nippa JH, Alexander RH, Folse R (1971) Pulse wave velocity in human veins. J Appl Physiol 30, 558–563.

Ohta M, Iwamoto K, Minagawa Y, Tatsumura M, Maeda K (1984) Studies on transitory FHR increase and the variation of fetal movements with the use of CW ultrasonic Doppler fetal actgraph. Proc 44th Mtg Jap Soc Ultrason Med 737–738.

Okamoto T, Tabata Y, Maruyama Y, Aoki T, Kakiuchi T, Nakahasi H, Sakazaki T, Yamashita M, Nakata M (1984) An application of pulse Doppler instrument for the stones at the lower part ureter; report 1. Proc 45th Mtg Jap Soc Ultrason Med, pp 593–594.

Paulus DA (1981) Noninvasive blood pressure measurement. Med Instrum 15, 91–94.

Perri AJ, Rose J, Feldman AE, Parker J, Karafin L, Kendall AR (1978) An evaluation of the role of the doppler stethoscope and the testicular scan in the diagnosis of torsion of the spermatic cord. Invest Urol 15, 275–277.

Pinkerton JA (1979) Intraoperative Doppler localization of intestinal arteriovenous malformation. Surgery 85, 472–474.

Pinnella JW, Spira M, Erk Y, Freed D, Hartely CJ (1982) Direct microvascular monitoring with implantable ultrasonic Doppler probes. J Microsurg 3, 217–221.

Powell MR, Thoma W, Fust HD, Cabarrou P (1983) Gas phase formation and doppler monitoring during decompression with elevated oxygen. Undersea Biomed Res 10, 217–224.

Roberts JD, Jones BM, Greenhaigh RM (1986) An implanted ultrasound Doppler probe for microvascular monitoring: an experimental study. Br J Plast Surg 39, 118–124.

Roederer GO, Folcarelli PH, Dixon RD, Baumann FG, Riles TS, Imparato AM (1986) Combined use of OPG-Gee and duplex scanning in a clinical setting. Bruit 10, 22–24.

Rosenfeld EZ, Beer G, Spitzer-Rahat S, Weinreb A (1982) On the correlation of acoustical sounds and ultrasound velocimetry in the subclavian steal syndrome. J Cardiovasc Surg (Torino) 23, 494–500.

Rossazza C, Pourcelot L (1984) The Doppler effect in ophthalmology. Annee Ther Clin Ophtalmol 35, 117–139.

Roversi GD, Canussio V, Cabibbe GG, Beaussart JL (1973) The use of ultrasonics (Doppler effect) in the study of fetal cardiac activity: possibilities and limitations. Minerva Ginecol 25, 448–453.

Rozman J (1979) Ultrasonic bubble detection during hemodialysis. Abstracts WFUMB 79, Scimed, Tokyo, p 399.

Rubissow GJ, Mackay RS (1974) Decompression study and control using ultrasonics. Aerospace Med 45, 473–478.

Sagar KB, Rhyne TL, Greenfield LJ (1983) Echocardiographic tissue characterization and range-gated Doppler ultrasound for the diagnosis of pulmonary embolism. Circulation 67, 365–370.

Sapov LA, Iunkin IP, Volkov LK, Menshikov VV (1976) Study of the potentials for ultrasonic location of gas bubbles for the control of decompression in divers. Voen Med Zh 6, 65–67.

Saumet JL (1983) Diagnosis and physiopathology interest by ultrasonic velocimetry of the ulcero-mutilating acropathy (UMA). Ultrasonics 4, 31–36.

Schoonderwaldt HC, Cho Chia Yuen G, Colon EJ, et al (1981) The presection value of the Doppler LP test for shunt-therapy in patients with normal pressure hydrocephalus. J Neurol 225, 15–24.

Semb BKH, Pedersen T, Hatteland K, Storstein L, Lilleaasen P (1982) Doppler ultrasound estimation of bubble removal by various arterial line filters during extracorporeal circulation. Scand J Thorac Cardiovasc Surg 16, 55–62.

Smith DF, Foley WS (1982) Real-time ultrasound and pulsed Doppler evaluation of the retroplacental clear area. J Clin Ultrasound 10, 215–219.

Steer PJ (1977) Monitoring in labour. Brit J Hosp Med 15, 219–225.

Stromberg B, Hammarlund K, Oberg PA, Sedin G (1983) Transepidermal water loss in newborn infants. 9. The relationship between skin blood flow and evaporation rate in full term infants nursed in a warm environment. Acta Paediatr Scand 72, 729–733.

Suzuki I, Satomura S (1958) A study on the pulsation wave of eye pressure with ultrasonic interference method (1). Acta Soc Ophthamol Jap 62, 1698–1701.

Thierfelder C, Magnus S, Pardemann G, Vogel S (1978) Functional vascular diagnostics of the dental pulp, a new application field for the ultrasonic Doppler procedure. Dtsch Gesundheitsw 33, 1105–1109.

Thierfelder C (1979) The blood supply of the single tooth and its functional change caused by prosthetic therapy. Stomatol DDR 29, 174–177.

Tsuchida S, Yamaguchi O (1981) Urodynamics. Diagnostic ultrasound in urology and nephrology (Eds H Watanabe, JH Holmes, HH Holm, BB Goldberg), pp 221–227, Ogaku-Shoin, New York.

Utsu M, Sakakibara S, Ishida T, Chiba Y, Hasegawa T (1983) Dynamics of tracheal fluid flow in the human fetus, studied with pulsed Doppler ultrasound. Nippon Sanka Fujinka Gakkai Zasshi 35, 2017–2018.

Vucicevic ZM, Russo L (1977) Importance and accuracy of Doppler in diagnosis of amaurosis fugax. Ultrasound in medicine 3B (Eds D White, R Brown), pp 1393–1394, Plenum Press, New York.

Wagner A, Szreter T, Powalowski T (1978) Use of Doppler ultrasonic method for detection of upper airways strictures. Otolaryngol Pol 32, 181–187.

Waltemath CL, Preuss DD (1971) Determination of blood pressure in low-flow states by the Doppler technique. Anesthesiology 34, 77–79.

Wells PNT, Halliwell M, Skidmore R, Webb AJ, Woodcock JP (1977) Tumor detection by ultrasonic Doppler blood-flow signals. Ultrasonics 15, 231–232.

Woodcock JP (1976) Ultrasonic assessment of arterial thickening. Ultrasonics 14, 99.

Wyatt AP, Ratnavel K, Loxton GE (1973) The technique and possible application of supra-orbital artery blood-pressure estimation. Br J Surg 60, 741–743.

Wei-Qi W, Lin-xin Y (1982) A double beam Doppler ultrasound method for quantitative blood flow velocity measurement. Ultrasound Med Biol 8, 421–425.

Yoshitoshi Y, Machii K, Sekiguchi H, Mishina Y, Ohta S, Hanaoki Y, Kohashi Y, Shimizu S, Kuno H (1966) Doppler measurement of mitral valve and ventricle wall velocities. Ultrasonics 4, 27–28.

Yukioka H, Takekawa S, Nishimura K, Fujimori M (1981) A study of noninvasive method for measuring central venous pressure during general anesthesia. Anesth Analg (Cleveland) 60, 901–903.

Appendix

SPECIAL FUNCTIONS ARISING FROM WOMERSLEY'S THEORY

M'_{10} and ε'_{10} which appear in eqn 2.15 are complex functions of the non-dimensional parameter α (defined in eqn 2.16) and may be written:

$$M'_{10}(\alpha) = (1 + h_{10}^2 - 2h_{10} \cos \delta_{10})^{\frac{1}{2}} \qquad \text{A.1}$$

and

$$\varepsilon'_{10}(\alpha) = \tan^{-1} \left(\frac{h_{10} \sin \delta_{10}}{1 - h_{10} \cos \delta_{10}} \right) \qquad \text{A.2}$$

where

$$h_{10} = \frac{2}{\alpha} \left[\frac{(\text{Ber } 1)^2(\alpha) + (\text{Bei } 1)^2(\alpha)}{\text{Ber}^2(\alpha) + \text{Bei}^2(\alpha)} \right]^{\frac{1}{2}} \qquad \text{A.3}$$

and

$$\delta_{10} = \frac{3\pi}{4} - \tan^{-1} \left[\frac{(\text{Bei } 1)(\alpha)}{(\text{Ber } 1)(\alpha)} \right] + \tan^{-1} \left[\frac{\text{Bei}(\alpha)}{\text{Ber}(\alpha)} \right] \qquad \text{A.4}$$

The Kelvin functions Ber and Bei are the real and imaginary parts of a complex Bessel function of order zero, while (Ber 1) and (Bei 1) are the real and imaginary parts of a complex Bessel function of order one. Summations for the calculation of each of these functions have been given by Abramowitz and Stegun (1965) Handbook of Mathematical Functions, Dover Publications, New York.

The functions ψ which appears in eqn 2.17 is also a function of α and may be written:

$$\psi = \left(\frac{\tau J_0(\tau) - \tau J_0(y\tau)}{\tau J_0(\tau) - 2J_1(\tau)} \right) \qquad \text{A.5}$$

where J_0 and J_1 are Bessel functions of the first kind and orders 0 and 1 respectively, and $\tau = \alpha i^{3/2}$.

SUBJECT INDEX